Oil, Gas and Petrochemical: Processing, Production and Management

Oil, Gas and Petrochemical: Processing, Production and Management

Edited by **Jane Urry**

SYRAWOOD
PUBLISHING HOUSE

New York

Published by Syrawood Publishing House,
750 Third Avenue, 9th Floor,
New York, NY 10017, USA
www.syrawoodpublishinghouse.com

Oil, Gas and Petrochemical: Processing, Production and Management
Edited by Jane Urry

© 2016 Syrawood Publishing House

International Standard Book Number: 978-1-68286-070-0 (Hardback)

Contents

Preface

Different approaches, evaluations, techniques and advanced studies on processing and production of oil, gas and petrochemicals have been included in this book. Diverse range of topics such as uses of petrochemicals, their production and refinement, natural gas, etc. have been covered in this text in a comprehensive manner. With state-of-the-art inputs by acclaimed experts of this field, this book targets students and professionals alike. It will help new researchers by foregrounding their knowledge in this branch.

This book has been the outcome of endless efforts put in by authors and researchers on various issues and topics within the field. The book is a comprehensive collection of significant researches that are addressed in a variety of chapters. It will surely enhance the knowledge of the field among readers across the globe.

It gives us an immense pleasure to thank our researchers and authors for their efforts to submit their piece of writing before the deadlines. Finally in the end, I would like to thank my family and colleagues who have been a great source of inspiration and support.

Editor

Investigation of Underground Sour Gas Storage in a Depleted Gas Reservoir

R. Azin[1,2]*, R. Malakooti[2], A. Helalizadeh[3] and M. Zirrahi[4]

1 Department of Chemical Engineering, School of Engineering, Persian Gulf University, Bushehr 7516913817 - Iran
2 Persian Gulf Science and Technology Park, Bushehr - Iran
3 Department of Petroleum Engineering, Petroleum University of Technology, Ahwaz - Iran
4 Department of Chemical & Petroleum Engineering, Schulich School of Engineering, University of Calgary, Calgary, Alberta, Canada T2N 1N4 - Canada
e-mail: reza.azin@pgu.ac.ir - reza.malakooti@pet.hw.ac.uk - helalizadeh@yahoo.com - mzirrahi@ucalgary.ca

* Corresponding author

Résumé — **Étude de stockage souterrain de gaz naturel acide dans un réservoir épuisé** — Le Stockage Souterrain de Gaz (SSG) implique le stockage de grandes quantités de gaz naturel pour satisfaire la demande en gaz naturel des secteurs domestiques, commerciaux ou industriels. Le stockage du gaz naturel acide peut être avantageux du point de vue économique car il réduit les coûts de traitement et augmente le potentiel d'exploitation de réservoirs partagés.

Cet article étudie la faisabilité du SSG dans un gisement iranien fracturé et épuisé de gaz à condensat. La simulation compositionnelle est utilisée pour construire le modèle dynamique de réservoir, développer la phase d'adaptation de l'histoire d'exploitation et établir les cycles d'Injection/Soutirage (I/S). Un flux de gaz peu corrosif et trois flux de gaz acide de compositions différentes ont été testés pour stockage dans le réservoir pendant une période estivale.

Les résultats de simulation montrent que la présence de sulfure d'hydrogène et de dioxyde de carbone dans le flux de gaz injecté améliore la production de condensats. L'Amélioration de Production de Condensats (APC), définie comme le pourcentage d'augmentation de la récupération de condensats grâce à l'injection de gaz acide par rapport à l'injection de gaz peu corrosif, a été calculée pour les différentes compositions du gaz stocké. Le Taux de Contenance en Condensats (TCC), défini comme le rapport entre la production de condensats dans le gaz acide soutiré et celle dans le méthane soutiré, a été également évalué pour les différentes compositions de gaz stocké. Les résultats ont montré que l'APC a des taux plus élevés pendant les premiers cycles, mais qu'elle diminue dans les derniers. Le TCC est plus élevé pour le gaz acide comparativement au gaz peu corrosif. De plus, le pouvoir calorifique du gaz produit, calculé pour tous les cycles d'injection/soutirage, est comparé avec celui du gaz injecté. Il est apparu que la différence de pouvoir calorifique entre le gaz produit et le gaz injecté augmente avec la teneur en sulfure d'hydrogène et dioxyde de carbone du gaz injecté. En outre, il est aussi apparu que le réservoir possède, à la fin des cycles d'injection/soutirage, une pression plus basse dans le cas du stockage souterrain de gaz acide que dans celui de gaz peu corrosif. La présence des composants acides fait baisser le coefficient z du flux de gaz injecté et induit une différence plus faible entre le coefficient z du gaz injecté et celui du fluide de réservoir.

Abstract — **Investigation of Underground Sour Gas Storage in a Depleted Gas Reservoir** — *Underground Gas Storage (UGS) involves storage of large quantities of natural gas to support the natural gas demand in domestic, commercial and industrial areas. Storage of sour gas can be advantageous from economic standpoint, as it reduces treatment costs and increases the potential of* *production from shared reservoirs. This paper investigates feasibility of UGS in one of Iranian depleted fractured gas condensate reservoirs. Compositional simulation was employed to build dynamic reservoir model, develop the history matching phase of the reservoir and construct Injection/Withdrawal (I/W) cycles. One sweet gas stream and three sour gas streams with different compositions were tested for storage into reservoir during summer season.*

Results of simulation showed that presence of H_2S and CO_2 in the injected gas stream improves condensate production. Condensate Production Enhancement (CPE), defined as the percentage of condensate recovery increase due to sour gas injection relative to the sweet gas injection, was calculated for different compositions of storage gas. Also, Condensate Holding Ratio (CHR), defined as the ratio of condensate in the withdrawn sour gas to that in the withdrawn CH_4, was estimated for different storage gas compositions. Results showed that CPE has a higher rate in earlier cycles and declines at later cycles. CHR is higher for sour gas storage compared to sweet gas. Furthermore, heating value of produced gas was calculated in all I/W cycles and compared with heating value of injected gas. It was indicated that difference between heating value of produced and injected gas increases with increase of H_2S and CO_2 content of the injected gas.

Also, it was found that the reservoir has lower pressure rise at the end of I/W cycles in the case of underground sour gas storage compared to sweet gas storage. The presence of acid gas components decreases the z-factor of injected gas stream resulting in smaller difference between z-factors of injected gas and reservoir fluid.

Abbreviations

CHR   Condensate Holding Ratio
CPE   Condensate Production Enhancement
EOS   Equation of State
I/W   Injection/Withdrawal
PR   Peng-Robinson
PVT   Pressure Volume Temperature
UGS   Underground Gas Storage

Nomenclature

Np_{so} Cumulative condensate production with sour gas injection (m³)

Np_{sw} Cumulative condensate production with sweet gas injection (m³)

$K_{C_{5+}}^{pure}$ K-value of condensate (C_{5+}) in injected pure methane (dimensionless)

$K_{C_{5+}}^{sour}$ K-value of condensate (C_{5+}) in injected sour gas (dimensionless)

P_{pc} Pseudo-critical pressure (Pa)

P'_{pc} Pseudo-critical pressure adjusted for nonhydrocarbon components (Pa)

T_{pc} Pseudo-critical temperature (°C)

T'_{pc} Pseudo-critical temperature adjusted for nonhydro-carbon components (°C)

$x_{C_{5+}}^{pure}$ Liquid mole fraction of condensate (C_{5+}) in injected pure methane (dimensionless)

$x_{C_{5+}}^{sour}$ Liquid mole fraction of condensate (C_{5+}) in injected sour gas (dimensionless)

y_{H_2S} Mole fraction of component H_2S (dimensionless)

$y_{C_{5+}}^{CH_4}$ Vapor mole fraction of condensate (C_{5+}) in injected pure methane (dimensionless)

$y_{C_{5+}}^{inj}$ Vapor mole fraction of condensate (C_{5+}) in injected gas (dimensionless)

$y_{C_{5+}}^{pure}$ Vapor mole fraction of condensate (C_{5+}) in injected pure methane (dimensionless)

$y_{C_{5+}}^{sour}$ Vapor mole fraction of condensate (C_{5+}) in injected sour gas (dimensionless)

ε Pseudo-critical temperature adjustment factor for nonhydrocarbon components (°C)

INTRODUCTION

Underground Gas Storage (UGS) is the unique efficient process to store large quantities of natural gas in an underground inventory. This technology has been adopted in many countries to match the constant supply from long-distance pipelines to the variable demand of markets. Depleted gas reservoirs represent the best candidates for UGS, in which natural gas is injected into the reservoir in the low-demand season, while it is withdrawn from the reservoir in the high-demand season [1]. Previous studies [2-4] focused on underground sweet gas storage in depleted gas reservoirs. In addition to sweet gas injection, sour gas, a blend of natural gas, hydrogen sulfide (H_2S) and carbon dioxide (CO_2) streams, can be recognized as another option for the purpose of UGS.

Underground sweet gas storage causes double costs in the industry. First, the sour gas stream is sweetened before injection into the reservoir; second, the withdrawn gas is usually contaminated with the fluid remaining in the reservoir and needs treatment before use. On the other hand, underground sour gas storage is a more economic option for UGS projects because the sweetening and treating process is performed once. This advantage brings about strategic management for shared gas reservoirs which contain sour gas. In this manner, the rate of gas withdrawal would be higher from the shared reservoirs since operators do not need to construct treatment facilities before reinjection of the produced gas.

When gas is stored in depleted gas condensate reservoirs, it can enhance the condensate recovery through pressure maintenance and gas cycling. Injection of dry gas increases the dew point pressure and helps revaporization of heavy hydrocarbon fractions into the gas phase [5]. Among all gases, pure CO_2 and separator gases containing H_2S and CO_2 are the most effective injection streams to minimize the condensate drop-out in near-wellbore zones [6]. Also, the presence of CO_2 and H_2S in the gas phase causes water content and condensate holding to increase. Decrease in the compressibility factor is another impact of acid gases present in the injection gas [7]. Based on compositional simulation, CO_2 is found to be an efficient dry gas stream that can be injected into the retrograde reservoir to improve condensate recovery [8]. Changes in the wetting behavior of reservoir rock and interfacial tension between the storage gas and reservoir fluid are other important issues that arise during UGS. Melean et al. [9] investigated the influence of gas/condensate interfacial tension and wetting behavior of condensate on the porous substrate. They found that CO_2 is more advantageous than CH_4 and N_2 in reducing capillary forces and promoting the spreading of the condensate phase on water.

Sour natural gas mixtures result in problems arising mainly in two ways: increased potential for corrosion in wellbore and surface facilities and increased tendency for hydrate formation at elevated pressures [10]. These issues imply that all aspects of sour gas mixtures including their advantages and disadvantages must be considered whether or not underground sour gas storage is an economic and efficient process.

The main purpose of this work was to analyze the influence of CO_2 and H_2S on the gas condensate reservoir behavior when subjected to underground sour gas storage. Injection/ Withdrawal scenarios with different CO_2 and H_2S compositions were studied to find out the changes in the condensate recovery, heating value and reservoir pressure in the case of sour gas storage.

1 METHODOLOGY

1.1 Reservoir Summary

An Iranian depleted fractured gas reservoir was nominated to study the possibility of underground sour gas storage.

The reservoir's characteristics and production history information have been addressed in previous studies by Azin et al. [2] and Jodeyri Entezari et al. [3].

Figures 1 and 2 show the relative permeability curves of water, condensate and gas phases present in the reservoir. The capillary pressure curve between water and condensate phases versus water saturation is shown in Figure 3. The capillary pressure between gas and liquid phases was assumed to be zero. There was no active aquifer for this reservoir.

The condensate saturation in blocks is very low and does not exceed more than 0.001 during long-term production.

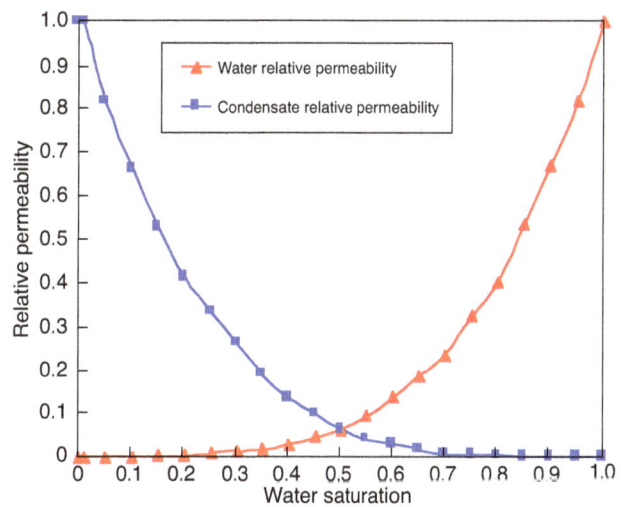

Figure 1

Typical relative permeability curves of water and condensate phases.

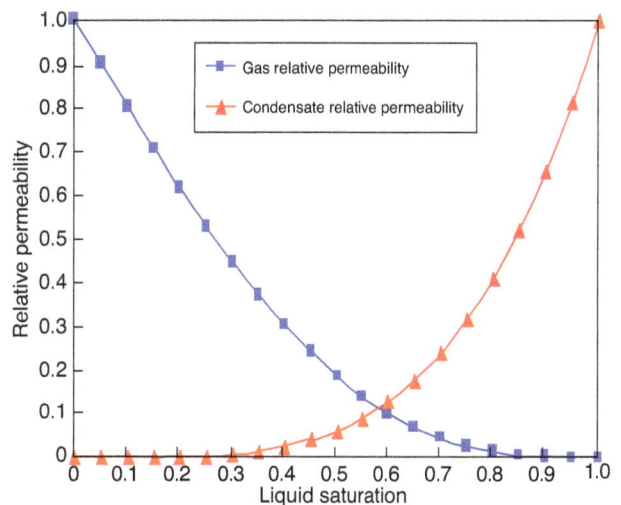

Figure 2

Typical relative permeability curves of gas and condensate phases.

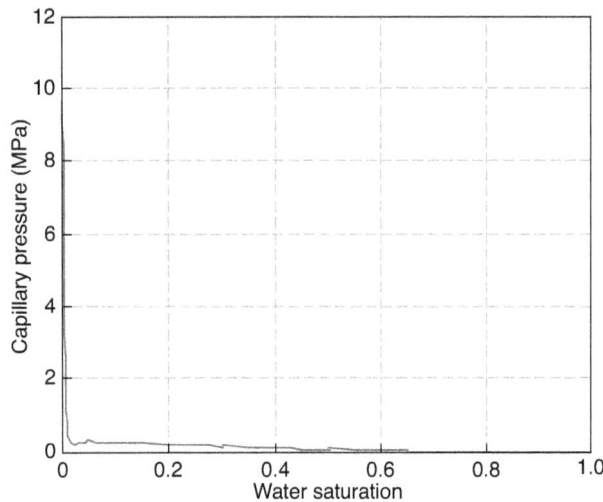

Figure 3

Capillary pressure curve *versus* water saturation.

Figure 2 shows that the liquid saturation should be about 0.25 in order to flow. This necessitates a condensate saturation of about 0.05, which is difficult to achieve in reservoir simulation. This observation confirms that condensate presents within the reservoir as an immobile phase and only gas is considered to be a mobile phase. Difference in the condensate components' chemical potential in gas and liquid phases leads to condensate evaporation into the gas phase. Then condensate components are separated from the withdrawn gas in the surface facilities.

The mechanism of fluid flow between the matrix and fracture throughout successive Injection/Withdrawal cycles can be summarized as: the injected gas stream diffuses into the matrix blocks and reaches equilibrium with immobile condensate components. Thus, the gas phase becomes rich as a result of this equilibrium and condensate component evaporation during injection and shut-in periods. Enriched gas diffuses from the matrix into the fracture and then is withdrawn from the reservoir in the production phase. Finally, condensate components are removed from stream production because of pressure reduction in the surface facilities.

1.2 Reservoir Simulation Model

Azin *et al.* [2] designed a simulation model in Eclipse software [11] to study the possibility of UGS in the candidate reservoir. To do that, petrophysics and geological information was used in order to set up a static model of the reservoir. The fractured carbonate gas reservoir with bottom water was discretized to yield grid block dimensions of approximately 111 × 41. To build a dual porosity model, the

simulation zone was subdivided into 34 simulation layers in the vertical direction, with 17 layers belonging to fractures.

The reservoir fluid model was prepared using the PVTi module in Eclipse software [11]. Components of the initial reservoir fluid were lumped into 7 pseudo-component groups and are given in Table 1. Pressure Volume Temperature (PVT) calculations were performed with the 3-parameter Peng-Robinson Equation of State (EOS). The Lohrenz-Bray-Clark (LBC) viscosity correlation was utilized to obtain the viscosity of gas and liquid [11]. History matching was accomplished as a final stage of simulation model construction and a reasonable match was achieved for gas and condensate production rates and reservoir pressure [2].

TABLE 1

Initial reservoir fluid composition

Components	Observed composition
C_1, N_2	0.92097
C_2, CO_2	0.05361
C_3, nC_4	0.01715
iC_4, nC_5	0.00265
FC_6	0.00188
C_7, C_{11}	0.00338
C_{13+}	0.00036

1.3 I/W Gas Storage Cycle Simulation

I/W gas storage cycles were developed on a completed simulation model obtained by Azin *et al.* [2]. The strategy used to make the storage cycles was similar to that considered in the studies of Malakooti and Azin [4]. Regarding annual gas consumption during cold months, [16] exhibited total consumption of 68×10^9 m^3 natural gas in Iran from November 1st in 2008 up to March 31st in 2009. Hence, it was decided to withdraw 5.1×10^9 m^3 gas in each cycle, which provided 7.5% of total natural gas consumption during five cold months. To reach this objective, the injection rate was set to 327.7 m^3/s from April 15th to October 15th; then, the production rate was equal to 393.3 m^3/s from November 1st to March 31st of next year. For this reservoir with a given target rate, the base pressure was found to be about 15.86 MPa [4]. Abandonment reservoir pressure before I/W cycles can start is called base pressure.

At optimum conditions [4], four new vertical wells were defined at different parts of the reservoir in addition to the one existing vertical well to make the depletion phase faster and increase the reservoir capacity to produce such gas volume. Figure 4 represents the well productivity index trend during the depletion phase and successive I/W cycles. All wells were supposed to shut in from October 16th to October

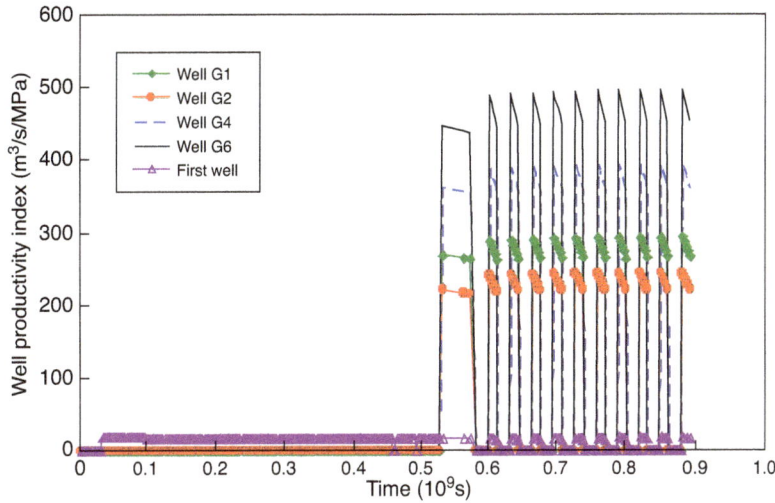

Figure 4

Well productivity index trend during depletion phase and successive I/W cycles.

TABLE 2

Composition of the injected gas streams

Name	C_1, N_2	C_2, CO_2	C_3, nC_4	iC_4, nC_5	FC_6	C_7, C_{11}	C_{13+}	H_2S
Sweet gas	0.975	0.0246	0.0004	0	0	0	0	0
Sour gas (1)	0.955116	0.034498	0.006978	0.000045	0.000001	0	0.000012	0.00335
Sour gas (2)	0.906784	0.07742	0.008613	0.000047	0.000001	0	0.000013	0.007122
Sour gas (3)	0.851784	0.12742	0.008613	0.000047	0.000001	0	0.000013	0.012122

30th during each I/W cycle. To examine the influence of sour gas streams on the performance of the UGS reservoir, four injection gas streams (one sweet gas stream and three sour gas streams) were used. The compositions of the storage gas streams are given in Table 2. The sweet gas stream was mainly composed of methane without any trace of CO_2 and H_2S, while the sour gas streams contained CO_2 and H_2S with different compositions.

2 RESULTS AND DISCUSSION

2.1 Effect of H_2S and CO_2 Impurities on Condensate Evaporation

The performance of underground sweet gas injection with different I/W schemes was studied previously [2-4]. It was shown that injection of sweet gas into a depleted gas condensate reservoir will increase the reservoir pressure in successive I/W cycles and cause the trapped condensate to revaporize. The results of this study indicate that condensate production tends to increase when sour gas is injected into the reservoir, as shown in Figure 5. Condensate Production Enhancement (CPE), defined by Equation (1), reveals the effect of sour compounds on the condensate production:

$$CPE = \frac{Np_{so} - Np_{sw}}{Np_{sw}} \tag{1}$$

where:

Np_{so}: Cumulative condensate production with sour gas injection,

Np_{sw}: Cumulative condensate production with sweet gas injection.

Figure 6 depicts the CPE trend *versus* successive I/W cycles for different sour gas stream compositions. For instance, at the end of the 10th cycle, CPE will be 3, 5 and 7 percent when the mole percent of ($CO_2 + H_2S$) in injected streams is 0.85, 2.8 and 8.3%, respectively. This figure also indicates that the rate of CPE is not uniform; the rate is higher at initial stages and decreases in later cycles, where the remaining condensate trapped in the porous media becomes more difficult to extract by injected gas.

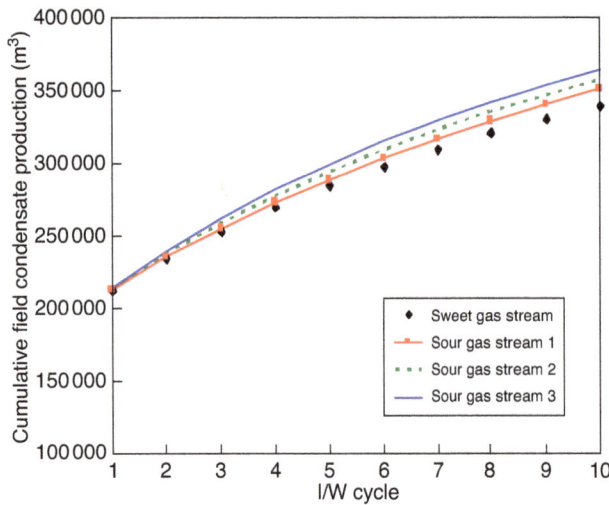

Figure 5

Field condensate production total in the case of different injected gas streams.

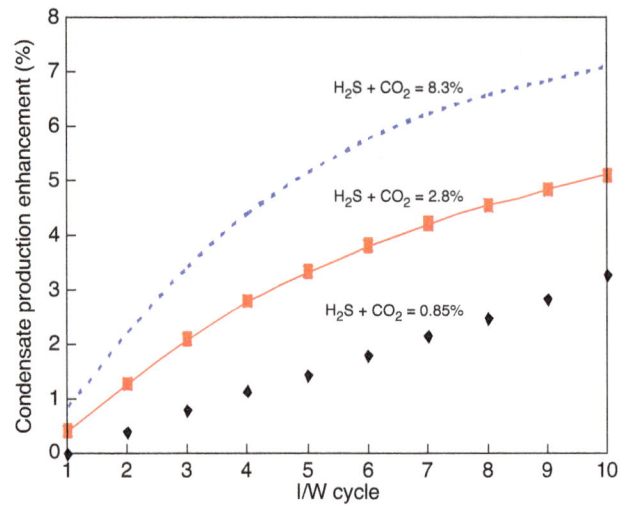

Figure 6

Condensate production enhancement caused by various mole fraction of CO_2 and H_2S.

According to Figures 5 and 6, the presence of CO_2 and H_2S in the injection gas increases the tendency of heavy hydrocarbons to vaporize and produce during the withdrawal stage. One reason for this phenomenon is that CO_2 and H_2S molecules have a potential to make London dispersion forces with heavy hydrocarbons. The oxygen and sulfur atoms of CO_2 and H_2S are electronegative and both molecules have a polar structure. On the other hand, heavy hydrocarbons have a high chance that their molecular electron density will not be evenly distributed [17]. This uneven distribution causes a temporary multi-pole state in heavy hydrocarbons, leading to attraction by CO_2 and H_2S molecules in the gas phase. The chance of distributing in electron density will be stronger by presenting CO_2 and H_2S and longer chains of carbons in hydrocarbon molecules. So, heavy hydrocarbons will have a greater tendency to revaporize and produce a phase when CO_2 and H_2S exist in storage gas.

To show this effect, the Condensate Holding Ratio (CHR) of injected gas is defined as the ratio of condensate in injected sour gas in equilibrium with reservoir fluid to that in pure CH_4 (as a base component for storage gas) in equilibrium with reservoir fluid. The CHR can be defined as Equation (2):

$$CHR = \frac{y_{C_{5+}}^{inj}}{y_{C_{5+}}^{CH_4}} \quad (2)$$

The CHR can be interpreted as a function of the condensate equilibrium ratio, called the K-value in terms of classical thermodynamics. The K-value of condensate (C_{5+}) in the mixture of injected gas is defined as Equations (3) and (4) for storage of sour and pure methane, respectively:

$$K_{C_{5+}}^{sour} = \frac{y_{C_{5+}}^{sour}}{x_{C_{5+}}^{sour}} \quad (3)$$

$$K_{C_{5+}}^{pure} = \frac{y_{C_{5+}}^{pure}}{x_{C_{5+}}^{pure}} \quad (4)$$

Also, the mole fraction of condensate in the gas phase for each state is obtained by rearrangement of Equations (3) and (4):

$$y_{C_{5+}}^{sour} = K_{C_{5+}}^{sour} x_{C_{5+}}^{sour} \quad (5)$$

$$y_{C_{5+}}^{pure} = K_{C_{5+}}^{pure} x_{C_{5+}}^{pure} \quad (6)$$

By substituting Equations (5) and (6) in (2), the Condensate Holding Ratio is obtained, as follows:

$$CHR = \frac{K_{C_{5+}}^{sour} x_{C_{5+}}^{sour}}{K_{C_{5+}}^{pure} x_{C_{5+}}^{pure}} \quad (7)$$

C_{5+} is the major constituent of the reservoir liquid phase and accounts for more than 85mole%, regardless of the type of storage injection gas. Therefore, changes in the mole fraction of C_{5+} in sour gas injection and pure methane injection can be assumed to be negligible. By this assumption Equation (7) changes as:

$$CHR = \frac{K_{C_{5+}}^{sour}}{K_{C_{5+}}^{pure}} \quad (8)$$

Values of the CHR and K-value of condensate (C_{5+}) for several types of storage gas were calculated by using the Peng-Robinson Equation of State (PR-EOS) and are shown in Figures 7 and 8. Figure 7 shows that the presence of CO_2 and H_2S in injected gas increases the K-value of the condensate. Also, it can be deduced from these figures that the effect of H_2S is superior to CO_2. A similar trend is observed in the CHR, as shown in Figure 8. This figure also

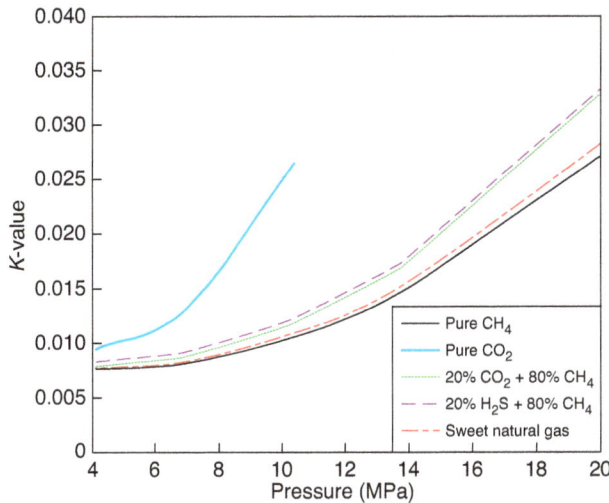

Figure 7

K-value of condensate (C_{5+}) for different types of storage gas.

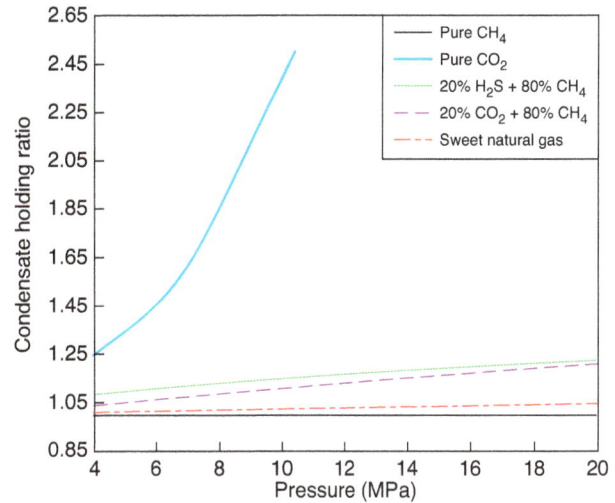

Figure 8

CHR for different types of storage gas at 77.22°C.

shows that injection of pure CO_2 increases the condensate production drastically more than other streams. Previous studies [6, 9] confirmed this trend.

Melean *et al.* [9] studied the effects of thermodynamic conditions on the interfacial properties. Relevant parameters which they studied are dimensionless Bond number and capillary number to relative interfacial properties to gravity and viscous forces, respectively. Their study shows that CO_2 is more effective in lowering gas/oil IFT, followed by CH_4 and N_2. Therefore, the condensate phase loses its tendency to remain on the surface of the water phase that often covers the porous rock. As a result, heavier hydrocarbon is ready to vaporize and produce with the gas phase during the withdrawal stage of successive UGS cycles.

2.2 Effect of H_2S & CO_2 Impurities on the Heating Value

One of the most important factors in UGS projects is the study of the difference in the heating value of injected and produced gas streams during Injection/ Withdrawal cycles. To achieve this purpose, the heating value of produced gas was calculated in all I/W cycles and compared with the heating value of injected gas. The heating value of a gas stream is calculated by multiplying the heating value of each component into its mole fraction. The heating value of defined components such as CH_4, C_2H_6 and C_3H_8 are obtained from the GPSA book [12]. Other components are assumed to be paraffinic components and their heating values are calculated by interpolating between defined and available paraffinic component heating values with molecular weight in the range near to the heavy components.

To study the effect of the presence of H_2S and CO_2 in injected gas on the heating value of produced gas, the ratio of difference between the produced and injected gas streams' heating value to the injected gas's heating value are calculated. As shown in Figure 9, the difference between the heating value of produced and injected gas increases with the increase in H_2S and CO_2 content of the injected gas. A gas stream with a 10% H_2S mole fraction has a higher difference between the heating value of injected and produced gas compared with other gas streams. This figure also shows that the heating value of produced gas increases during production in each I/W cycle because the heavy component mole fraction in the production gas stream increases with pressure decline. On the other hand, because of the decrease in total heavy components within the depleted reservoir during I/W cycles, production of heavy components decreases; so, the heating value of produced gas decreases.

Also, Figure 10 compares the heating values for different streams of injection and production in which the heating value of sweet streams is higher than that of sour streams.

2.3 Effect of H_2S and CO_2 Impurities on Pressure Rise

As the reservoir is volumetric with no active aquifer, the difference between the z-factor of injected and reservoir fluid was found to be the main cause for pressure rise during successive I/W cycles [2]. The vaporizing-gas displacement process derives from the fact that the injected dry gas with a lower specific gravity and higher z-factor compared with the reservoir fluid is enriched by vaporizing intermediate and heavy components from the condensate liquid. As a result, residual reservoir fluid becomes leaner and its z-factor starts

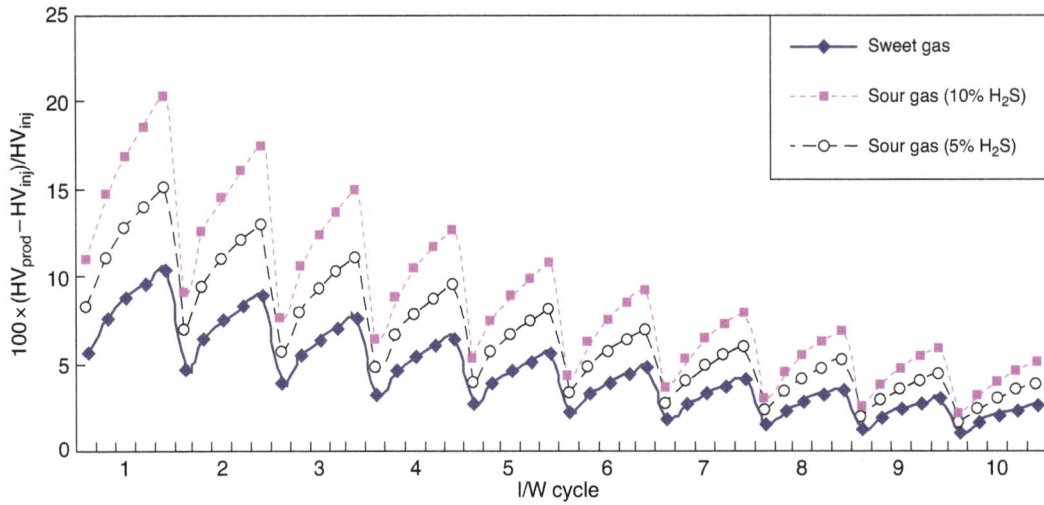

Figure 9

Ratio of difference between heating value of produced and injected gas streams to heating value of injected gas stream in each I/W cycle.

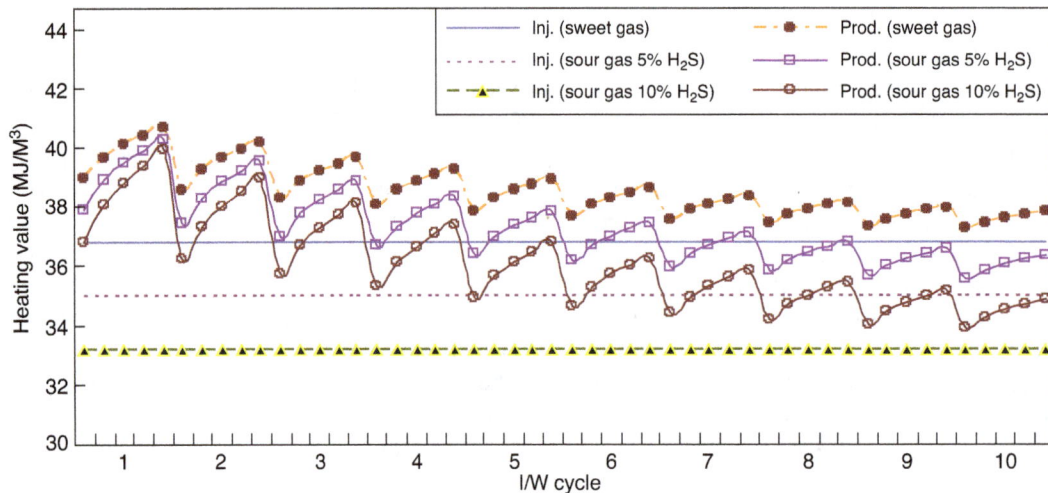

Figure 10

Heating value of injection and production streams *versus* successive I/W cycles.

to increase. Subsequently, reservoir pressure will slowly rise at the end of I/W cycles [2]. The z-factor of injection and reservoir fluid was calculated through numerical representation of the Standing and Katz [13] z-factor graph offered by Dranchuk and Abou-Kassem equations [14].

It is well known that existence of CO_2 and H_2S leads to large errors in the evaluation of the compressibility factor. The remedy to this problem is to adjust the pseudo-critical properties to account for the unusual behavior of these components. The equations used for this adjustment are [15]:

$$T'_{pc} = T_{pc} - \varepsilon \tag{9}$$

$$P'_{pc} = \frac{P_{pc}T'_{pc}}{T_{pc} + y_{H_2S}(1 - y_{H_2S})\varepsilon} \tag{10}$$

T'_{pc} and P'_{pc} are then used to calculate T_{pr} and P_{pr} for a sour gas stream. The pseudo-critical temperature adjustment factor ε is obtained using a graphical procedure defined by Wichert and Royan [15]. The reduction of the compressibility z-factor caused by the presence of CO_2 and H_2S is shown in Figure 11. As the acid gas content of the injection stream increases, the z-factor of the injection stream decreases at pressures of interest.

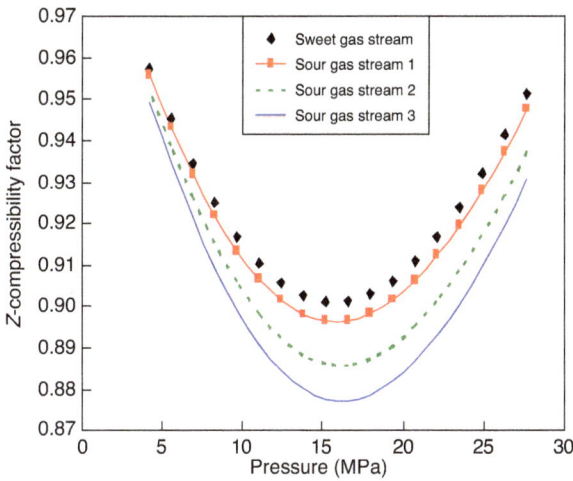

Figure 11

Compressibility z-factor trend *versus* pressure for different gas streams.

The results reveal that the UGS reservoir will experience a lower pressure rise at the end of I/W cycles when the mole fraction of CO_2 and H_2S increases, because the difference between the z-factors of injected and reservoir fluids is smaller. However, the reservoir pressure differences among the sweet gas and three sour gas streams are small because of a small amount of CO_2 and H_2S in the gas streams. This fact is illustrated in Figures 12 and 13. Figure 12 represents that maximum and minimum of operating pressure during each injection/withdrawal cycle take place between 16 MPa and less than 21 MPa as it was expected beforehand since base pressure and initial reservoir pressure for this specific reservoir were 15.86 and 21.37 MPa respectively. Reservoir pressure difference among different streams can be observed more accurately in Figure 13, in which only pressure at the end of I/W cycles has been plotted *versus* time. Figure 14 summarizes calculation of changes in the z-factor of injected

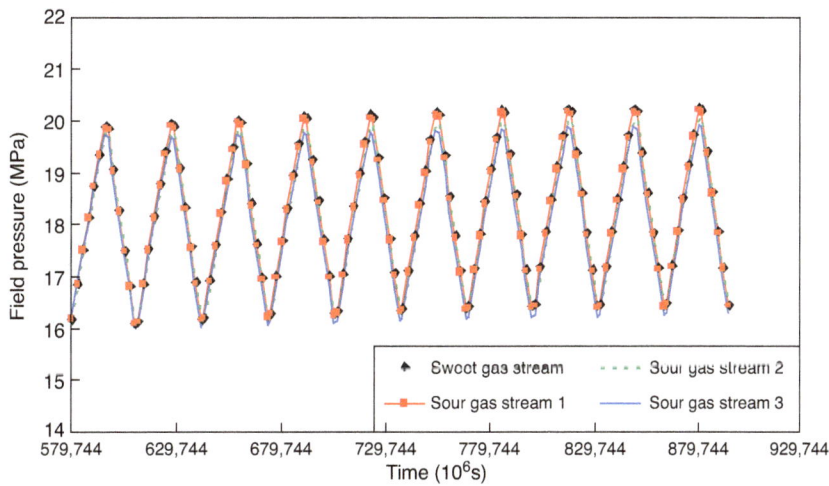

Figure 12

Pressure trend *versus* time for different injected streams during successive I/W cycles.

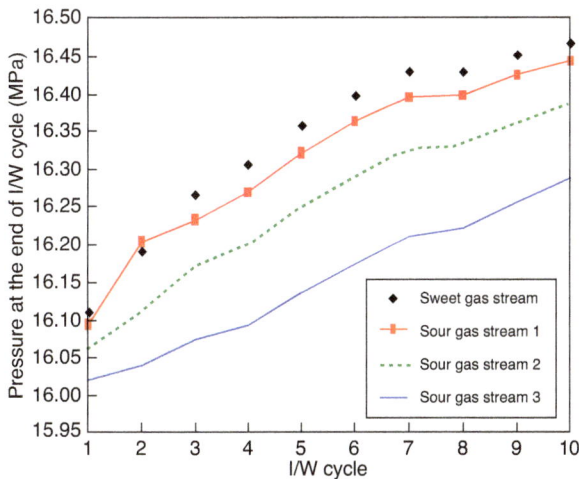

Figure 13

Pressure rise at the end of successive I/W cycle for different compositions of injected gas streams.

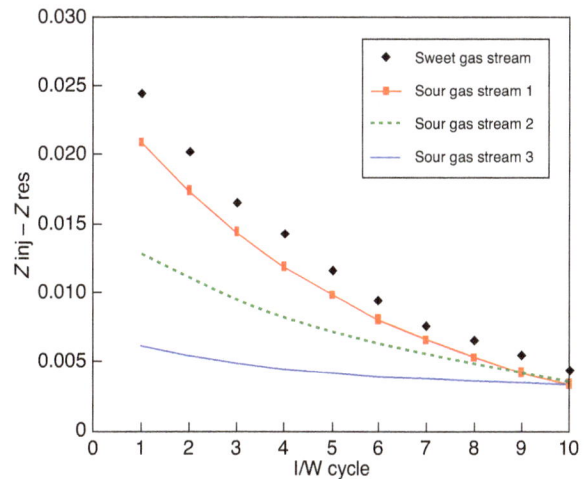

Figure 14

Difference between compressibility z-factor of injected gas and reservoir fluids during I/W storage cycles.

gas and reservoir fluid at different pressures for all cycles. The difference between the compressibility z-factor of injected gas and reservoir fluids declines during storage cycles because the composition of the reservoir fluid approaches that of the injected fluid.

CONCLUSIONS

– Condensate Production Enhancement (CPE) has a higher rate in earlier cycles and declines in later UGS cycles;

– the Condensate Holding Ratio (CHR) is higher in sour gas injection compared with sweet gas injection, resulting in higher condensate recovery;

– as the composition of CO_2 and H_2S increases in the injected gas stream, the difference between the heating value of produced and injected gas tends to increase;

– because of heavy component production at the surface, the heating value of produced gas increases in each I/W cycle. Meanwhile, the reduction in the amount of remaining heavy hydrocarbons within the reservoir causes the heating value of produced gas to be higher at earlier stages of UGS compared with later ones;

– the presence of CO_2 and H_2S components in the injection gas stream decreases the compressibility z-factor of the injection stream and causes less pressure rise than that without any trace of CO_2 and H_2S;

– the difference between the compressibility z-factor of injected gas and reservoir fluids has a smaller value in the case of sour gas streams.

REFERENCES

1 Tureyen O.I., Karaalioglu H., Satman A. (2000) Effect of the Wellbore Conditions on the Performance of Underground Gas-Storage Reservoirs, SPE 59737, *SPE/CERI Gas Technology Symposium*, Calgary, Alberta, Canada, 3-5 April.

2 Azin R., Nasiri A., Jodeyri Entezari A. (2008) Investigation of Underground Gas Storage in a Partially Depleted Gas Reservoir, *J. Oil Gas Sci. Technol.* **63**, 691-703.

3 Jodeyri Entezari A., Azin R., Nasiri A., Bahrami H. (2010) Investigation of Underground Gas Storage in a Partially Depleted Naturally Fractured Gas Reservoir, *Iran. J. Chem. Chem. Eng.* **29**, 1, 103-110.

4 Malakooti R., Azin R. (2011) Optimization of Underground Gas Storage in a Partially Depleted Gas Reservoir, *J. Pet. Sci. Technol.* **29**, 824-836.

5 Adel H., Tiab D., Zhu T. (2006) Effect of Gas Recycling on the Enhancement of Condensate Recovery, Case Study: Hassi R' Mel South Field, Algeria, SPE 104040, *First International Conference and Exhibition*, Cancun, Mexico, 31 August-2 September.

6 Zaitsev I.Y., Dmitrievsky S.A., Norvik H., Yufin P.A., Bolotnik D.N., Sarkisov G.G., Schepkina N.E. (1996) Compositional Modeling and PVT Analysis of Pressure Maintenance Effect in Gas Condensate Field: Comparative Study, SPE 36923, *SPE European Petroleum Conference*, Milan, Italy, 22-24 October.

7 Clark M.A. (1999) Experimentally Obtained Saturated Water Content, Phase Behavior and Density of Acid Gas Mixtures, *MSc Thesis*, University of Calgary, Canada.

8 Seto C.J., Jessen K., Orr Jr F.M. (2003) Compositional Streamline Simulation of Field Scale Condensate Vaporization by Gas Injection, SPE 79690, *SPE Reservoir Simulation Symposium*, Houston, Texas, USA, 3-5 February.

9 Melean Y., Bureau N., Broseta D. (2001) Interfacial Effects in Gas-Condensate Recovery and Gas Injection Processes, SPE 71495, *SPE Annual Technical Conference and Exhibition*, New Orleans, Louisiana, USA, 30 September-3 October.

10 Wichert E., Aziz K. (1972) Calculate Zs for Sour Gases, *Hydrol. Process.* **51**, 119-122.

11 Schlumberger, *Eclipse Reference Manual*, Houston, TX, Schlumberger Information Solution, 2005.

12 *Physical Properties*, Engineering Data Book, Gas Processors Suppliers Association, Tulsa, Okvtv, 1998.

13 Standing M.B., Katz D.L. (1942) Density of Natural Gases, *Trans. AIME* **146**, 140-149.

14 Dranchuk P.M., Abou-Kassem J.H. (1975) Calculation of z-Factors for Natural Gases Using Equations of State, *J. Can. Pet. Technol.* **14**, 34-36.

15 Wichert E., Royan T. (1997) Acid Gas Injection Eliminates Sulfur Recovery Expense, *J. Oil Gas* **95**, 17, 67-72.

16 http://www.NIGC.ir, 23/07/2009.

17 http://en.wikipedia.org/wiki/London_dispersion_force, 2/09/2010.

Method for Simulation and Optimization of Underground Gas Storage Performance

K. Wojdan[1], B. Ruszczycki[2*], D. Michalk[3] and K. Swirski[1]

[1] Institute of Heat Engineering, Faculty of Power and Aeronautical Engineering, Warsaw University of Technology,
Nowowiejska 21/25 00-665 Warszawa - Poland
[2] Transition Technologies S.A., Pawia 55, 01-030 Warszawa - Poland
[3] Statoil Deutschland, Kavernenanlage Etzel, Beim Postweg 2, 26446 Friedeburg - Germany
e-mail: b.ruszczycki@tt.com.pl

* Corresponding author

Résumé — Méthode de simulation et d'optimisation des capacités de stockage de gaz souterrain —
La simulation et l'identification correctes des capacités de débit d'une unité de stockage de gaz ne
sont envisageables que si l'on considére les limites non linéaires propres aux activités souterraines
et une stratégie optimale (ensemble de régles de décision) appliquée au mode opératoire spécifique
aux centrales. Le fonctionnement efficace d'une installation de stockage représente un vrai défi,
étant donné la complexité des limites liées à la mécanique des roches en milieu souterrain ainsi
que les autres limites techniques imposées aux différents dispositifs opérant dans l'unité. Cet
article a pour vocation de préparer le terrain afin de relever ces défis en définissant plusieurs
stratégies opérationnelles, à savoir un ensemble d'actions qui constituent un mécanisme de
haut niveau prédéfini permettant une activité rentable et efficace de l'unité. Nous décrirons de
façon détaillée un exemple de stratégie spécifique au travers de résultats de simulation
représentatifs et nous exposerons une procédure d'optimisation destinée à maximiser les
capacités opérationnelles des unités. Nous examinerons le modéle général des unités et la
nature des limites liées à la mécanique des roches, et nous analyserons l'élaboration de
l'algorithme informatique.

*Abstract — Method for Simulation and Optimization of Underground Gas Storage Performance —
Proper simulation and identification of the flow potential of a gas storage plant is only possible if the
nonlinear limits related to cavern operation and an optimal strategy (set of decision rules) related to
the plant operation mode are considered. An efficient operation of a storage plant is a challenging
task due to the complexity of cavern rock mechanical restrictions, as well as other technical restrictions imposed on different plant devices. The scope of this paper is to cater for these challenges by
defining a set of different operational strategies, i.e., sets of actions which constitute the predefined
high level mechanism that allows for an economic and efficient plant operation. In this paper, one
specific strategy example is described in detail. Specifically, we give simulation results and outline
an optimization procedure designed to maximize the plant's performance capabilities. The general
plant model and the form of rock mechanical restrictions of cavern operation are reviewed, and
the construction of the computational algorithm is analyzed.*

LIST OF ABBREVIATIONS AND SYMBOLS

AGDs   Above Ground Devices
BPR   Bottom Pressure
LLUOR   Lower Limit of Unrestricted Operation
SOE   Storage Operation Expert method
p_{min}, p_{max} Minimum and maximum allowable pressure inside the cavern
p_{LL} Lower limit of unrestricted operation pressure
p_R Recreation pressure

INTRODUCTION

Gas storage facilities, based either on underground salt caverns or depleted gas fields, are the key element in an industrial gas grid. Operation of the cavern plant, which is used not only for seasonal gas storing, but also for balancing on the daily or weekly scale, is a complex task. At any time point an operator can undertake many different actions, therefore the proper computational tool is required. The aim of our article is to build a framework that comprises all elements involved in the plant operation; having the model of caverns and above ground devices (compressors, preheaters, cyclon and glycol separators, etc.) does not in itself give a straightforward answer about how to operate the plant efficiently. The presented framework is intended to be used in order to choose the most optimal way of plant operation and to simulate the future gas storage states. The operational needs of industrial gas storage are determined by gas market demands, however secure plant operation is a major factor to be considered while fulfilling these demands. Due to the complex form of several different requirements, one of the most important parameters of a gas storage facility, *i.e.*, maximal hourly injection and withdrawal gas flow, is sometimes artificially undervalued. All factors limiting the accuracy of the forecast of injection and withdrawal flow potential result in a quite conservative plant operation. The lack of an accurate prognosis therefore obstructs an efficient operation of the plant within the safety margins. The factors contributing to the difficulty of the forecasting task include:

– nonlinear characteristics of caverns [1-3] and above ground facilities (including compressors);
– complex nonlinear cavern operation constraints [4-7];
– the possibility of violating geological cavern constraints which may lead to the temporary closure of the cavern [8-10];
– scale of the gas storage facility which can include several caverns and several compressors [11];

– long term time horizon (some time dependent cavern operation restrictions have a one year cycle);
– quick cycling of operation mode [12].

The most complex form of rock-mechanical restrictions is connected with the operation of the cavern in the low-pressure region, the analysis of such a case is discussed in [13-15]. From the perspective of integration of all elements into the overall framework, the challenging task is to include the time-dependent form of rock mechanics restrictions.

In this paper, the Storage Operation Expert (SOE) method is presented. It takes into consideration the thermodynamic model of gas caverns, model of above ground facilities, nonlinear constraints of rock mechanics imposed on the cavern operation and a predefined strategy for utilization of caverns. This method can be used to support plant operation in terms of technical and commercial concerns. Depending on simulation inputs, this method provides information about:

– realistic hourly plant withdrawal and injection limits for the flexible time horizon. This information can be easily converted into nomination limits available to storage users (based on individual storage users curves). To calculate hourly plant limits one should define an infinite (for injection) or negative infinite (for withdrawal) gas flow demand (refer *Sect. 3.2.*);
– state of rock mechanics timers that are active when a single cavern pressure is in a restricted range related with rules and limits of rock mechanics. Based on this information an operator can see how many days a cavern was operating in the restricted range. Specifically, the operator is informed about how many days he has to safely get the cavern out of the restricted range and how many days he can still safely operate in restricted ranges;
– optimal realization of aggregated business nominations for the plant. This optimal realization is a set of trajectories for each cavern in the plant. Each trajectory provides information about the gas flow associated with each cavern for every step of simulation (*e.g.* every hour of plant operation). Moreover, an operator can obtain the information about trajectories of wellhead pressure, temperature and rock mechanic timers. In order to calculate this information, one should set the gas flow demand to the accepted aggregated business nominations trajectory. Furthermore, our method can be used to perform "WHAT-IF" simulations which allow operators to test the plant for different gas demand scenarios, different plant operation strategies, different availability and configuration of Above Ground Devices (AGDs) and different availability and configuration of gas caverns. The

presented solution performs simulation with the resolution of one hour or one day. The simulator calculates an optimal decision trajectory for each cavern in the storage according to the following equation:

$$Decision\ (k,t) = \{connected/disconnected,\ flow\}$$

where k denotes the cavern number and t is the time (in units of simulation steps). This decision trajectory determines the time evolution of calculated storage and cavern states. The cavern state is defined by:
- wellhead pressure, volume and temperature;
- timers and other information related with the rock mechanics limits, *i.e.*, how long a cavern can operate below p_{LL} (lower limit of unrestricted operation) pressure level, or how long a cavern still has to wait above p_R, recreation pressure level (see *Sect. 2*);
- connected/disconnected status.

The storage state is defined by:
- hourly or daily volume increment;
- mode (injection with compressors, withdrawal with compressors, free flow injection or free flow withdrawal);
- availability of caverns and AGDs;
- configuration of AGDs (*e.g.* two out of three compressors running).

1 STORAGE PLANT MODEL

The plant model used in the simulations (*Fig. 1*) consists of five main parts:
1. system pipe, which connects a set of AGDs with the gas grid. The pressure in the system pipe is the same as the pressure in the gas grid;
2. set of AGDs. This module represents all AGDs which take part in the withdrawal or injection action;
3. field pipe which connects caverns with the set of AGDs. It is assumed that: during withdrawal, pressure in the field pipe is δp_{wdt} (for example $\delta p_{wdt} = 2$ bar) lower than the lowest pressure among caverns connected to the field pipe. During injection, the pressure in the field pipe is δp_{inj} (for example $\delta p_{inj} = 3$ bar) larger than the highest pressure among caverns connected to the field pipe;
4. pressure control valves. Individual caverns may be disconnected or connected to the field pipe. When they are connected, a dedicated controller controls the gas flow to and from the caverns by acting on the control valves. Due to the control operation range and safety regulations, it is assumed that the difference between the maximum and minimum pressure in the set of connected caverns cannot exceed p_{window} (for example $p_{window} = 50$ bar);
5. gas caverns are modeled with an advanced thermodynamic simulator. The presented method uses this simulator to calculate wellhead pressure, temperature and normalized volume response for a given flow.

An objective of the method used in the AGDs model, is to identify for a given set of input parameters, the "bottlenecks" of the plant which restrict the maximal flow, and to find the device configuration which provides the maximal flow capabilities. The gas flow in or out of the storage may be natural or forced by a set of compressors depending on the pressure difference between the field pipe and the system pipe (*Fig. 1*). Therefore, the model of the AGDs includes four storage operation modes: injection with compressors, withdrawal with compressors, free flow injection and free flow withdrawal. Depending on the mode, a model of a different set of devices is used to calculate the maximal withdrawal or injection rate. This output value is calculated for a given set of input parameters and device availability which can be imported from the service and maintenance plan. The four main inputs are: field pipe pressure, system pipe pressure, gas mole weight and gas heat capacity. Moreover, in the injection mode case, the gas temperature in the system pipe is needed, while in the withdrawal mode case, the gas temperature in the field pipe is needed. The external input parameters are defined *a priori* as prediction trajectories, based on some expert knowledge related with an expectation of future gas parameters and system pipe pressure. The length of these trajectories is the same as the time horizon of the simulation.

The field pipe pressure is calculated in each computational step based on information about storage mode, and a set of caverns which are connected to the field pipe. In the withdrawal mode case, the gas temperature in the

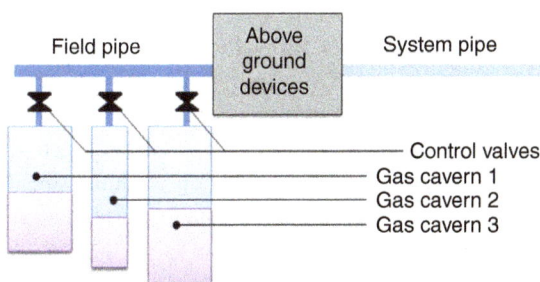

Figure 1

Simplified model of the plant.

field pipe is calculated as the weighted average of wellhead temperature of caverns that are connected to the field pipe.

The cavern model is used to calculate the wellhead pressure, temperature and volume response for the given gas flow. Such a model includes the equation of state of gas, gas chemical composition, cavern geometry and the rock mechanical cavern response. The particular, the cavern model used in this simulator was supplied by *ESK Gmbh Germany*, several other cavern models are also available, such as NOMIX by *GreyLogix*, CavInfo Software Suite by *SOCON*, WinKaga by *Chemkop*, SCGS Toolbox by *Technical Toolboxes, Inc.*

2 LIMITATIONS OF CAVERN OPERATION

The main limitations imposed on operation of a single cavern are:
- the maximum injection and withdrawal rate,
- the range of acceptable cavern pressure,
- the maximum allowed pressure change rate during gas injection or withdrawal.

These constrains are simple and easy to handle. However, there is also another class of more complex, and possibly time-dependent, cavern operation constraints, which originate from geological restrictions related with rock mechanics, *e.g.* maximum duration of operation at low pressure levels [4]. Violation of such constraints may result in local failures of the rock mass at the cavern wall, leading to the temporary closure of cavern operation, which is an essential loss from the gas storage plant perspective. The detailed form of those restrictions may be different for different plants. One should notice that even a single plant might have caverns associated with different classes of such constraints. The main differences between the classes of caverns are due to their different geometrical properties, such as depth of the cavern location, or the value of v/s ratio. In the example presented in this paper, there is a real life assertion that the plant consists of two classes of caverns. There are 7 class A and 5 class B caverns (the difference between the rock-mechanics restriction for these two cavern classes is discussed in the remainder of this section). The real number of caverns in the plant where the presented method had been tested was different than the number presented here, but the information about the real configuration is protected and cannot be published. The differences and consequences of the rock mechanical recommendations for different cavern properties are discussed in [16]. Usage of cavern simulator

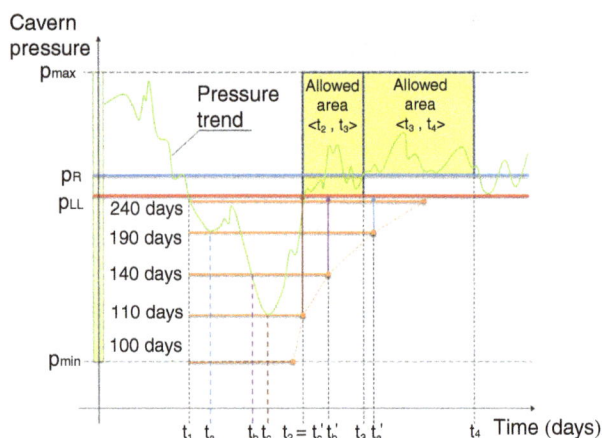

Figure 2

Rock mechanics limitations associated with class A caverns.

potentialities in analysis of short-term and long-term mechanical and thermodynamical cavern behavior is discussed in [17].

Figure 2 illustrates real life rock mechanic limitations associated with class A caverns. In the initial stage (before time instance t_1) the pressure was in the range $\langle p_{LL}, p_{max} \rangle$. Operating below or at p_{LL} limit causes undesired cavern convergence. Thus reaching or decreasing cavern pressure below this limit is related with a future necessity to maintain the cavern pressure in the range $\langle p_R, max \rangle$ for some defined time period. The p_{LL} limit is reached at time t_1 when a dedicated LLUOR cavern timer is started (initial timer state is zero). The system remembers the minimal value of cavern pressure recorded in the time period $\langle t_1, \text{current time} \rangle$. The lowest recorded pressure value determines the maximal allowed time for operating below p_{LL} limit (*Fig. 2*, orange horizontal lines). In our example, this maximal allowed time is in a range from 100 (when minimal pressure is reached) to 240 days (in case when pressure falls slightly below p_{LL} level). When the LLUOR cavern timer reaches the maximal allowed time, the pressure value must remain in the range of $\langle p_{LL}, p_{max} \rangle$. The nonlinear relationship between the minimum registered pressure and the time limit, is set individually for each cavern and has been approximated by a polynomial of degree sixth. In Figure 2, this time limit, t_a', t_b' and t_c' is determined by the cavern state at t_a, t_b and t_c respectively.

For each class A cavern an additional time period (t_2, t_3) is defined. In this time period, the pressure value can be in the range of $\langle p_{LL}, p_{max} \rangle$. At t_2, the LLUOR cavern timer is frozen. The OFFSET cavern timer starts ticking only if the cavern pressure was outside the restricted range in the last possible day (when LLUOR cavern timer equals the maximal allowed time for staying below p_{LL} level).

It is worth considering the case in which the cavern pressure exits from the pressure range $\langle p_{min}, p_{LL} \rangle$ before the "maximum allowed time" for staying below p_{LL} has expired. In this case, the timer that keeps track of the operation below p_{LL} is frozen, and all calculated future time points are shifted forward by the length of the time spent above p_{LL}. The OFFSET timer is not ticking, the t_3 time is defined in a such way that the $\langle t_3, t_4 \rangle$ period has some defined length (for example 80 days) and the $\langle t_1, t_4 \rangle$ period is not longer than one year.

In the whole time period $\langle t_3, t_4 \rangle$, the pressure value must be in the range $\langle p_R, max \rangle$. All timers are set to zero at time t_4. The cavern cycle is over and a new cycle is started. When the pressure value reaches the p_{LL} limit for the first time in a new cycle, the LLUOR cavern timer is started, and the whole procedure (that is based on described rules) is repeated.

Figure 3 illustrates real life rock mechanics limitations associated with class B caverns. The cavern pressure is not restricted by geological constraints until it hits the minimal allowed value. At time instance t_1, the LLUOR timer and the BPR (Bottom Pressure) timer start ticking. The BPR timer counts the days when cavern pressure was at the minimal pressure or close to the minimal pressure. When a cavern reaches minimum pressure the LLUOR timer begins to count the days that cavern pressure is below the p_{LL} level. The maximum value for this

timer is defined for each cavern (for example 80 days). The operator can keep the pressure below the p_{LL} level only for a limited time after it hits the minimal pressure. At time t_2, the cavern pressure becomes higher than minimum pressure plus some margin (ϵ), thus the BPR timer is frozen. At t_3, the pressure exits the restricted range. The LLUOR timer is frozen and the recreation pressure timer starts ticking. From this moment, the pressure has to be in the range $\langle p_{LL}, p_{max} \rangle$ for the time period $\langle t_3, t_4 \rangle$. The length of this time period is equal to the length of the $\langle t_1, t_2 \rangle$ time period. Thus the recreation pressure timer has to reach the value of the BPR timer. At t_4, all the timers are set to zero. In the case that the pressure leaves the range $\langle min, min + \epsilon \rangle$ and returns to this range, the length of the $\langle t_3, t_4 \rangle$ time period should be the same as the total time of staying at, or close to the minimal pressure.

3 CAVERN OPERATION STRATEGY

The goal of this section is to describe a strategy, *i.e.*, the predefined high level decision mechanism for operating caverns. There may be more than one strategy defined for the plant, but only one strategy can be used in a single simulation. For a given state of the gas storage plant, a high level strategy defines a set of actions that can be executed. Each action has a unique priority assigned, which depends on the cost and rock mechanics factors related with the action. In each simulation step, the action which results in the total gas flow that is as close as possible to the gas flow demand is chosen. If there are more desirable actions which result in the same total gas flow, then an action with the highest priority level will be used. Each action has a unique priority level and serves the following three functions:

- defines the set of active caverns that can take part in gas injection or withdrawal process;
- defines the pressure target for each active cavern;
- limits the set of active caverns to those which fit the pressure window p_{window} (*Sect. 1*), the bottom (top) limit of the pressure window is affixed to the lowest (highest) cavern pressure from the set of active caverns in case of injection (withdrawal) mode.

In the optimization algorithm, the set of active caverns (returned by the action) is limited in order to find a minimal set of caverns which provides the highest gas flow. This optimal correction is computed by solving an optimization task (*Sect. 4*) and it is not a part of the strategy definition. In other words, the action of the strategy defines a framework for the optimization task. This framework limits the number of caverns that are used by the optimization task. This limitation takes

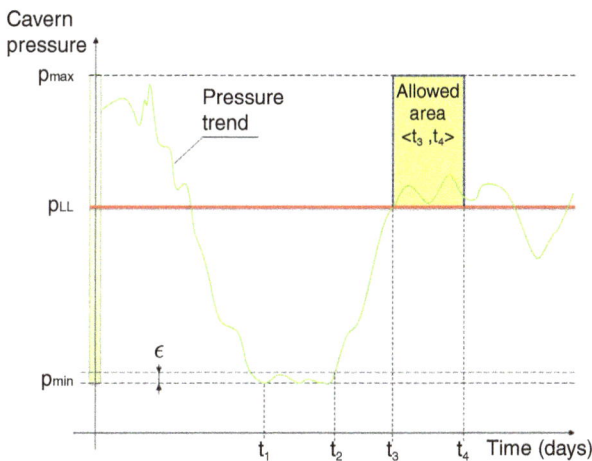

Figure 3

Rock mechanics limitations associated with class B caverns.

TABLE 1

The "strategy book". Decision (or decision combination) selection depends on states of class A and class B caverns

		SB$_1$		SB$_2$		Priority
SA$_1$	DA$_1$	DB$_1$	DA$_1$		High
	DA$_1$	DB$_2$	DA$_1$	DB$_2$	Low
SA$_2$		DB$_1$		DB$_2$	High
		DB$_2$	DA$_2$		Medium
	DA$_2$				Low

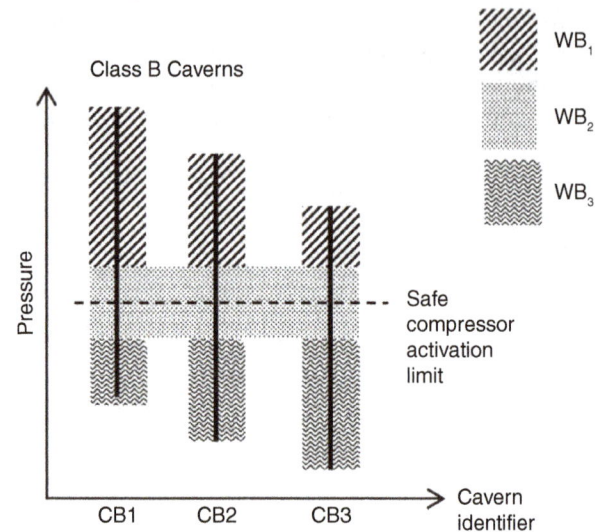

Figure 4

Default strategy, withdrawal: selected pressure ranges for B class caverns.

into consideration plant architecture and physical limits, some expert knowledge related with cavern operation and a mechanism that protects the plant against switching caverns on and off due to caverns thermodynamic effects.

3.1 An Example of a Real Life Strategy

In this section, an example of the real life strategy is described. It consists of three substrategies: withdrawal, standard injection and recreation injection substrategy. Each strategy consists of the following two operations:
- assigning the discrete system state based on current system parameters,
- associating the system state with the set of potential actions by means of the "strategy book", *i.e.*, the rules for decision selection (*Tab. 1*).

In order to handle cavern pressure thermodynamic effects and to avoid oscillations of cavern enable/disable decisions, the following mechanisms were introduced:
1. safety pressure margins on the following parameters:
   - minimal and maximal cavern pressure,
   - compressor activation level (please refer to safe compressor activation level in *Fig. 4*),
   - p_{LL} pressure level (please refer to safe LLUOR level in *Fig. 5*),
   - p_R level,
   - pressure window size (*Sect. 1*).
2. deadband pressure, which is used to identify an active pressure range in each cavern (*Fig. 4* and *5*). Figure 6 describes the idea of the deadband pressure.

The selected pressure ranges for the withdrawal substrategy for class A and class B caverns (with different rock mechanics rules) are presented in Figure 5 and Figure 4 respectively. The withdrawal substrategy is executed if the gas flow demand for the current step

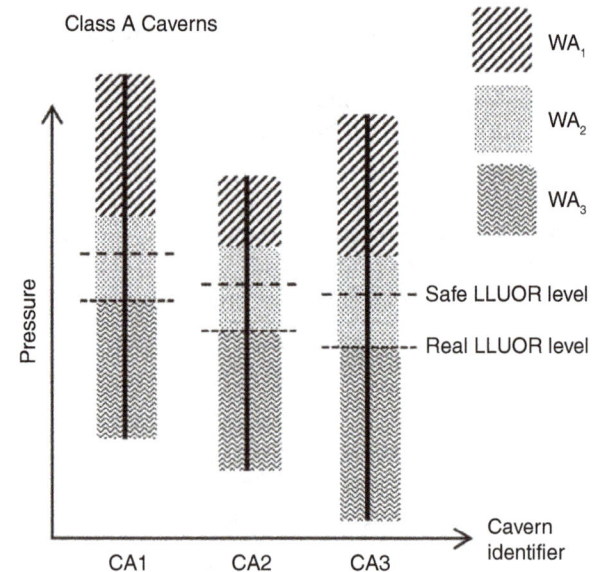

Figure 5

Default strategy, withdrawal: selected pressure ranges for A class caverns.

requires the withdrawal mode and no get out action is required. The get out action is required if at least one cavern in the plant requires the injection action in order to fulfill geological constraints related to rock mechanics rules (*Sect. 2*).

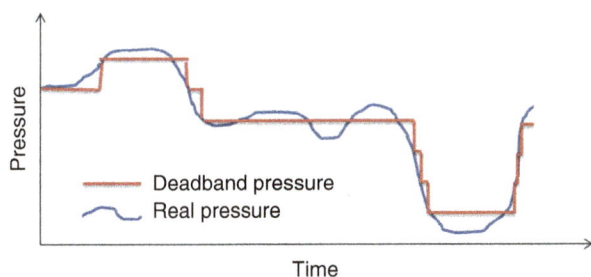

Figure 6

Deadband pressure.

The set of states and actions has been defined based on selected pressure ranges from Figures 4 and 5.

States for Class A Caverns

State SA_1. If at least one class A cavern is in the WA_1 pressure range, then the entire set of class A caverns is in the SA_1 state. The following decision is possible in the SA_1 state:

- decision DA_1: select an optimal subset of all class A caverns that are above safe p_{LL} level and get them to safe p_{LL} level. Thus the entire set of active caverns consists of class A caverns that are above safe p_{LL} level.

State SA_2. If all class A caverns are in the WA_2 or WA_3 range, then the set of class A caverns is in the SA_2 state. The following decision is possible in the SA_2 state:

- decision DA_2: select an optimal subset of all class A caverns and get them to safe minimal pressure. Thus the set of active caverns consists entirely of class A caverns.

States for Class B Caverns

State SB_1. If at least one class B cavern is in the pressure range, then the entire set of class B caverns is in the state. The following decisions are possible in the SB_1 state:

- decision DB_1: select an optimal subset of all class B caverns that are above safe compressor activation level and get them to the safe compressor activation level. Thus the set of active caverns consists of class B caverns that are above safe compressor activation level;
- decision DB_2: select an optimal subset of all class B caverns and get them to the safe minimal pressure. Thus the set of active caverns consists entirely of class B caverns.

State SB_2. If all class B caverns are in the WB_2 or WB_3 range, then the set of class B caverns is in the SB_2 state. The following decision is possible in the SB_2 state:

- decision DB_2: Select an optimal subset of all class B caverns and get them to the safe minimal pressure. Thus the set of active caverns consists entirely of class B caverns.

State identification is based on pressure ranges instead of single pressure levels in order to handle rapid changes of pressure related with thermodynamic effects. Such pressure changes are consequences of gas temperature changes resulting from connecting and disconnecting a cavern from the field pipe.

Table 1 presents the "strategy book" which is a set of decision rules. The selection of a decision depends on the state of class A and class B caverns. For instance, if class B caverns are in state SB_2, and class A caverns are in state SA_1, then the algorithm will first try to execute the DA_1 action only. If the withdrawal rate for simulated action will be lower than the current demand, then the algorithm will try to use simultaneously both DA_1 and DB_2 actions, i.e., it will select an optimal subset from the set of active caverns that consists of all class A caverns that are above the safe p_{LL} level and all class B caverns. The second combination has a lower priority than using solely the DA_1 decision, but it will be applied if the withdrawal rate provided by this combination is higher. The flow will be spread among selected caverns in proportion to their flow limits.

The standard injection and recreation injection sub strategy will be described in a more general form. The framework for defining those substrategies of the default strategy, is the same as for withdrawal mode. The injection substrategy is executed whenever the gas flow demand for a current step requires an injection mode and no get out action is required. The main idea behind this substrategy can be summarized by the following rules:

- if there are any class A caverns with pressures in restricted ranges, use such caverns with the highest priority and get them out of restricted ranges;
- if there are any class B caverns that reached the minimum pressure and are still below p_{LL} level, use such caverns with the medium priority and get them out of the restricted range;
- utilization of all other caverns has low priority.

A combination of the above rules is possible.

The recreation injection substrategy is the last strategy discussed here. It is required if at least one cavern needs to get out from the restricted pressure range in order to fulfill rock mechanics rules. Such a necessity is calculated for each cavern automatically. In the case when the get out action is required, the mode of the plant is set to injection. The plant mode change time (for example 45 minutes) is taken into consideration by the simulation algorithm. This paper does not include a detailed description of this strategy, but the main idea can be summarized by the following rules:

High priority rule: if there are any class A or class B caverns that need to get out the restricted range, then set them as active caverns. The pressure target is set in such a way that the algorithm will try to get them out of the restricted range, plus some safety margin that will cover a thermodynamic pressure drop after the get out action is done. This safety margin protects the plant from oscillating between injection and withdrawal modes.

Low priority rule: the set of active caverns consists of all caverns that need to get out of the restricted range and includes all other caverns, starting from the lowest pressure cavern that fits into the allowable pressure window (*Sect. 1*). By construction of the algorithm, the cavern with the lowest pressure from the set of caverns that require a get out action, is always included. However, the cavern gas flow sharing algorithm is different for caverns that require a get out action. The calculation algorithm will try to assign the maximum flow to such caverns, and the remaining flow is spread between all other caverns. The low priority rule was introduced in order to cover the case in which the get out action has been assigned to a small number of caverns, but the injection gas flow demand is high. In such a case, the algorithm might not be able to reach the gas flow demand by applying a high priority rule only.

3.2 Simulator Inputs

Main inputs of the simulator are presented below:
- scenario of the gas flow demand. One should define a trajectory of withdrawal demand or injection demand for each step of simulation. If the goal is to calculate the maximum possible withdrawal or injection rate on the assumed time horizon (for example 7 days), then this input should be set to infinity in the case of injection mode, or negative infinity in the case of withdrawal mode. To calculate an optimal gas flow for aggregated business nominations, this input should be the same as the aggregated business nominations;
- prognosis of physical gas parameters, including system pipe pressure, gas mole weight and gas heat capacity;
- scenario of caverns and AGDs availability within the simulation horizon. This information can be taken from the service and maintenance plan;
- the initial state of the plant. There is a possibility to set the initial pressure for each cavern manually, or the

thermodynamic state of each cavern can be automatically taken from the real plant (there is an assumption that the thermodynamic model of caverns is always synchronized with the real plant). The initial state is also defined by the initial state of rock mechanics rules (*Sect. 2*). One can set the state of each rock mechanics timer manually, or it can be automatically recalculated based on the historical plant data.

3.3 Computational Algorithm

The presented method starts the simulation with a particular plant state which has been loaded during the initialization of the algorithm. The plant state comprises the caverns operation history, current configuration and availability of the above ground devices The algorithm chooses the proper action and simulates plant future states with assumed time resolution (*e.g.* one hour), taking as a further input the gas nomination trajectory. In each step of the main loop the proper action is chosen, the caverns simulation is performed, and the caverns operation history is updated. At the beginning of each step, we identify a discrete plant state according to predefined strategy book. Formally, it is a mapping from the continuous space of variables describing the current pressures in the system, pressure levels of compressor activations and restriction on cavern operations, onto the set of discrete space variables according to which we specify the tentative set of caverns, according to Section 3.1. At this stage, we generate a series of possible action attempts, each attempt gives us a tentative list of "active" and "disabled" caverns. At this stage, we enter into the subloop where the optimization procedure (*Sect. 4*) is performed for each attempt. At each step of simulation, the model of above ground devices has to be executed several times in order to calculate the maximal possible flow for each probed configuration. The ADG's model contains internal optimization procedures as there are also possible different configurations of devices included in AGD's (for example the serial or parallel connection of compressors). The following work-flow describes in details the procedure in the main loop:

A1. Compute the storage mode. First, the algorithm checks if there is a need to start the get out action in order to avoid violating the rock mechanics constraints. If there is such a need then the plant mode is set to "injection". This calculation is based on static pressure-to-volume caverns characteristics and assumed pessimistic estimation of the injection gas flow rate provided by AGDs. If no get out action is required, then the plant mode is determined by the input gas demand trajectory;

A2. Compute compressor activation pressure level. Compressor activation pressure level is the boundary level of gas pressure in the field pipe that is required to start or stop the compressor. It is calculated based on the storage mode and the value of a system pipe gas pressure (storage input pipe);

A3. Compute strategy. First, the caverns availability (based on the service and maintenance plan) is checked. Next, the current plant state is matched with the set of states defined by the strategy. By construction, this matching is always possible and unique. The set of possible actions is returned (*Tab. 1*);

A4. Simulate each action and choose the one that provides the gas flow that is as close as possible to the gas demand. In case of more than one "winning" action, chose the one with the highest priority level. Priorities are set *a priori* based on economic and safety factors.

Let us have a closer look at the A4 procedure. For each simulated action, the following procedures are performed:

A4.1. Compute the optimal flow that can be provided by the plant for the given action. This procedure uses the AGDs model and information about cavern flow limits to calculate the maximal possible injection or withdrawal gas flow, and determines an optimal subset of active caverns which provides the highest possible injection or withdrawal gas flow. The optimization task is described in Section 4;

A4.2. Compute caverns shares. The share of each selected cavern in the gas injection/withdrawal action is calculated. The set of selected caverns is given by the A4.1 procedure. By default, the shares are proportional to the caverns gas flow limits. The sharing method is different in the case of the get out action. In such a case, our method first wants to use the potential of caverns for which the get out action is set, so it tries to assign the whole gas flow to such caverns. If 100% of the potential of such caverns is used, and still there is some gas flow "unassigned", then the algorithm spreads this flow among all other selected caverns proportionally to their flow limits;

A4.3. Assign flows to caverns based on shares and limit the flow demand to the cavern flow limit. If the demand flow assigned to the particular cavern (based on cavern shares) is higher than the flow limit, then it has to be limited. In the default sharing method, the ratio between the flow demand and the flow limit is the same for all caverns. If the limiting of the demand flow occurs, it affects all selected caverns in the same way;

A4.4. Simulate the cavern thermodynamic response once the "winning" action was chosen. At the initialization of the algorithm, the preceding operational history of all caverns is loaded. At the end of each step, the cavern thermodynamic response of wellhead pressure, volume and temperature for the given flow, is simulated based on the thermodynamic model of the cavern. The cavern states are not only calculated for selected caverns (with assigned flow) but also for all other caverns (zero flow). The cavern operational history is updated after each step of simulation.

4 OPTIMAL SELECTION OF CAVERNS

The potential maximal flow of a plant is limited by two factors: the sum of gas flow limits of utilized caverns and the possible flow provided by AGDs. The flow provided by AGDs depends mainly on the difference between the system pipe and the field pipe pressure. The field pipe pressure is dictated by the set of utilized caverns. Thus increasing the number of connected caverns increases the caverns flow potential, but on the other hand increasing it above some boundary value might have a negative effect on the flow provided by the AGDs. In order to determine whether, there is a need to set a "red flag" signaling that the gas should be injected into the cavern initiating the get out action, we assume the worst case scenario, *i.e.*, the minimal gas flow that can be delivered by AGDs and also the possibility that all caverns operating in the restricted pressure range have to simultaneously execute the get out action. The available gas flow is divided among all the caverns performing the get out action, taking into account restrictions such as the maximum allowed pressure change rate during the injection. The general optimization task is defined by finding the following maximum:

$$\max_{x} \{f(x) - \alpha b(x)\} \tag{1}$$

$$f(x) \equiv \min\{m(x), g(p_{field}(x))\} \tag{2}$$

with constraint:

$$r \leq s(x) \leq \max(r, \text{roundUp}(c/m(x)) \tag{3}$$

where:

− x is the set of all selected caverns. This set must at least cover all caverns from an active set that needs to execute a *get out action* in order to fulfill rock mechanics rules (only if such *get out action* is required);

− $f(x)$ is the maximal possible plant flow as a function of selected caverns;

– $m(x)$ is the sum of maximal flows provided by selected caverns. The flow provided by each cavern is calculated by choosing a lower value from two possibilities: one is the cavern flow limit, scaled to the size of the simulation step; the second value is the total gas volume which can be still injected or withdrawn from the cavern. For instance, if the gas pressure is close to the minimal allowed pressure in the withdrawal mode, the possible gas flow is low;

– $p_{field}(x)$ is the field pipe pressure determined by a set of selected caverns and the plant mode, see Section 1;

– $g(p_{field})$ is an estimation of the flow limit of AGDs (best case scenario) in function of field pipe pressure. The AGDs model is used to calculate the maximal possible flow for current p_{field};

– $s(x)$ is the number of caverns in the x set;

– c is the injection or withdrawal gas flow demand for the current simulation step;

– r is the number of caverns from an active caverns set that needs to execute a get out action in order to fulfill rock mechanics rules;

– α is the penalty coefficient;

– $b(x) = 0$ if the configuration of utilized caverns is the same as in the previous step, otherwise $b(x) = 1$.

The penalty protects the plant from frequently switching cavern statuses. If the penalty factor is given by $\alpha = h_a$, then the optimizer will change the set of the selected caverns only if such an action will raise the value of the optimized function by more than h_a. Otherwise, the optimizer would prefer to stay with the current caverns configuration. By default, h_a is equal to the average gas flow rate of a single cavern.

The penalty coefficient can be set to zero according to a penalty reset period which is denoted by τ, for example $\tau = 24$ hours. Therefore, there is a high probability that the cavern cluster might be changed every τ hours. The specific hour is controlled by another variable which is denoted by τ_{offset}.

In the first stage of solving, an optimization task of the subset of caverns which can operate in a free flow mode is isolated from the set of all active caverns (the optimizer input). The optimization task is solved separately for this subset of input caverns with compressors turned off, and for the entire set of input caverns with compressors turned on. The action which results in the largest value of the flow is chosen. The value of the function $f(x)$ is calculated in the following way:

– compute the maximal flow rate that can be realized by the AGDs for the current mode (withdrawal free flow, withdrawal compressor flow, injection free flow, injection compressor flow), based on:

  – the calculated decision about compressor utilization,

  – the information about AGDs availability (*i.e.* from the service and maintenance plan),

  – the assumed trajectory of gas physical parameters (*i.e.* gas pressure in a transport line),

  – the computed field pipe temperature (based on wellhead gas temperature from the set of active caverns) in case of withdrawal mode;

– scale the computed flow rate to the simulation time resolution;

– if the maximal possible flow is higher than the gas flow demand for the current simulation step, then the maximal gas flow is reduced to the assumed gas demand. If there is a get out action required then the maximal gas flow provided by the plant is calculated in the following way:

– let u be a sum of all injection flow limits related with caverns that need to get out from restricted ranges (according to rock mechanics limitations);

– let d be a gas injection demand for a current step;

– let z be a maximal gas flow provided by the AGDs;

– final flow is a $min(z, max(u, d))$.

The calculation of an optimal subset of active caverns can take a long time. To test each subset of active caverns, the algorithm has to execute the model of the AGDs. However, in many cases we can assume that if the difference between field pipe pressure and system pipe rises, the withdrawal flow rises and the injection flow drops. With such an assumption the optimization task can be solved in the following way:

– calculate the flow of the plant for the set of selected caverns in the previous simulation step (only if such set is allowed by strategy and rock mechanics constraints). In such a case, there is no penalty;

– sort all the active caverns in such a way that the first element of the sorted list is a cavern with the lowest (highest) pressure in case of an injection (withdrawal) mode;

– in the first iteration calculate, the function $f(x)$, where x consists of the first element of the sorted list;

– in all subsequent iterations of the optimization task:

  – add the next cavern from the sorted list to the set x,

  – if the set x is the same as the set of selected caverns from the previous step, skip this iteration ("no penalty" solution was already computed);

– iterations are stopped if the value of the function $f(x)$ does not increase in the next iteration. If the increase is lower than the adjustable parameter δ the iterations are also stopped.

Please note that if the penalty value is constant (the set of selected caverns form the previous simulation step is excluded from calculations), the function $f(x)$ is convex under the presented assumptions. Thus the calculation of an optimal subset of caverns can be even speeded up in case of a large caverns set by applying one of many discrete optimization methods dedicated to convex functions.

The presented algorithm calculates the optimal subset of active caverns that provides the highest flow. In case of more than one "winning" subset, the algorithm will choose the subset with the lowest number of caverns.

5 SIMULATION RESULTS

The long term (three months) simulation of the plant model was executed in order to show the method decisions and trajectories of the plant parameters. The step size was 1 hour. In the presented illustrative example, gas flow nominations were set in the following way:
- in steps 0-39, the withdrawal flow demand was 2.5 million normal m^3/h, which exceeded the maximal plant flow, therefore, the actual flow was determined by the optimization procedure;
- in steps 40-838, the withdrawal flow demand was between 0.2-0.5 million normal m^3/h. Supplying this demand was within the capabilities of the plant;
- in steps 839-1 456, gas injection was required, the nominations were 0.2-0.5 million normal m^3/h. Such an injection was within the capabilities of the plant;
- in steps 1 457-2 160, gas injection of 2.5 million normal m^3/h was required, which exceeded the maximal flow of the plant, therefore the actual flow was determined by the optimization procedure.

For the purpose of demonstrative interpretation of simulator actions, the gas pressure in the system line was kept constant. In the initial state, the amount of gas in each cavern was sufficiently large to allow the cavern pressure to be well above the p_{LL} level (*Fig. 2* and *Fig. 3*). The resulting pressure trajectories are shown in Figure 7.

In the first stage, the gas free flow is possible (withdrawal without the need to turn compressors on). The optimizer procedure chooses the set of caverns that provides the maximal flow of the plant. Note that not all available caverns are used, the flow is restricted by the AGDs operational limits (*Sect. 5*). When the differences between the pressures in the cavern pool become large enough to induce the switch of the configuration, the caverns with the lowest pressure are turned off and the caverns that were inactive are turned on. Around step 100, the cavern pressure values approach the compressor activation level. The optimizer performs more frequent switches in order to maintain the configuration that allows the free flow. After step 229 (point a in *Fig. 7*), free flow is no longer possible and the compressor has to be turned on. From step 259 (point b in *Fig. 7*), only class B caverns are used, which is dictated by high priority action for states SA_2 and SB_2 according to the "strategy book" (*Tab. 1*). After step 321 (point c in *Fig. 7*), the current configuration does not provide the demanded flow, therefore the set of class A caverns is selected (medium priority action) which provides the demanded flow. At step 421 (point d in *Fig. 7*), the flow demand has decreased from 0.45 to 0.2 million normal m^3/h, therefore the high

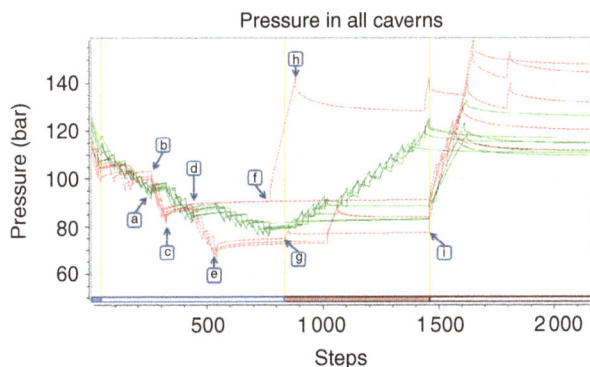

Figure 7

Pressure trajectories of class A caverns (green solid line) and class B caverns (red dashed line). The explanation of performed actions (a-h) and specifications of different gas demand nominations (rectangles above x-axis and vertical orange lines) are presented in Section 5.

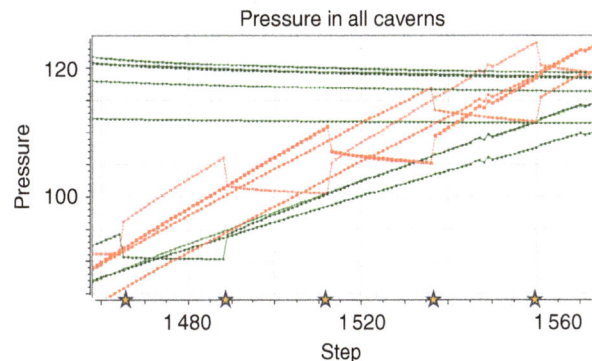

Figure 8

The magnification of pressure trajectories during the injection action. The configuration switching occurs every τ (24 hours) which is denoted by stars on x-axis.

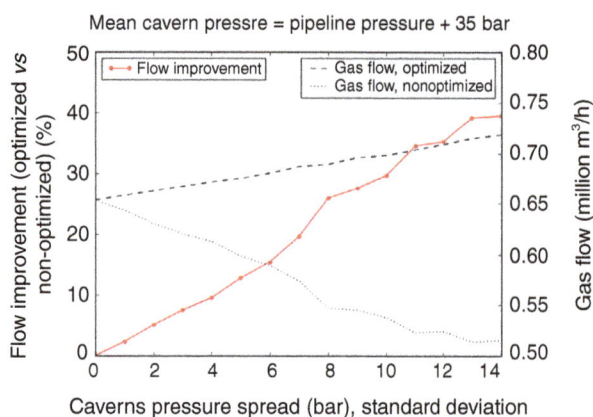

Figure 9

The gain in the maximum flow (between optimized and non-optimized configuration) depends significantly on the pressure spread in the caverns set.

priority action is chosen again as it covers 100% of the flow demand. This action is kept until step 543 (point e in *Fig. 7*), when the medium priority action is chosen again. At step 771 (point f in *Fig. 7*), one of the class B caverns starts the get out action, as is dictated by the state of the LLUOR timer and the current pressure in the cavern. The plant mode is then changed to injection in spite of the withdrawal demand. The entire flow is directed into the cavern performing the get out action. From step 839 (point g in *Fig. 7*), gas injection is required (the gas demand varies from 0.2 to 0.5 million normal m^3/h), therefore additional caverns are connected. At step 882 (point h in *Fig. 7*), the get out action is completed and the plant continues to fulfill the injection requirements. At step 1 457 (point i in *Fig. 7*), the injection demand rises above the plant's available potential, therefore the optimal configuration to achieve the maximal plant flow is used. The injection into each cavern takes place until the pressure in the cavern reaches the maximal allowable pressure value, with configuration switching occurring every τ hours, where $\tau = 24$ (*Fig. 8*). Figure 9 illustrates the improvement of the plant's flow when the optimal configuration of caverns is used instead of all available caverns connected. This improvement depends obviously on several different parameters, we present it as a function of the pressure spread in the caverns set (represented by pressure standard deviation). The mean value of the pressure values was kept constant. The flow's improvement due to the optimizer action rises as a function of the pressure spread in the set of active caverns up to when the standard deviation of the spread was about 15 bar and the

pressure values cover the entire allowable pressure window.

CONCLUSIONS

The presented method constitutes a framework which bridges models of plant elements with the operational rules. The main achievement is an integration of different computational layers into an independent fully functional tool. The presented method defines an optimal strategy (decision rules) for a plant operation. Such an optimal strategy significantly increases the gas storage potential without any change to infrastructure. The increased capacity results in more dynamical responses to changes in natural gas demand, and hence the gas availability for the end user. This method allows for computation of realistic injection and withdrawal plant flow limits that can be transformed to the storage user's nomination limits. Moreover, SOE can be used for calculating an optimal realization of aggregated business nominations for the plant. Proper simulation and identification of the plant's potential is only possible if we consider important nonlinear limits related with cavern operation and the optimal strategy (set of decision rules) which controls the plant operation. It should be pointed out that, in comparison to other gas storage plant simulators, the SOE method takes into account time-dependent recommendations, resulting from rock mechanics, that are related to gas cavern operation. These rules vary for different types of caverns. SOE monitors the state of each cavern following these recommendations. The simulator automatically detects the necessity to execute an injection that will protect a given cavern from violating geological and operational constraints, therefore SOE takes responsibility for meeting the complex nonlinear geological constraints. Optimizing the operation of depleted or partially depleted gas field reservoirs is in principle a different issue [18], however the general framework presented here might be adapted also for that case. The SOE method has been initially tested in an underground gas storage plant located in Germany where it assists the daily operation.

REFERENCES

1 Karimi-Jafari M., Bérest P., Brouard B. (2006) Some Aspects of the Transient Behaviour of Salt Caverns, *SMRI Fall Conference*, Rapid City, South Dakota, 1-4 Oct.

2 Clerc-Renaud A., Dubois D. (1980) Long-term Operation of Underground Storage in Salt, *Fifth International Symposium on Salt-Northen Ohio Geological Society*, Vleveland, 29 May-1 June.

3 Bérest P., Brouard B., Karimi-Jafaria M., Sambeek L.V. (2007) Transient behavior of salt caverns-Interpretation of mechanical integrity tests, *Int. J. Rock Mech. Min. Sci.* **44**, 5, 767-786.

4 Lux K.-H., Dresen R. (2012) Design of salt caverns for high frequency cycling of storage gas, in *Mechanical Behavior of Salt VII*, P. Bérest (ed.),Taylor & Francis Group, London.

5 Thompson M., Davison M., Rasmussen H. (2009) Natural gas storage valuation and optimization: A real options application, *Naval Res. Logistics* **56**, 3, 226-238.

6 Smirnov V.I., Khachaturyan N.S., Shafarenko E.M. (1996) Mathematical and Physical Modelling While Underground Storage Stability Assessment, *Int. J. Rock Mec. Min. Sci. Geomech. Abstracts* **35**, 4, 566-568.

7 Rokahr R.B., Hauck R., Staudtmeister K., Zander-Schiebenhöfer D. (2000) The results of the pressure build-up test in the brine filled cavern Etzel K102, *Proc. SMRI Fall Meeting*, San Antonio, pp. 89-103.

8 Bérest P., Nguyen Minh D. (1981) Stability of Cavities in Rocksalt, *International Society for Rock Mechanics International Symposium*, Tokyo, 21-24 Sept.

9 Bérest P., Brouard B. (2011) Creep closure rate of a shallow gas cavern, *SMRI Spring 2011 Technical Conference*, Galveston, Texas.

10 Bérest P., Brouard B. (2003) Safety of Salt Caverns Used for Underground Storage, *Oil Gas Sci. Technol.* **58**, 361-384.

11 Flanigan O. (1995) *Underground Gas Storage Facilities: Design and Implementation*, Gulf Professional Publishing, Houston.

12 Karimi-Jafari M., Gatelier N., Brouard B., Bérest P., Djizanne H. (2011) Multi-Cycle Gas Storage in Salt Caverns, *Solution Mining Research Institute Fall (2011) Technical Conference*, York, 3-4 Oct.

13 Lux K.H., Rokahr R.B. (1980) Some remarks on theoretical simulation models in salt cavern construction with regard to the evaluation of discharge measurements, *Proc. Int. Symp. RockStore*, Bergman M. (ed.), Pergamon, pp. 329-335.

14 Schmidt U., Staudtmeister K. (1989) Determining minimum permissible operating pressure for a gas cavern using the finite element method, *Proc. Int. Conf. on Storage of Gases in Rock Caverns*, Nilsen B., Olsen J. (eds), Balkema, pp. 103-113.

15 DeVries K.L., Nieland J.D. (1999) Feasibility study for lowering the minimum gas pressure in solution-mined caverns based on geomechanical analyses of creep-induced damage and healing, *Proc. SMRI Spring Meeting*, Las Vegas, pp. 153-182.

16 Leuger B., Staudtmeister K., Zapf D. (2012) The thermo-mechanical behavior of a gas storage cavern during high frequency loading, in *Mechanical Behavior of Salt VII*, P. Bérest (ed.),Taylor & Francis Group, London.

17 Brouard B., Karimi-Jafari M., Bérest P., Frangi A. (2006) Using Locas Software to Better Understand the Behavior of Salt Caverns, *Solution Mining Research Institute, Spring 2006 Technical Meeting*, Brussels, 30 April-3 May.

18 Azin R., Nasiri A., Jodeyri Entezari A. (2008) Underground Gas Storage in a Partially Depleted Gas Reservoir, *Oil Gas Sci. Technol.* **63**, 691-703.

3

A Step Forward to Closing the Loop between Static and Dynamic Reservoir Modeling

M. Cancelliere*, D. Viberti and F. Verga

*Politecnico di Torino, Department of Environment, Land and Infrastructure Engineering,
24, Corso Duca Degli Abruzzi, 10129 Torino - Italy
e-mail: michel.cancelliere@polito.it - dario.viberti@polito.it - francesca.verga@polito.it*

* Corresponding author

Résumé — **Un pas en avant pour boucler la boucle entre la modélisation de réservoir statique et dynamique** — La tendance actuelle pour la calibration du modèle dynamique est de trouver plusieurs modèles calibrés au lieu d'un unique ensemble de paramètres qui permettent de reproduire les données historiques. Actuellement, plusieurs des flux de travail impliquant des techniques assistées par le calage d'historique, en particulier celles basées sur les optimiseurs heuristiques ou sur la recherche directe, ont l'avantage de conduire à un certain nombre de modèles calibrés et par conséquence vont partiellement résoudre le problème de la non-unicité des solutions. L'importance de parvenir à des solutions multiples est que les modèles calibrés peuvent être utilisés pour une quantification réelle de l'incertitude qui pèse sur les prévisions de production, qui représentent la base de l'analyse des risques soit techniques soit économiques. Dans cet article, l'importance d'intégrer les incertitudes géologiques dans une étude de réservoir est démontrée. Un flux de travail est présenté comprenant l'analyse de l'incertitude associée à la distribution spatiale des faciès de milieux de sédimentation fluvial dans la calibration des modèles dynamiques simulés numériquement et, par conséquence, les prévisions de production. La première étape dans le flux de travail a été de générer un ensemble de réalisations de faciès à partir de différents modèles conceptuels. Après la phase de modélisation des faciès, les propriétés pétrophysiques ont été attribuées aux domaines de simulation. Ensuite, chaque réalisation de distribution des faciès a été calibrée séparément. Des techniques d'assimilation de données ont été utilisées pour calibrer les modèles dans un laps de temps raisonnable. Les résultats ont montré que même l'adoption d'un modèle conceptuel pour la distribution des faciès clairement représentatifs de la géométrie interne du réservoir ne peut pas garantir des résultats fiables en termes de prévisions de production. En outre, les résultats ont également montré que les réalisations qui semblent tout à fait acceptables après étalonnage n'étaient pas représentatives de la véritable configuration interne du gisement et elles ont fourni des prévisions de production erronées ; au contraire, quelques réalisations qui ne présentent pas un bon ajustement des données de production pourraient prédire de façon fiable le comportement du gisement. Ainsi, il a été confirmé que l'approche statistique est le seul moyen de réduire l'incertitude inhérente à la modélisation de gisement et devrait être adoptée comme le standard dans les études de gisement.

Abstract — A Step Forward to Closing the Loop between Static and Dynamic Reservoir Modeling — The current trend for history matching is to find multiple calibrated models instead of a single set of model parameters that match the historical data. The advantage of several current workflows involving assisted history matching techniques, particularly those based on heuristic optimizers or direct search, is that they lead to a number of calibrated models that partially address the problem of the non-uniqueness of the solutions. The importance of achieving multiple solutions is that calibrated models can be used for a true quantification of the uncertainty affecting the production forecasts, which represent the basis for technical and economic risk analysis.

In this paper, the importance of incorporating the geological uncertainties in a reservoir study is demonstrated. A workflow, which includes the analysis of the uncertainty associated with the facies distribution for a fluvial depositional environment in the calibration of the numerical dynamic models and, consequently, in the production forecast, is presented. The first step in the workflow was to generate a set of facies realizations starting from different conceptual models. After facies modeling, the petrophysical properties were assigned to the simulation domains. Then, each facies realization was calibrated separately by varying permeability and porosity fields. Data assimilation techniques were used to calibrate the models in a reasonable span of time. Results showed that even the adoption of a conceptual model for facies distribution clearly representative of the reservoir internal geometry might not guarantee reliable results in terms of production forecast. Furthermore, results also showed that realizations which seem fully acceptable after calibration were not representative of the true reservoir internal configuration and provided wrong production forecasts; conversely, realizations which did not show a good fit of the production data could reliably predict the reservoir behavior. Thus a statistical approach was confirmed to be the only way to reduce the uncertainty inherent to reservoir modeling and should be adopted as a standard in reservoir studies.

LIST OF SYMBOLS

φ	Porosity
k	Effective permeability (m^2)
k_{abs}	Absolute permeability (m^2)
k_r	Relative permeability
y_k^t	True model state vector
C_{pp}	Error covariance matrix
AGM	Adaptive Gaussian Mixture
$EnKF$	Ensemble Kalman Filter
$FOPT$	Field cumulative oil production
$FWPT$	Field cumulative water production
OF	Objective Function
$WBHP$	Well Bottom Hole Pressure (Pa)
$WOPR$	Well Oil Production Rate (m^3/day)
$WGOR$	Well Gas Oil Ratio
$WWCT$	Well water cut
h	Bandwidth parameter

INTRODUCTION

The main goal of a reservoir study is the integration of a large number of data, expressing all the available geological, geophysical, petrophysical and engineering information, into a model that, ideally, should describe with proper accuracy the structural and petrophysical properties of the reservoir (Cosentino, 2001; Benetatos and Viberti, 2010). In general, the ability of reproducing the historical dynamic behavior of a field is taken as an indicator that the model is a reliable representation of the real system so that it can be used for prediction purposes. Therefore, considerable effort is devoted to the calibration process, also known as history matching, in the belief that a calibrated model will provide reliable estimates of hydrocarbon recovery. The traditional manual history matching workflow consists in a sequential approach that begins with the matching of global or field parameters, *e.g.* field pressure, and follows with the adjustment of individual flow units or layers, to finally match the well and near-wellbore data (Mattax and Dalton, 1990). This approach has proved to be very flexible because engineers can tune the values of the reservoir parameters based on their own experience as well as good judgment; however, its disadvantage is being an extremely slow trial and error procedure. Furthermore, a good fit for the production data does not necessarily give a good estimation of the parameters of the reservoir and this might lead to errors in the prediction of the reservoir performance (Tavassoli *et al.*, 2004). Additionally, an assessment of the overall uncertainty related to the calculated reserves and the associated production profiles is often required. When a single reservoir model is available a number of simulations are performed varying the input parameters and evaluating discrepancies from a reference scenario. At this point, a great number of production profiles are simulated and the overall uncertainty can be evaluated. This task may diversify depending on the development stage of the field. For undeveloped fields, without a production history to match, the aim is to define a significant set of production profiles. Sensitivity studies, known as "risk analysis", are often carried out based on different reservoir realizations, *i.e.* distribution of static properties, to assess the uncertainty affecting the estimate of the hydrocarbon originally in place. However, when it comes to production forecasts, a reference case is selected among the reservoir realizations. Then, only one or few parameters of the dynamic model are varied at a time. The best approach is to concentrate on the parameters that are deemed to have a significant impact on the reservoir behavior; however, dependencies among the considered parameters exist and should be taken into account. Even if a production history exists, the evaluation of the uncertainty affecting the production forecast is a very difficult task. Having fixed the static and dynamic data of the reference scenario during the history matching phase, the only parameters that can be varied in the forecast

phase are the well count, location and completion and the facilities constraints. Nevertheless, because the history match is an ill-posed problem and the solution is not unique, the representativeness of uncertainty in the forecast phase is definitely not guaranteed. Several optimal solutions, or reservoir history matches, are needed to truly evaluate the uncertainty associated to a calculated production forecast and thus to the recoverable reserves (Selberg *et al.*, 2006). The possibility to obtain multiple calibrated models partially addresses the problem of the non-uniqueness of the solutions and at the same time the set of suboptimal models can provide a more reliable prediction of the reservoir performance, since the overall uncertainty is reduced by integrating geological and production data on a wide range of scenarios. To this end assisted history matching techniques can prove of great help (*Fig. 1*).

1 CLOSED LOOP SIMULATION

Historically, during a reservoir study the work performed by geoscientists in each discipline was not integrated. The results were handed over from discipline to

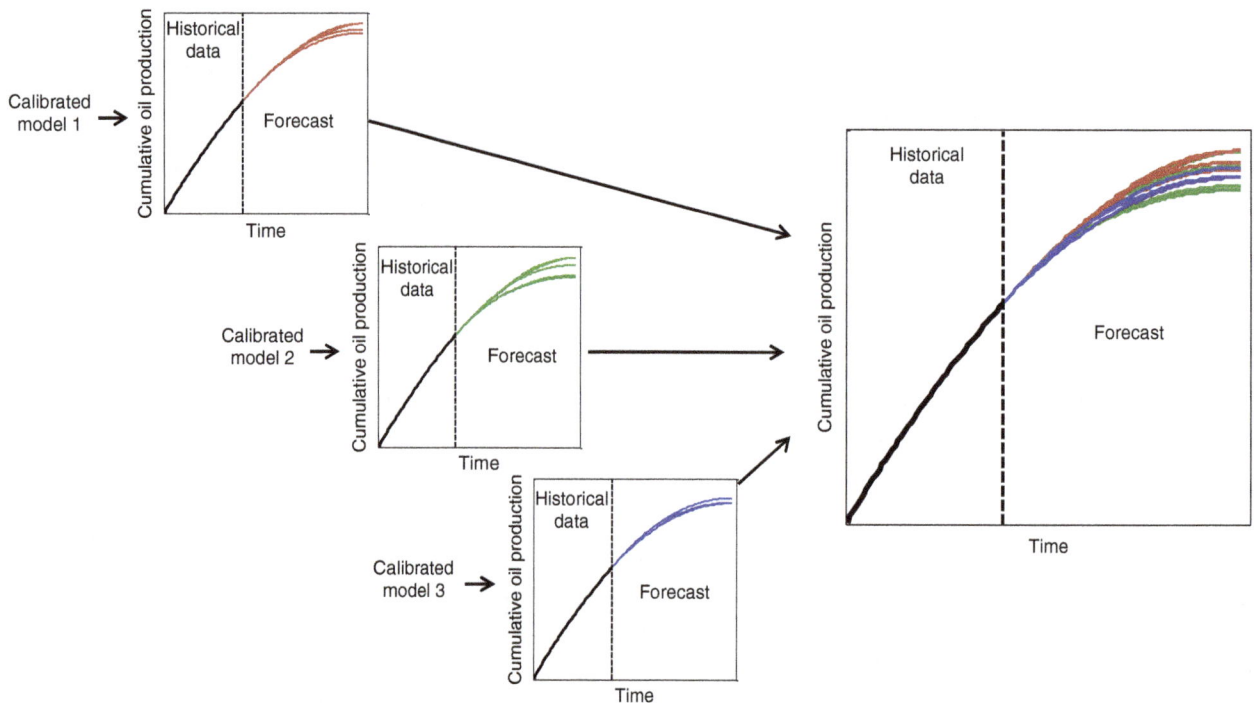

Figure 1

Uncertainty assessment using multiple calibrated models.

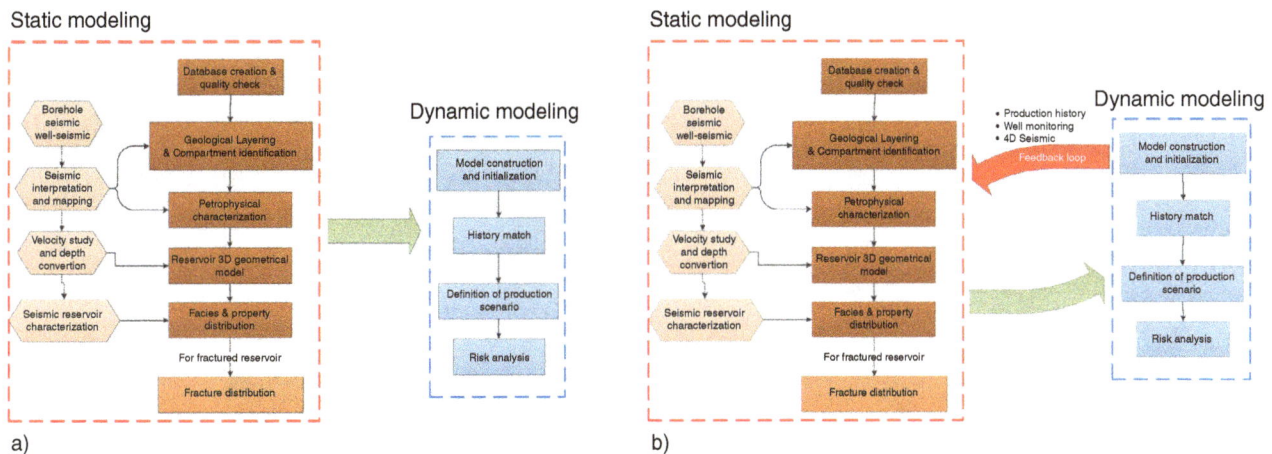

Figure 2

Traditional workflow a) *vs* closed loop workflow b) for reservoir modeling.

discipline without any active interconnection and with limited information exchange. Each discipline involved in the creation of the reservoir model had to provide data and highly accurate results in order to minimize the uncertainties during the construction of the reservoir model. This process was based on the claim that the independent accuracy of each discipline leads to an overall accuracy of the final model. However, this approach showed strong limitations in the case inconsistencies arose during the data processing, requiring a thorough and consistent re-evaluation of all the model parameters. In particular, the history matching workflow was limited to the calibration of the dynamic model by changing few parameters of a given static model according to a deterministic approach, so the workflow was unidirectional. As an example, local variations of permeability might be introduced in the model by the engineers regardless of the facies and the petrophysical properties distribution initially defined in the geological model, or variations of porosity could be adopted in order to pseudo-adjust a structural problem. However, in recent years, this trend has changed and the construction of the static model is more often conditioned by the dynamic data, closing the loop that had remained open for many years (*Fig. 2*).

This approach guaranties a closer and faster collaboration between the different disciplines involved in a reservoir study because the feedback coming from the history match can be potentially incorporated into the structural, static and dynamic model. In order to account for geological uncertainties and integrate dynamic data to geological model several algorithms of different nature have been proposed, such as: spectral decomposition (Reynolds *et al.*, 1996); pilot point

method (RamaRao *et al.*, 1995) gradual deformation method (Hu, 2000), probability perturbations methods (Caers, 2003) and spectral co-simulation perturbation (Le Ravalec-Dupin and Da Veiga, 2001). History matching is often performed on the basis of gradient-based methods or evolutionary strategies (Schulze-Riegert, *et al.*, 2001; Schulze-Riegert and Haase, 2003; Zitzler, 1999).

2 APPLICATION OF THE ENKF/AGM TO ADDRESS FACIES DISTRIBUTION UNCERTAINTY

A workflow, which includes the analysis of the uncertainty associated with the facies distribution of a fluvial depositional environment, is presented below. This type of depositional environment was selected because it is particularly representative of complex internal geometries which are found in several reservoirs worldwide. The Ensemble Kalman Filter (EnKF) modified according to Adaptive Gaussian Mixture (AGM) filter in order to loosen up the EnKF requirement of a Gaussian prior distribution of the petrophysical properties was used for data assimilation.

2.1 Model Description

A synthetic reservoir model was set up for the purpose of this research. The reservoir model contains $33 \times 53 \times 15$ (26 235) active gridblocks. The gridblocks are uniform in size with a horizontal dimension of 75×75 m^2. The reservoir is bounded by faults to the east, south, north and west. An anticline structure

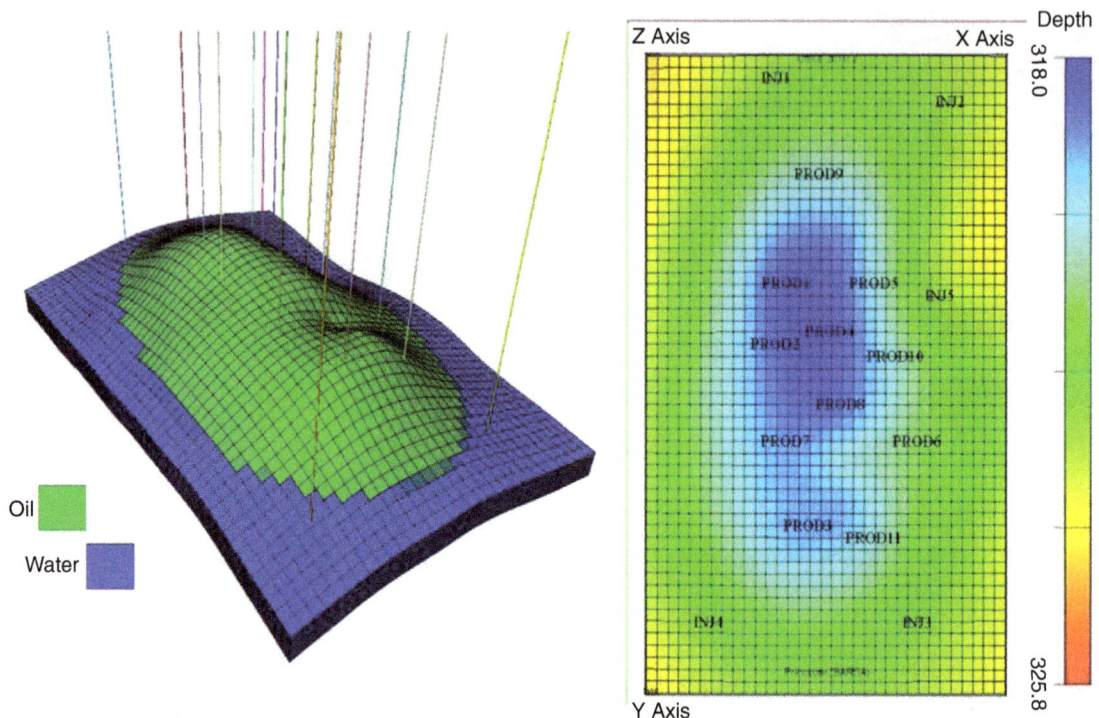

Figure 3

Water saturation and pressure field at the initial state.

completes the reservoir seal. No active aquifer is present. Undersaturated oil was considered, so no gas cap exists at the initial state. A total of 11 producing wells were drilled and most of them were located at the top of the structure. Water was injected through 5 wells for pressure support. The reservoir domain and well locations are shown in Figure 3.

Two different facies, channel sands and floodplain, typical of a fluvial depositional environment were modeled. The facies were distributed using an object modeling technique. The channel parameters were set as follows:
– a net-to-gross of 40% was set in all the domain;
– an orientation between 20 and 40 compass degrees;
– a triangular probability distribution was used to define the amplitude, wavelength, width and thickness of the channels. The values used for each parameter distribution are summarized in Table 1.

The facies modeling process was constrained to the well facies data. The facies distribution at each well is presented in Figure 4. The facies distribution in the 3D model varies in both horizontal and vertical directions, however, in the following, for simplicity's sake, only the first layer is shown, the remaining layers showing similar behavior.

TABLE 1

Object modeling parameters for the reference "true" case

Channel parameter	Min	Mean	Max
Amplitude (m)	400	500	550
Wavelength (m)	1 000	1 500	2 000
Width (m)	600	700	800
Thickness (m)	10	20	30

The petrophysical properties were distributed in the model using a sequential Gaussian simulation algorithm for each facies type. A spherical variogram and a normal probability distribution were used for porosity and log-permeability for both facies and for all layers simultaneously. The first step was to generate a porosity field for the channel and floodplain facies. The correlation length for both the major and minor directions of the variogram was set to 2 000 m for the flood plain facies and to 5 000 m for the channel sands. The mean values of porosity were set to 3% and 16% with 2% and 6% standard deviation for floodplain and channel sands, respectively (Fig. 5). Furthermore, in order to squeeze the tails to acceptable physical values, an exponential transformation was applied to both the fields.

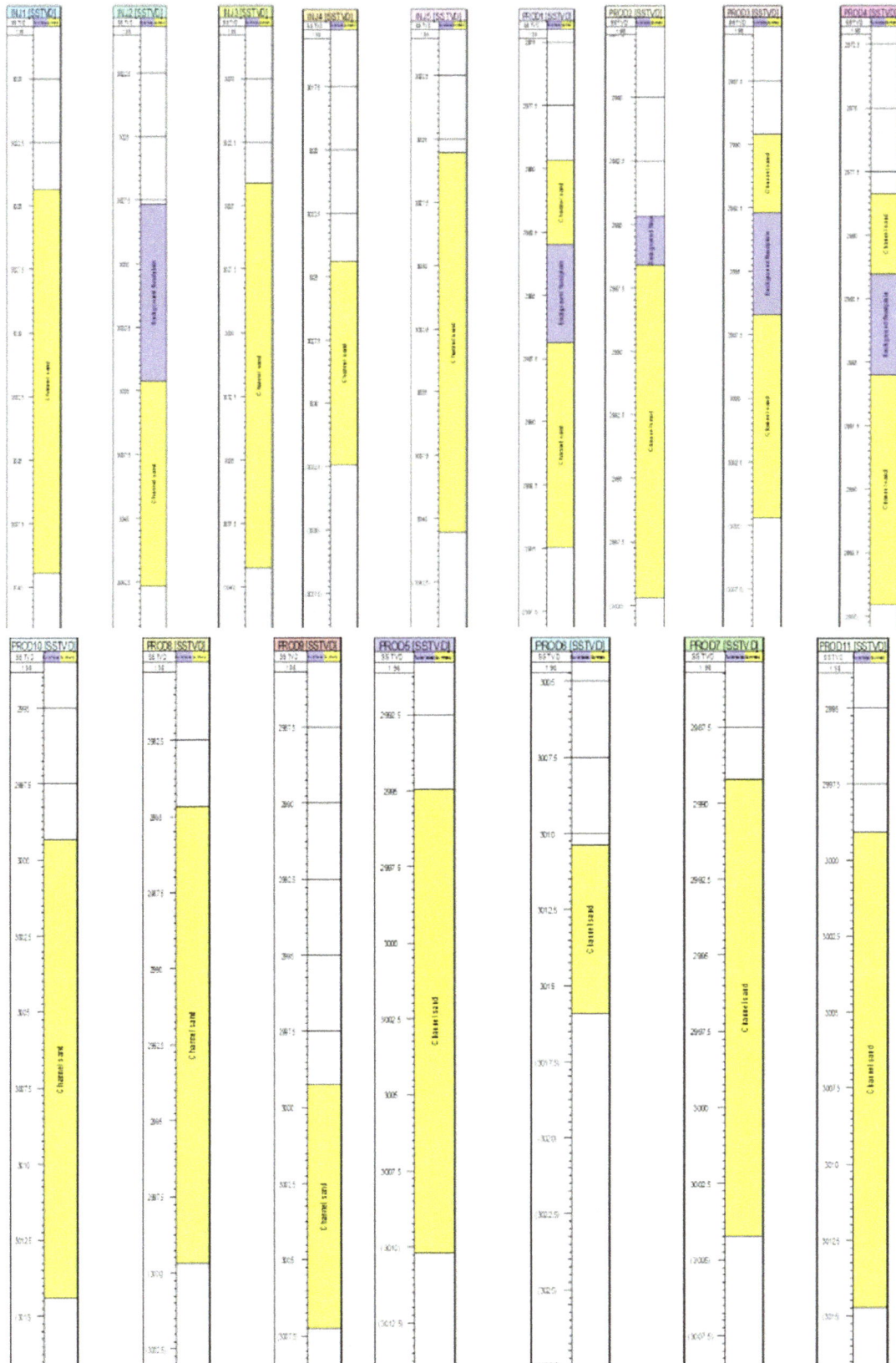

Figure 4

Facies data at the wells.

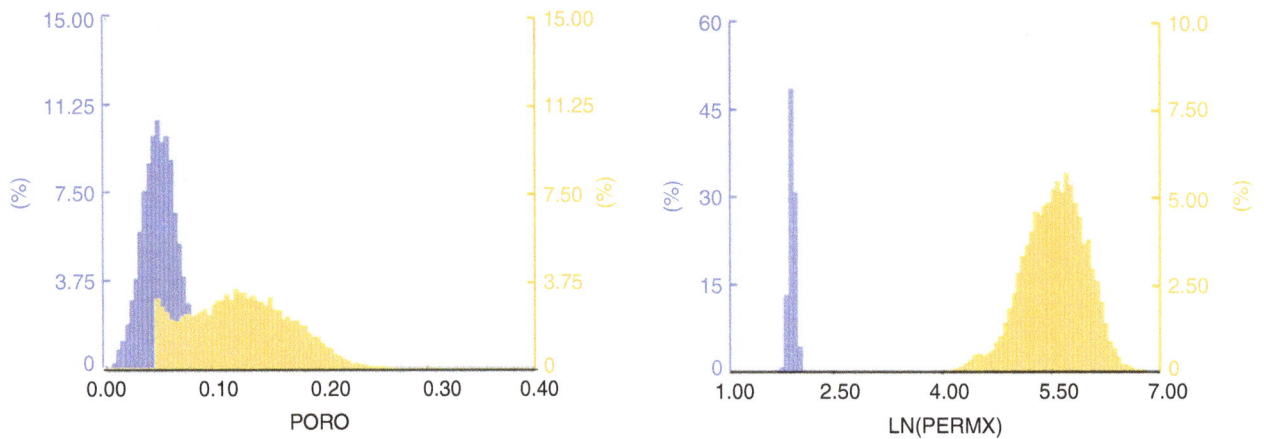

Figure 5

Porosity and permeability histograms for the channel and floodplain facies.

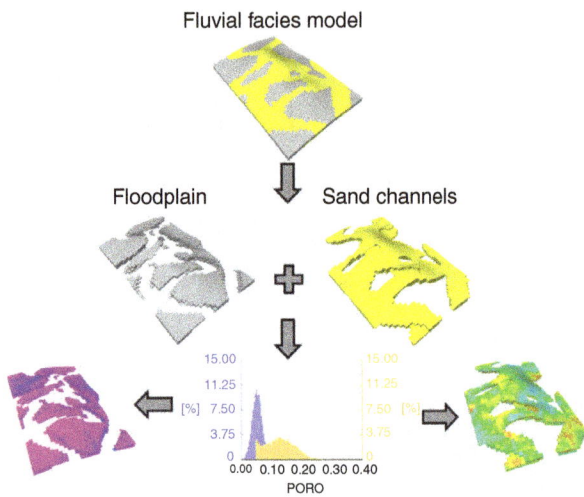

Figure 6

Facies to petrophysical properties modeling workflow.

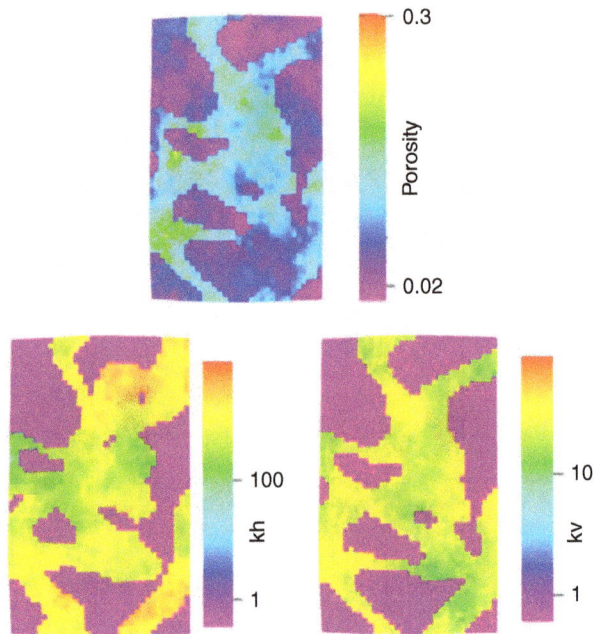

Figure 7

Porosity and permeability fields for layer 1 (true case).

The log-permeability (XY direction) field was generated using the sequential Gaussian co-simulation based upon cokriging algorithm (Deutsch, 2002) with a correlation factor to the porosity field of 0.8. The mean values of permeability were set to 7 mD and to 250 mD with a standard deviation of 0.05 and 0.4 in the logarithmic "space" for the floodplain and channel sands, respectively. The vertical permeability distribution was obtained by multiplying the horizontal permeability by the anisotropy factor, set equal to 0.1.

The facies modeling workflow and the procedure to assign the petrophysical properties are illustrated in Figure 6.

The petrophysical properties of the top layer as resulting from the facies distribution and characterization are shown in Figure 7.

The modeling workflow described above is intrinsically characterized by a high uncertainty level since the number of control points, *i.e.* well data, is too low in respect to the degree of freedom of the whole model. Thus the proposed History Matching (HM) workflow will involve not only the uncertainty related to the model petrophysical characterization, formalized as porosity and permeability distributions, but it will also involve different conceptual facies models and realizations in order to better integrate the facies/petrophysical modeling with HM through consolidated tools.

2.2 EnKF and AGM

A sequential data assimilation technique is a process which aims at estimating and predicting (analysis step) an unknown true state y_k^t of the system by integrating the forward model (system dynamics) F in time (k) using measurements, whenever available, to reinitialize the model before the integration continues. The model equation can be written as:

$$y_k^t = F(y_{k-1}^t) + q_{k-1} \qquad (1)$$

where q is the unknown model error over one time step.

The standard Kalman Filter is an optimal sequential data assimilation method for linear dynamics (*i.e.* F is a linear operator) under the hypothesis that the prior distributions of the state vector components and of the measurement errors are all Gaussian and unbiased, *i.e.* white noise is assumed (Evensen, 2009). Unfortunately, the majority of real world problems are highly nonlinear (and petroleum engineering problems belong to such category). In order to overcome the standard KF limitation, subsequent variants of the original scheme were proposed, such as the Ensemble Kalman Filter (EnKF) introduced by Evensen in 1994 (Evensen, 1994), which uses an ensemble representation of the solution states. Sets of ensemble realizations are generated using Monte–Carlo sampling for the initial state, model noise and measurement noise. Ensemble members are then forwarded in time by solving the nonlinear state equations and are analyzed by an approximate Kalman filter scheme. The ensemble covariance is used as an approximation of the true covariance, thus avoiding explicit evolution and storage of the covariance as well. The proposed method can therefore be used with realistic highly nonlinear models on large domains and it is also well suited for distributed computing environments. When applied to the history matching problem the ensemble Kalman filter is able to estimate a large number of model variables and to assimilate a wide range of data types and can be easily integrated with existing reservoir simulators.

In order to loosen up the EnKF requirement of a Gaussian prior distribution the Adaptive Gaussian Mixture filter (AGM) was introduced by Stordal *et al.* in 2011 (Stordal *et al.*, 2011, 2012; Valenstrand *et al.*, 2012). Gaussian mixture filters are based on the assumption that at each time-step k the p density $f(y_k|d_{1:k-1})$ can be approximated by a Gaussian kernel density estimator:

$$f(y_k|d_{1:k-1}) = \sum_{i=1}^{N_e} \omega_{k-1}^i \Phi(y_k - y_k^{f,i}, h^2 C_{yy,k}^{e,f}) \qquad (2)$$

where $\Phi(x - \mu, \Sigma)$ is a multivariate normal distribution with mean μ and covariance matrix Σ. $\{\omega_{k-1}^i\}_{i=1}^{N_e}$ are scalar weights which satisfy $\sum_{i=1}^{N_e} \omega_{k-1}^i = 1$. In the construction of the N_e Gaussian kernels a bandwidth parameter $h \in (0,1]$ is introduced. The choice of the bandwidth parameter h determines the magnitude of the Kalman filter update step. This is treated as a design parameter which can be beneficial to reduce the risk of filter divergence.

In the proposed scheme, Stordal assigns the same covariance matrix to each kernel. In this way, each ensemble member is updated linearly, proportionally to the EnKF update. This filter suffers from a degeneracy phenomenon and the most efficient way to get around it is resampling, *i.e.* drawing new particles according to the distribution of the ensemble and then reassigning them the uniform weights. In order to reduce weight collapse and consequent resampling, which is time consuming, an adaptive parameter α can be introduced, extending the scheme introduced in by Hoteit *et al.* (2008).

In this framework, the EnKf can be viewed as a mixture filter, obtained by using Gaussian kernels and forcing the weights to be uniform.

2.3 Methodology

The main goal of the case study discussed below was to include the uncertainty associated to the facies distribution, both in terms of conceptual models parameters and petrophysical characterization, in the calibration of the numerical model and, consequently, in the production forecast. The proposed workflow is presented in Figure 8.

The first step in the workflow was to generate a set of facies realizations starting from different conceptual models. Here a conceptual model is intended as a given set of geometrical parameters that are used to stochastically generate different facies realizations (*Fig. 9*). For example, straight channels oriented in the north-south

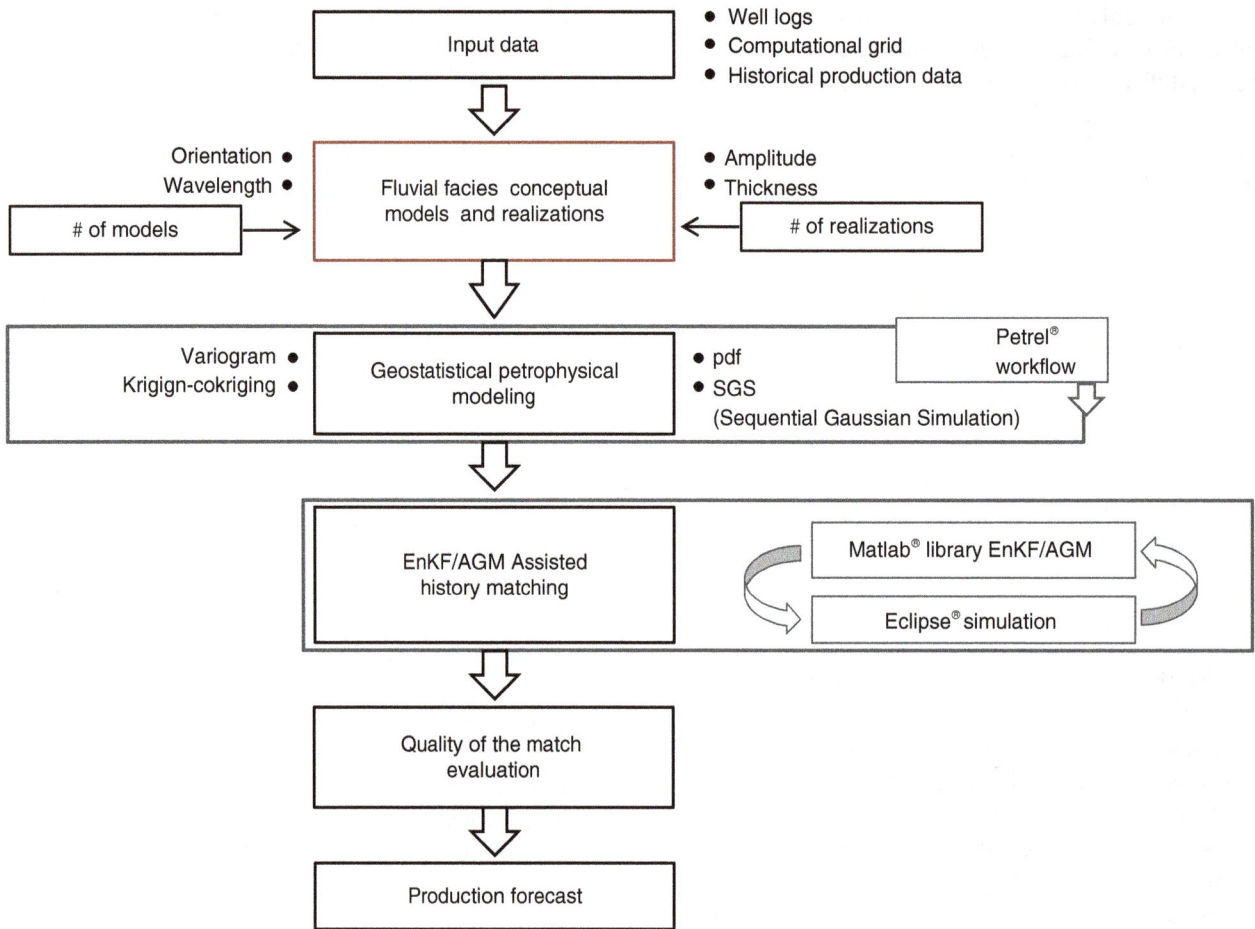

Figure 8

Assisted history matching workflow.

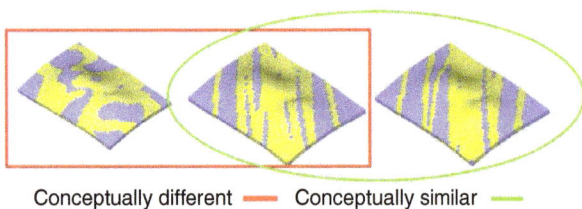

Figure 9

Conceptually different *vs* conceptually similar facies models.

direction are conceptually different from meandering channels oriented in the east-west direction.

After facies modeling the petrophysical properties were assigned to the simulation domain. The number of petrophysical realizations needed for each facies realization depends on the dimensions of the model and on the number of historical parameters to be matched. In the analyzed cases, porosity and log-permeability values have to be assessed for each cell of the domain. In general, no more than 200 ensemble members are required for a medium complexity problem in which the number of parameters – to be assimilated at each assimilation step – does not exceed 100. In this case, 60 ensemble members were needed. Then, each facies realization was calibrated separately with the aid of the AGM algorithm. The optimal value of the bandwidth parameter h is known to be problem dependent and the optimization requires a trial-and-error procedure. In this work, the bandwidth parameter h was set to 0.5, from a preprocessing analysis. After completion of all the assimilation steps, simulations were run again the final porosity and permeability fields. The quality of the match for each facies realization was computed using the following misfit function:

$$OF = \frac{1}{N_{obs}} \frac{1}{N_w} \sum_{obs=1}^{N_{obs}} \sum_{w=1}^{N_w} OF_{obs}^w \quad (3)$$

where the component OF_{obs}^w represents the contribution to the total objective function of the observation obs obtained at the well w:

$$OF_{obs}^w = \frac{1}{N_t} \frac{1}{\sigma_{obs,w}^2} \sum_{k=1}^{N_t} \left[\left(\bar{d}^k \right)_{obs}^w - \left(d_{true}^k \right)_{obs}^w \right]^2 \quad (4)$$

the term $\left(\bar{d}^k \right)_{obs}^w$ is the ensemble mean of the observation obs obtained at the well w at the time k, i.e.:

$$\left(\bar{d}^k \right)_{obs}^w = \frac{1}{N_e} \sum_{j=1}^{N_e} \left(d_j^k \right)_{obs}^w \quad (5)$$

and the coefficient $\sigma_{obs,w}^2$ is the variance of each observation at each well during the whole history matching:

$$\sigma_{obs,w}^2 = \frac{1}{N_t - 1} \sum_{k=1}^{N_t} \left[\left(d_{true}^k \right)_{obs}^w - \frac{\sum_{i=1}^{N_t} \left(d_{true}^i \right)_{obs}^w}{N_t} \right]^2 \quad (6)$$

In this way, the different scale sizes of the observations were nearly normalized.

Eventually, the posterior distribution of the field cumulative oil and water production were compared against the prior distributions.

2.4 Reservoir State Vector

The state vector contains the variables of interest. It describes the state of the dynamic system and represents its degrees of freedom. The state vector y_j^k mainly consists of three data types: static reservoir properties, dynamic reservoir properties and measured values:

$$y_j^k = \begin{bmatrix} v \\ g \\ d \end{bmatrix}_{j,k} \quad (7)$$

where j refers to the jth ensemble member and k is the time step index for time t_k.

In this case study, the static properties were represented by porosity and logarithmic transformation of the horizontal and vertical permeabilities for each gridblock, that is:

$$g_j^k = \left[\phi_1 \ldots \phi_N, \ln (k_x)_1 \ \ldots \ \ln (k_x)_N, \ln (k_z)_1 \ldots \right.$$
$$\left. \ln (k_z)_N \right]_{j,k}^T \quad (8)$$

The dynamic variables v_j^k were the pressure and water saturation values for each gridblock:

$$v_j^k = \left[p_{w1} \ldots p_{wN}, S_{w1} \ldots S_{wN}, S_{g1} \ \ldots \ S_{gN} \right]_{j,k}^T \quad (9)$$

The number of elements of the state vector corresponding to the observed data d_j^k can vary depending on the measurement availability at a given assimilation time step. Bottom hole pressures were observed at both producing and injector wells, while block pressure, water-cut and oil production rates were recorded at the production wells only:

$$d_j^k = [WBHP_1, \ldots, WBHP_{Nw}, WBP_1, \ldots, WBP_{Nw},$$
$$WWCT_1, \ldots, WWCT_{Npw}, WOPR_1, \ldots, WOPR_{Npw}]_{j,k}^T \quad (10)$$

The maximum length of the state vector is:

$$N_s = 5N + N_d = 131\,224 \text{ elements} \quad (11)$$

2.5 Initial Ensemble Models

Initial ensemble models were generated for the three different conceptual facies models. The geometric parameters used for the generation of the facies model are summarized in Table 2, 3 and 4, respectively. The first conceptual model (cases A) represents straight channels with orientation NE-SW, the second model (cases B) represents meandering channels with direction NE-SW and the third model (cases C) represents straight channels oriented in the N-S direction (Fig. 10).

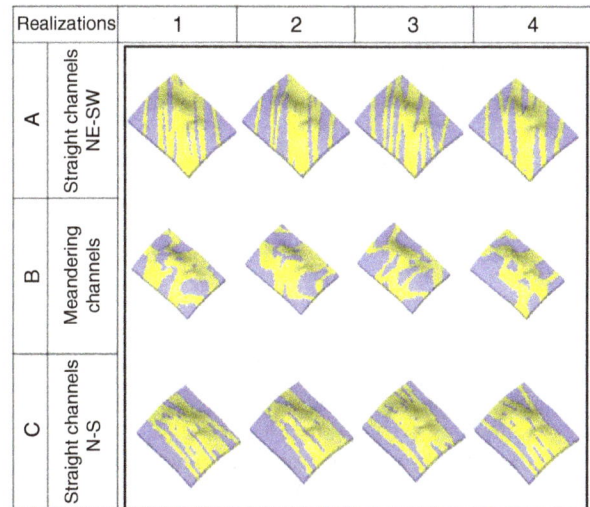

Figure 10

Set of facies realizations based on the different conceptual models.

TABLE 2

Object modeling properties. Case A

Channel parameter	Min	Mean	Max
Amplitude	0	0	0
Wavelength	1 000	1 500	2 000
Width	150	250	350
Thickness	20	20	20
Orientation	30	Uniform	50

TABLE 3

Object modeling properties. Case B

Channel parameter	Min	Mean	Max
Amplitude	400	500	550
Wavelength	1 000	1 500	2 000
Width	600	700	800
Thickness	10	20	30
Orientation	10	Uniform	40

TABLE 4

Object modeling properties. Case C

Channel parameter	Min	Mean	Max
Amplitude	20	20	20
Wavelength	1 000	1 500	2 000
Width	150	250	350
Thickness	20	20	20
Orientation	−10	Uniform	10

A total of 100 petrophysical realizations were created for each facies realization. The variogram parameters and probability density functions were equal to those used for the generation of the true case. The variogram parameters and the values of the probability density functions are summarized in Table 5 and Table 6, respectively.

Examples of petrophysical realizations from different facies models are shown in Figure 11.

TABLE 5

Variogram parameters

	Channel sand	Floodplain
Type of variogram	Spherical	Spherical
Corr. length (min direction)	2 000 m	5 000 m
Corr. length (major direction)	2 000 m	5 000 m
Corr. length (vertical)	10 m	10 m

TABLE 6

Facies petrophysical properties

	Channel sand	Floodplain
Type of distribution	Normal	Normal
Mean (porosity)	20%	5%
Standard deviation (porosity)	6%	2%
Mean ($k_{horizontal}$)	250 mD	7 mD
Standard deviation ($k_{horizontal}$)	0.4	0.05

2.6 Assimilation of Production Data

Two different scenarios were considered; in each scenario different geostatistical realizations (although generated with the same input parameters) and different production histories were used. The scope was that of evaluating the efficiency of the workflow under two different conditions. Scenario 1 represents a conservative situation in which the true value of the field cumulative oil production at the end of the forecast phase is very close to the medium value of the prior distribution. Conversely, in scenario 2 the prior field cumulative oil production was biased relative to the true case.

A total of twelve years of historical production data were considered available. Four of them were used for history matching; the remaining eight were used to check the accuracy of the forecast phase.

Four different types of observations were considered during the assimilation step: Well Bottom-Hole Pressure (WBHP), Well Block Pressure (WBP), well water cut (WWCT) and Well Oil Production Rate (WOPR). The well block pressure was used as an approximation of the well pressure measurements under static conditions, when the wells are shut-in. Although the wells were supposed to be produced

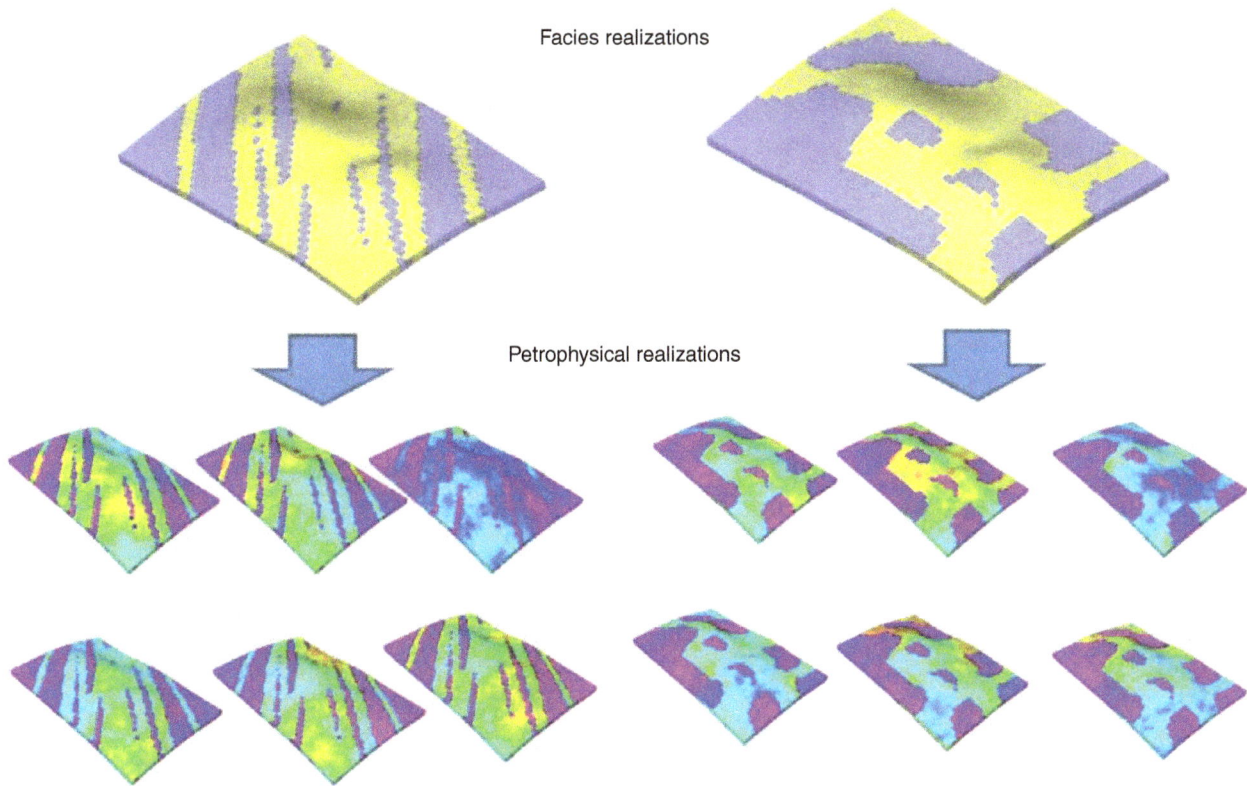

Figure 11

Examples of petrophysical realizations from different facies models.

at a constant rate, the actual WOPR values varied because in some simulation models wells were unable to respect the target value.

The data used for assimilation were 318 WBHP, 66 WBP (pseudo static pressure), 264 WCT and 264 WOPR values. All the data was subject to noise and the noise was assumed to be white. The imposed standard deviations are: 3 bar for pressure; 1% for water-cut; and 0.0001 sm^3/d for oil production rate.

During the history matching phase producers are set to honor the historical oil production rates, with a minimum operational BHP. For the forecast phase a constant field liquid rate constraint of 1150 m^3/day was imposed to the producers. The injector is set to replace the volume of fluid produced. The well production rate was reduced by a factor of 0.9 whenever the water cut exceed 0.9.

2.7 Facies Probability

The porosity mean was converted to a probability map that described the channel occurrence probability in each gridblock of the model. The transformation from mean porosity values to channel probability was expressed as a piecewise function (*Fig. 12*).

Defining ϕ_{ch} and ϕ_{fp}, the mean porosity values of channels and floodplain, respectively, and σ_{ch} and σ_{fp} their corresponding standard deviations, the transformation function was defined as:

$$
\begin{cases}
P_{ch} = 0 & \text{if } \phi \leq \phi_{fp} + \sigma_{fp} \\
P_{ch} = \frac{\phi - \phi_{fp} - \sigma_{fp}}{(\phi_{ch} - \sigma_{ch}) - (\phi_{fp} + \sigma_{fp})} & \text{if } \phi_{fp} + \sigma_{fp} < \phi < \phi_{ch} - \sigma_{ch} \\
P_{ch} = 1 & \text{if } \phi \geq \phi_{ch} - \sigma_{ch}
\end{cases}
\tag{12}
$$

Such map is needed to construct the channel probability map which helps assessing the presence of channels through its key parameters such as the mean value and standard deviation. An example will be shown in Figure 27 in Section 2.10.

2.8 Results: Scenario 1. History Matching Objective Function Evaluation

The mean values, the standard deviation and the ranking of the objective function of each facies realization are shown in the boxplot and table of Figure 13. The three first realizations in the OF ranking, *i.e.* sorted by increasing OF mean values, refer to three different conceptual facies models (C1, A3, B3). Therefore, acceptable matched models can be obtained independently from the facies conceptual model. Model A1 has the worst OF mean value and also a very wide distribution of the OF values from the different ensemble members. Conversely, models A3 and C1 show fairly good OF values for almost all their ensemble members.

The heat map relative to the assimilation phase for all the facies realizations is shown in Figure 14. A logarithmic color scale was used to qualitative distinguish different OF values. Results show that the imposed Well Oil Production Rates (WOPR) were respected fairly well in all the cases, which means that the

production wells did not reach the minimum bottom hole pressure. The quality of the water cut match varied depending on the well and facies realization. Production wells PROD1, PROD2, PROD7 and PROD8 showed a good match for almost all the cases. The C3 and B3 facies realizations showed a better overall

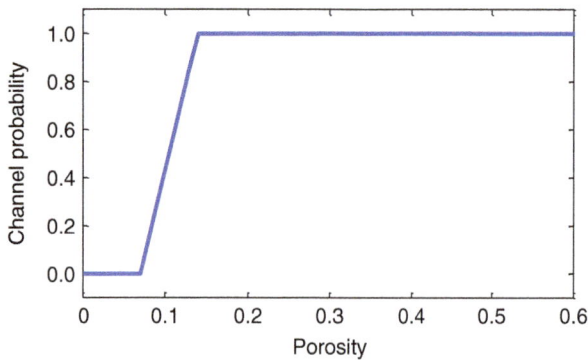

Figure 14

OF heat map of the assimilation phase.

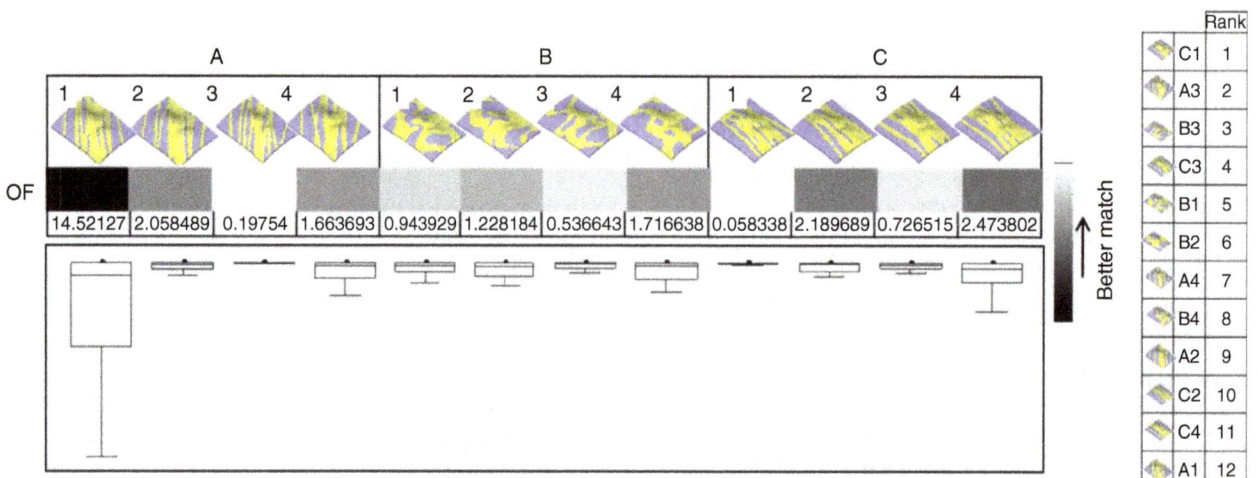

Figure 12

Channel probability *vs* porosity.

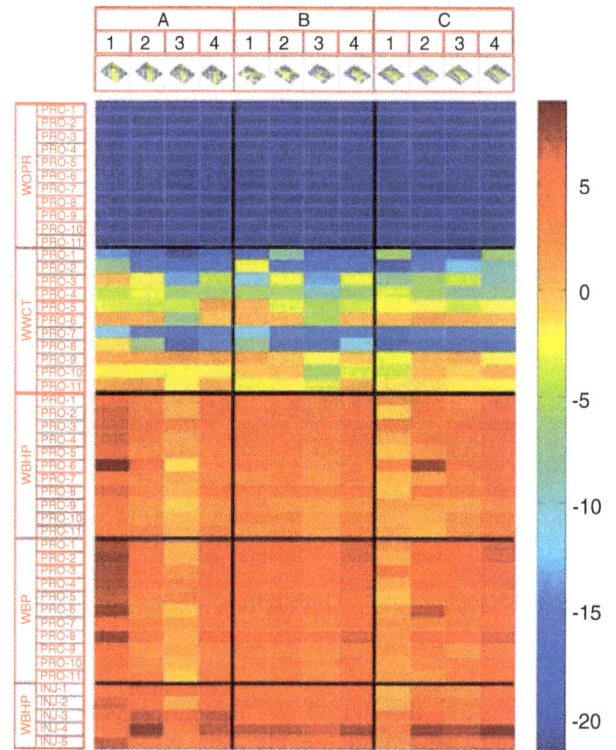

Figure 13

Ranking of the objective functions for each facies realization.

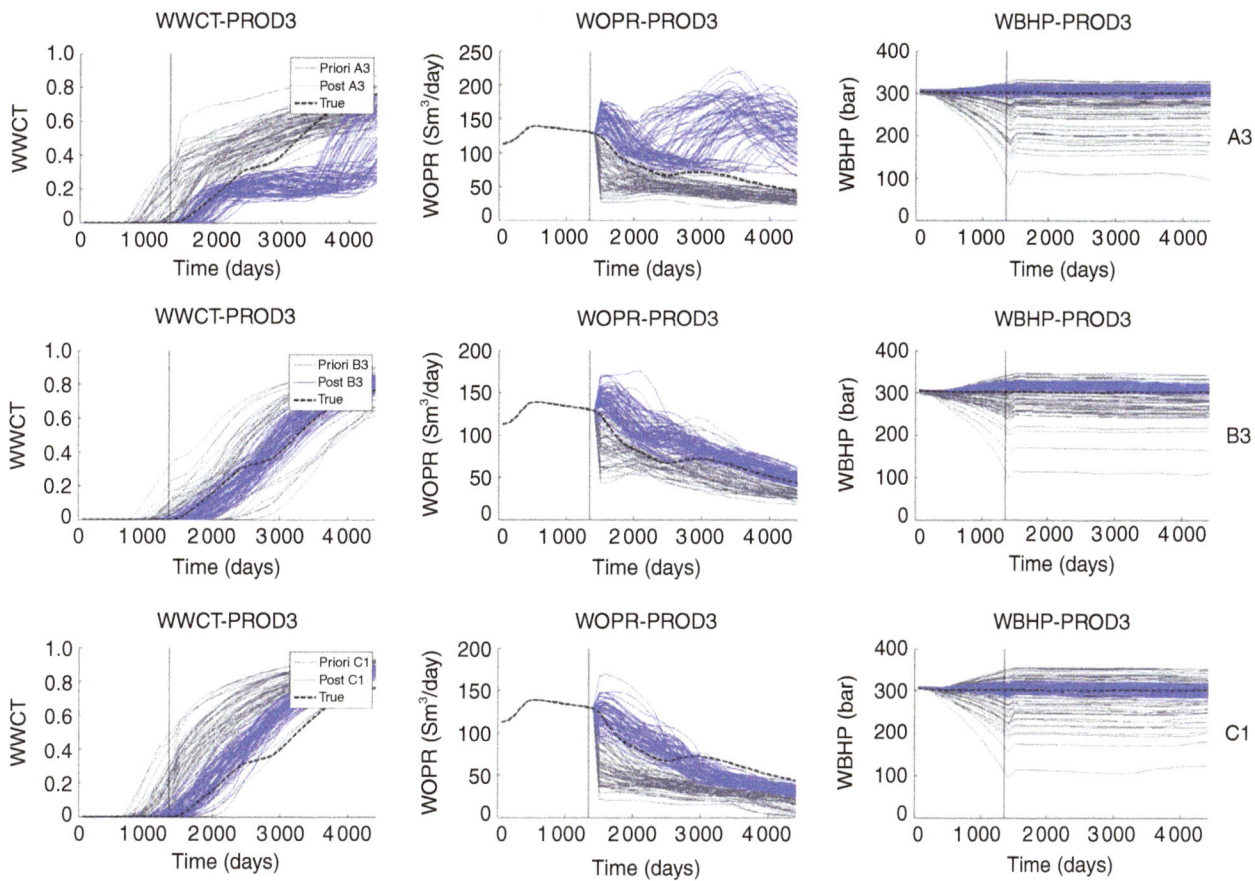

Figure 15

Prior and posterior observations *vs* time for well PROD3 for cases A3, B3 and C1.

match of the water cut than the other facies realizations. In general, the bottom hole pressure and well block pressure showed the worst OF values. In the column orientation, facies realization A3 and C1 showed a much more accurate response in terms of pressure than the other realizations.

2.9 Forecast Uncertainty

Results in terms of observed data (WOPR, WBHP, and WWCT) of the initial and calibrated models for well PROD3 are shown in Figure 15. The vertical line separates the assimilation and forecast periods. The true curve for each observed quantity is plotted as a reference. The gray and blue lines are the data obtained from the initial (prior) and final (posterior) ensembles, respectively. Case B3 provides a better match for WOPR and WWCT at well PROD3. Conversely, case B3 shows a less accurate reproduction

of the reservoir pressure if compared to cases A3 and B3. The calibrated models honor fairly well the WWCT during the assimilation period leading to a sensible reduction in the uncertainty compared to the initial models in the forecast phase. Similar behavior is observed for the WBHP.

The heat map associated to the forecast phase for all the facies realizations is shown in Figure 16. During the forecast phase, the total liquid rate was imposed, which means that the oil rate varies as a function of the water cut. Consequently, the values of the OF for the WOPR are well correlated to the OF values of WWCT and are much higher than in the assimilation phase. The realizations A3, B3, C1 and C3 show better pressure matches than the rest of the realizations.

The values of the OF, based on the cumulative oil production at the end of the forecast phase, for a progressively larger number of facies realizations are shown in

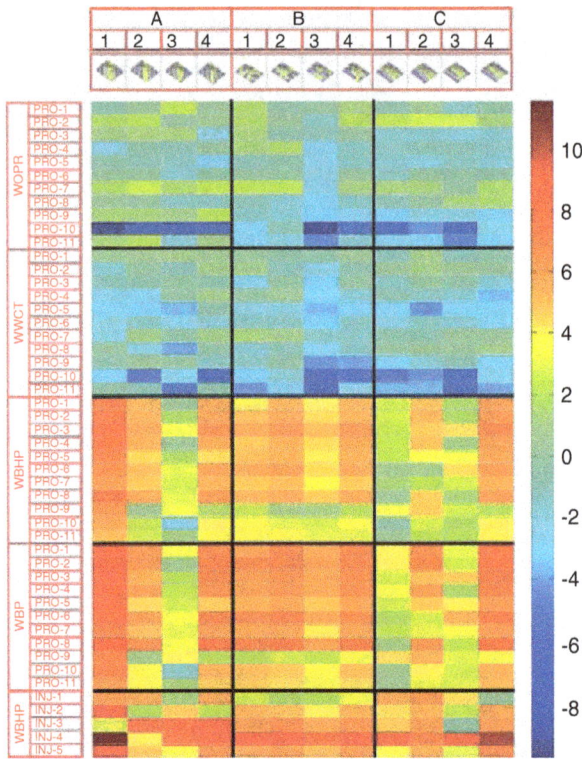

Figure 16

OF heat map of the forecast phase.

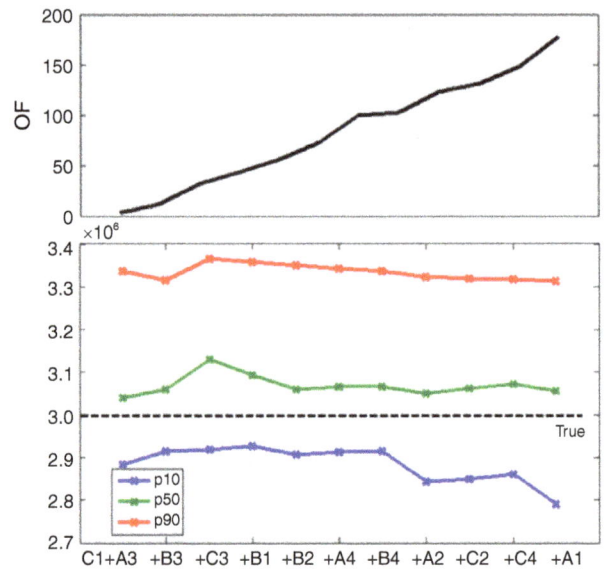

Figure 17

Cumulative oil production at the end of the forecast phase grouped OF ascending.

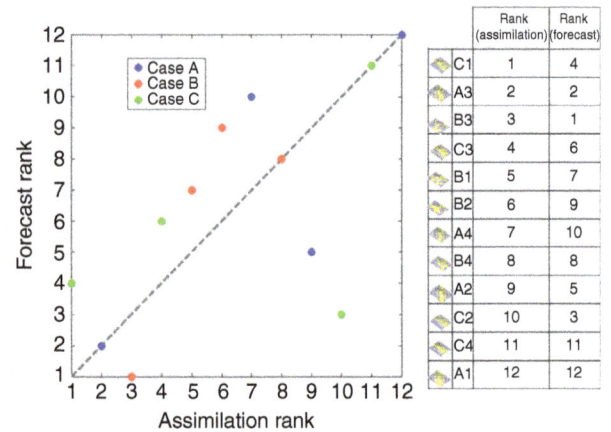

Figure 18

Forecast *vs* assimilation OF ranking.

Figure 17. The corresponding evolutions of the P90, P50 (or median) and P10 are indicated by a red, green and blue line, respectively. It is worth noting that the mean and P90 values remain almost constant independently of the considered realizations. In contrast, an overestimation of the WWCT values in the cases with high OF values leads to lower field cumulative oil values and, consequently, to lower P10 values.

The correlation between the OF rank of the facies realizations in the forecast phase and in the assimilation phase is shown in Figure 18. The facies realizations that lie on the diagonal line $x = y$ represent the realizations that did not change their ranking from one phase to the following. The realizations with a better rank in the forecast phase than in the assimilation phase are located in the upper part of the plot ($y > x$). Conversely, the realizations that undergo a ranking decrease in the forecast phase relative to the assimilation phase are located in the lower portion of the plot ($y < x$). This plot clearly shows the non-uniqueness of the solution because more than one reservoir configuration can provide a satisfactory reservoir response if compared with the observed behavior. Furthermore and even more importantly,

realizations which seem fully acceptable after calibration are not necessarily representative of the true reservoir internal configuration and provide wrong production forecasts; on the other hand, realizations which have a relatively low ranking after history matching can reliably predict the reservoir behavior.

The field cumulative oil production (FOPT) distribution obtained from the prior and posterior ensembles for each conceptual facies model is shown in Figure 19. The colored regions identify the [P10, P90] intervals and the dotted colored line indicates the mode (P50).

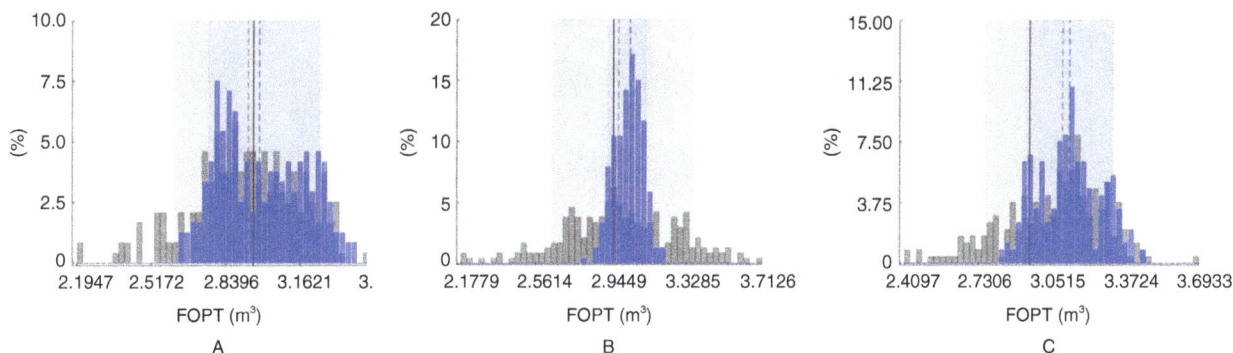

Figure 19

Prior (gray) *vs* posterior (blue) cumulative oil production distribution at the end of the forecast phase for each conceptual models group, A, B and C, respectively. Black solid line is the true value and dotted lines the mean values of prior and posterior distributions. Outlined regions represent, instead, the amplitude interval between the P10 to P90 of the corresponding distributions.

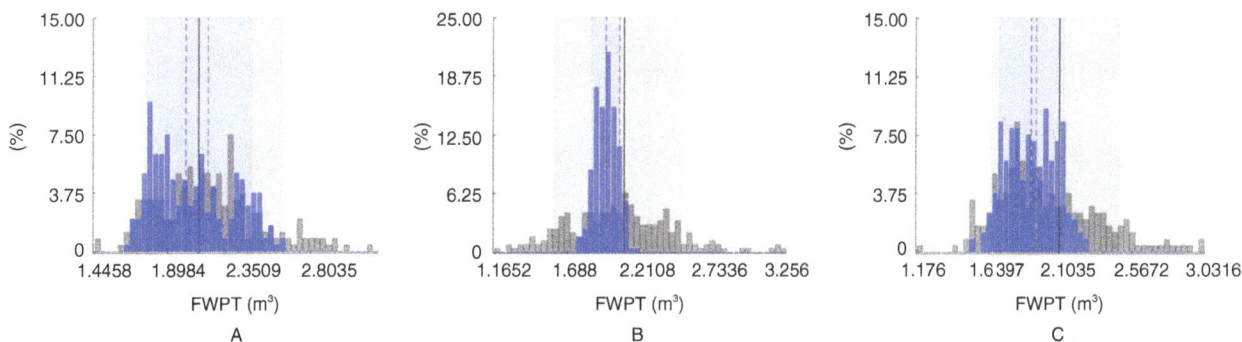

Figure 20

Prior (gray) *vs* posterior (blue) cumulative water production distribution at the end of the forecast phase for each conceptual models group A, B and C, respectively. Black solid line is the true value and dotted lines the mean values of prior and posterior distributions. Outlined regions represent, instead, the amplitude interval between the P10 to P90 of the corresponding distributions.

The continuous line represents the true value. The uncertainty of the cumulative oil production is efficiently reduced in all the cases, but it is worth noting that the facies models of type B show the best performance in reducing such uncertainty. In all models, the FOPT mean is very close to the true value for both the prior and posterior distributions.

The distribution of the FOPT is vital information for the evaluation of a development plan for a given reservoir. However, another important aspect is the field cumulative water production (FWPT), mainly because it has a great impact on the required surface equipment and facilities and, as a consequence, on the economic feasibility of a project. The FWPT distribution obtained from the prior and posterior ensembles for each conceptual facies model is shown in Figure 20. Similarly to the FOPT, the FWPT uncertainty shows an evident

reduction. The conceptual model B shows the larger reduction relative to the prior distribution. Conversely, model A is less effective to reduce the uncertainty.

2.10 Facies and Petrophysical Properties

The evolutions of the mean and standard deviation of the ensemble permeability for cases A3, B3 and C1 are shown in Figures 21, 22 and 23, respectively. The petrophysical properties generation was not constrained at the well locations, thus the corresponding standard deviation is greater than zero. The petrophysical distribution of model B3 seems to reproduce in a more realistic way the petrophysical properties of the true case.

Similarly, the evolutions of the mean and standard deviation of the ensemble porosity for cases A3, B3 and C1 are shown in Figure 24, 25 and 26, respectively.

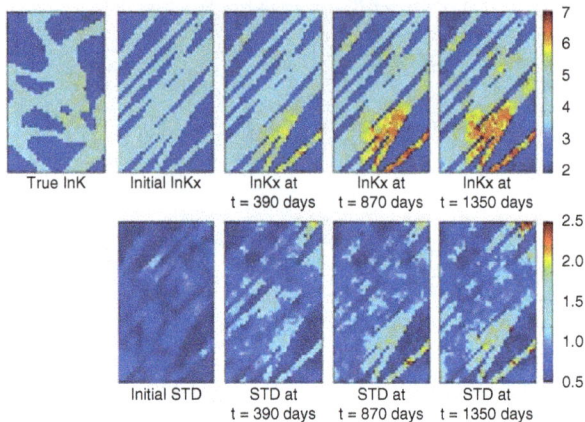

Figure 21

Comparison between the true ln(PERMx) field for layer 1 and ensemble mean in case A3 at different assimilation time (initial, at day 390, at day 870, at day 1 350).

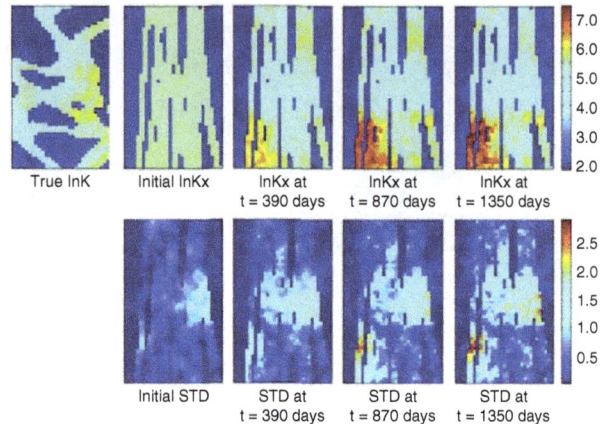

Figure 23

Comparison between the true ln(PERMx) field for layer 1 and ensemble mean in case C1 at different assimilation time (initial, at day 390, at day 870, at day 1 350).

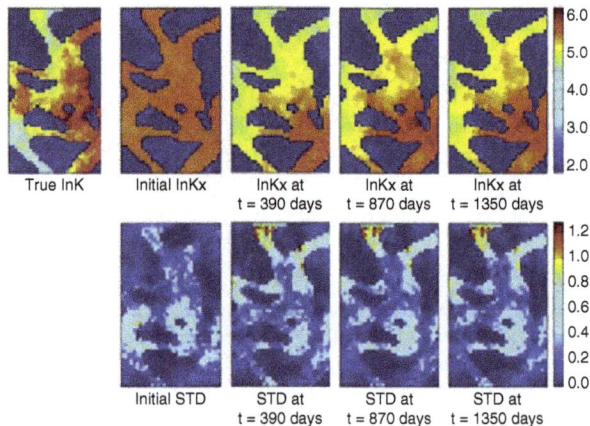

Figure 22

Comparison between the true ln(PERMx) field for layer 1 the ensemble mean in case B3 at different assimilation time (initial, at day 390, at day 870, at day 1 350).

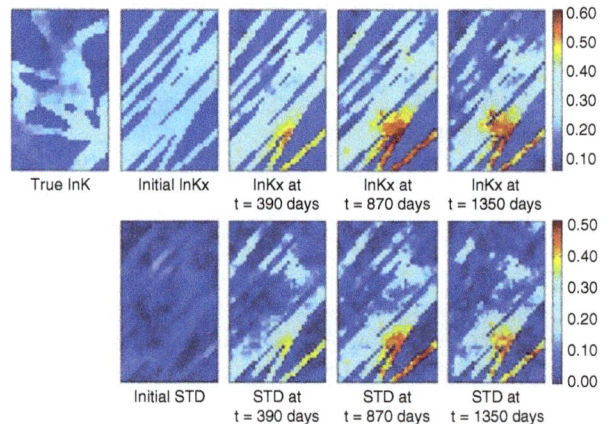

Figure 24

Comparison between the true porosity field for layer 1 and ensemble mean in case A3 at different assimilation time (initial, at day 390, at day 870, at day 1 350).

Case A3 shows unrealistic high values of porosity in the southern channel, which is a clear indication that a wider channel exists, *i.e.* more volume is needed in that zone, as confirmed by comparing it to the true case. Case B3 shows a fairly good match of the porosity field and lower values of standard deviation. Localized spots of higher values of standard deviation exist in zones with large porosities, indicating a higher uncertainty in these zones. Case B3, indeed, is characterized by the same facies model of the true case. As expected, the assimilation process results in the most satisfying petrophysical characterization even though the reference porosity and log-permeability distribution were not included in the ensemble (*Fig. 22, 25*).

The mean and standard deviations of the channel probability field are shown in Figure 27. Red zones in mean probability map represent high probabilities of channel occurrence and blue zones low probability of channel occurrence. This was obtained by applying Equation (12). The map seems to capture fairly well

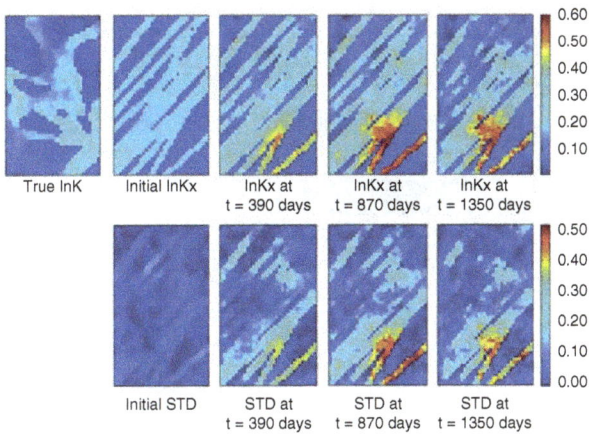

Figure 25

Comparison between the porosity field for layer 1 and ensemble mean in case **B3** at different assimilation time (initial, at day 390, at day 870, at day 1 350).

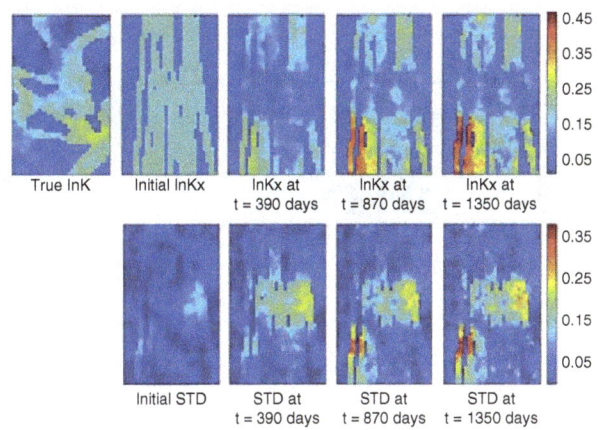

Figure 26

Comparison between the true porosity field for layer 1 and ensemble mean in case **C1** at different assimilation time (initial, at day 390, at day 870, at day 1 350).

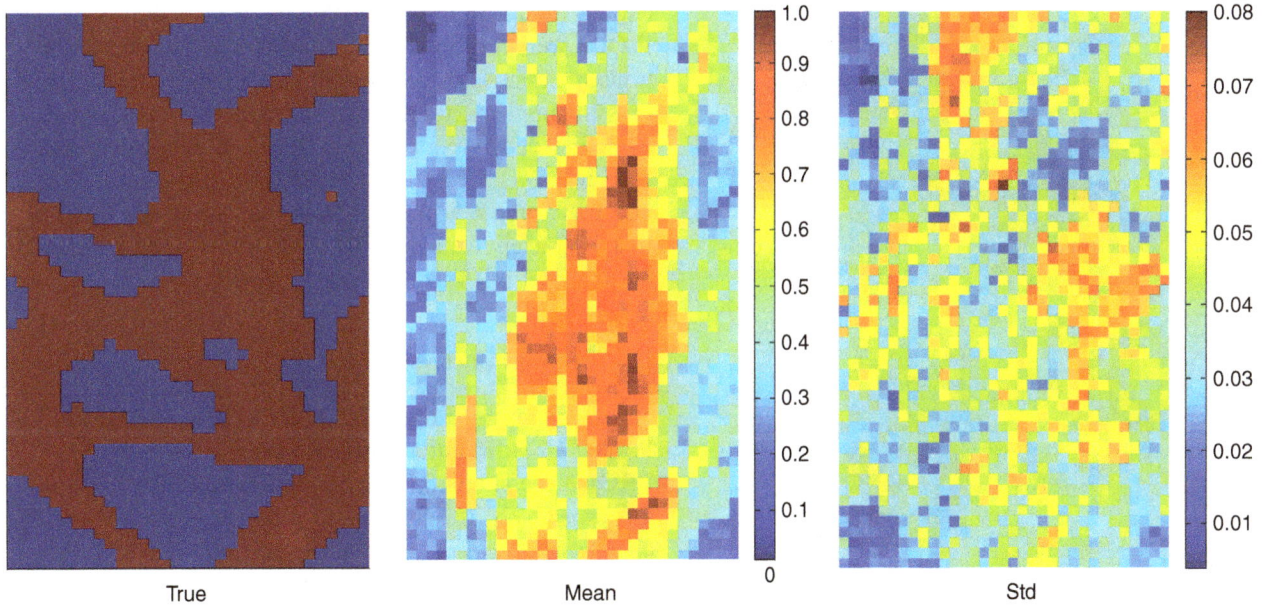

Figure 27

True, mean and standard deviation of the channel probability field considering all the cases.

the presence of channels in the central region of the reservoir, which is related to the presence of several wells providing the dynamic observations used to calibrate the models. The obtained facies probability map could be used as soft data for the generation of a new set of facies models in an iterative procedure so as to better represent the reservoir internal geometry. Eventually, the red areas of the standard deviation map show the zones of the reservoir where the information on the probability of channel occurrence is less reliable.

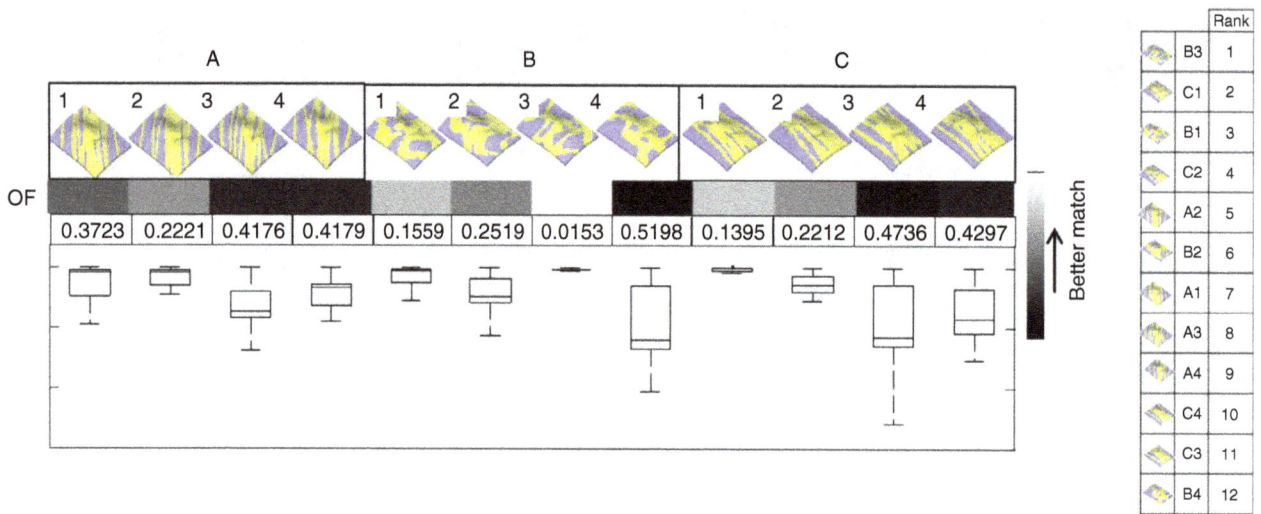

Figure 28

Objective function and ranking for each facies realization.

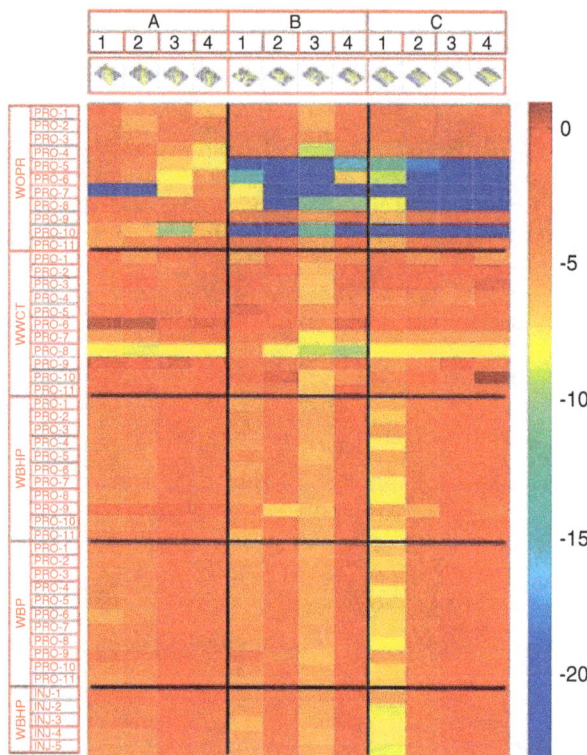

Figure 29

Heat map of the assimilation phase.

2.11 Results: Scenario 2. History Matching Objective Function Evaluation

The mean values, the standard deviation and the ranking of the objective function of each facies realization are shown in the boxplot and table of Figure 28.

The heat map relative to the assimilation phase for all the facies realization is shown in Figure 29. A logarithmic color scale was used to qualitative distinguish different OF values. Results show that not always the imposed WOPR were respected, which means that the production wells in some cases reached the minimum bottom hole pressure imposed at the wells. The water cut match quality varies depending on the well and facies realization. However, wells PROD8 and PROD6 present the best and the worst WWTC match, respectively. The C1, B3 and B1 facies realizations show a better overall match with respect to the other facies realizations.

2.12 Forecast Uncertainty

Results in terms of observed data (WOPR, WBHP, and WWCT) of the initial and calibrated models for well PROD3 are shown in Figure 30, for cases A2, B3 and C1 respectively. Results are shown for the models of each conceptual facies model with the best OF value, *i.e.* A2, B3 and C1. The vertical line separates the assimilation and forecast periods. The true curve for each observed data is plotted as a reference. The gray and blue lines are the data obtained from the initial (prior) and final (posterior) ensemble, respectively. Cases A2 and B3 match fairly well the historical water cut values both in assimilation and forecast phase. The WOPR have almost perfect matches in the assimilation phase and acceptable matches for the forecast phase in all the models except for well PROD2 in case C1. Cases B3 and C1 match accurately the pressures in the assimilation and forecast phases. Conversely, case A2 fails to reproduce the bottom hole pressure during the assimilation phase in both wells.

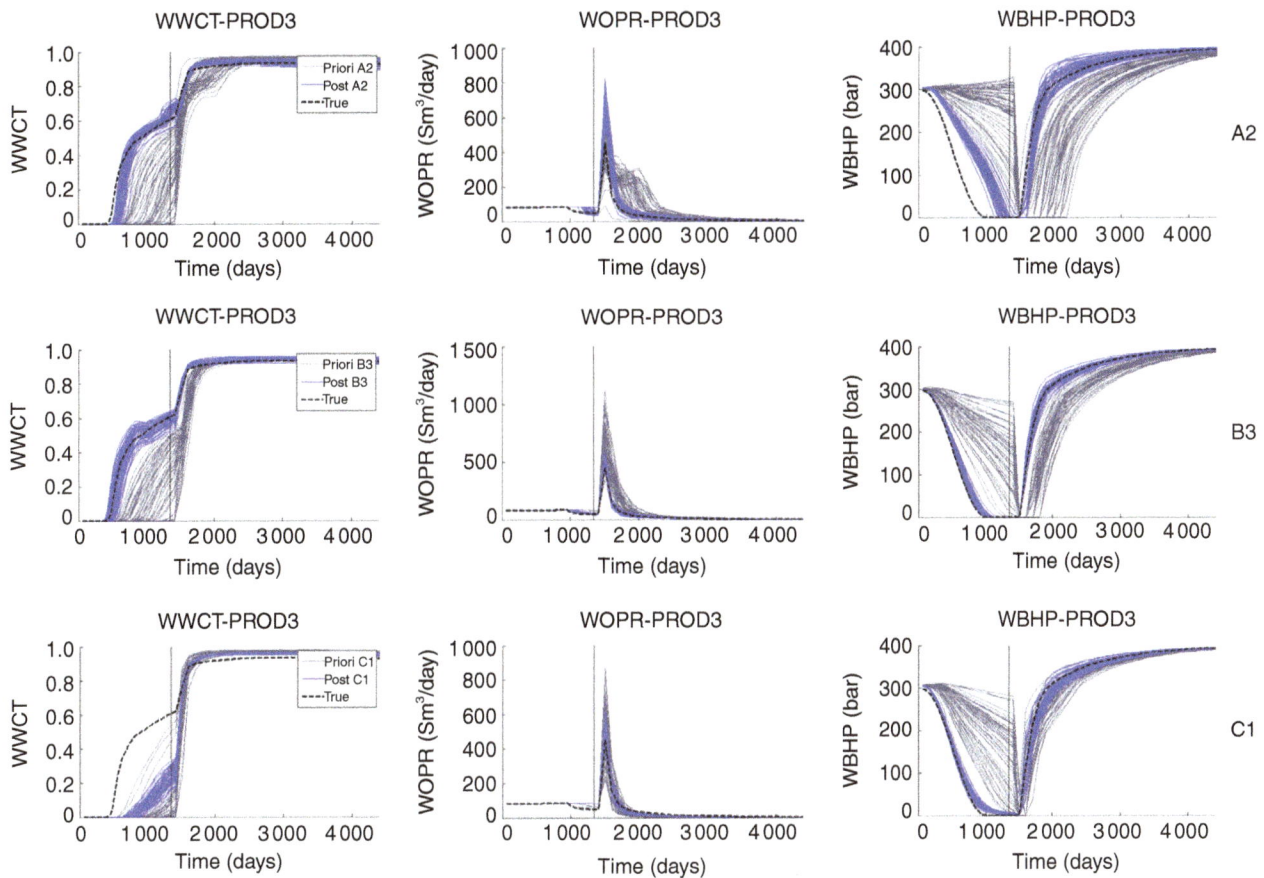

Figure 30

Prior and posterior observation *vs* time for well PROD3 for cases A2, B3 and C1.

The heat map associated to the forecast phase for all the facies realization is shown in Figure 31. Low values of WBHP for the injection wells are obtained; in these cases the injection pressure reached the maximum operational value so the wells change to a pressure controlled condition. The oil rates show higher mismatch than the rest of observations. In general, the realization B3 shows the lowest mismatch values.

The values of the OF, based on the cumulative oil production at the end of the forecast phase, for a progressively larger number of facies realizations are shown in Figure 32. The corresponding evolutions of the P90, P50 (or median) and P10 are indicated by a red, green and blue line, respectively. The FOPT uncertainty increases when cases of higher OF values are considered. Cases with lower OF values exhibit a FOPT distribution in which the true value lies outside the P10-P90 range. On the other hand, when all the available models are accounted for, a wider FOPT (P10-P90) range is obtained, but the true case more realistically lies within the P10-P90 range.

The correlation between the OF rank of the facies realizations in the forecast phase and in the assimilation phase is shown in Figure 33. The facies realizations that lie in the diagonal line $x = y$ represent the realizations that did not change their assimilation ranking from one phase to the following. The realization with a better rank in the forecast phase than in the assimilation phase are located in the upper part of the plot ($y > x$). Conversely, the realizations that experiment a ranking decreasing in the forecast phase with respect to the assimilation phase are located in the lower of the plot ($y < x$). In this scenario as well, realizations which seem fully acceptable after calibration are not necessarily representative of the true reservoir internal configuration and provide wrong production forecasts; on the other hand,

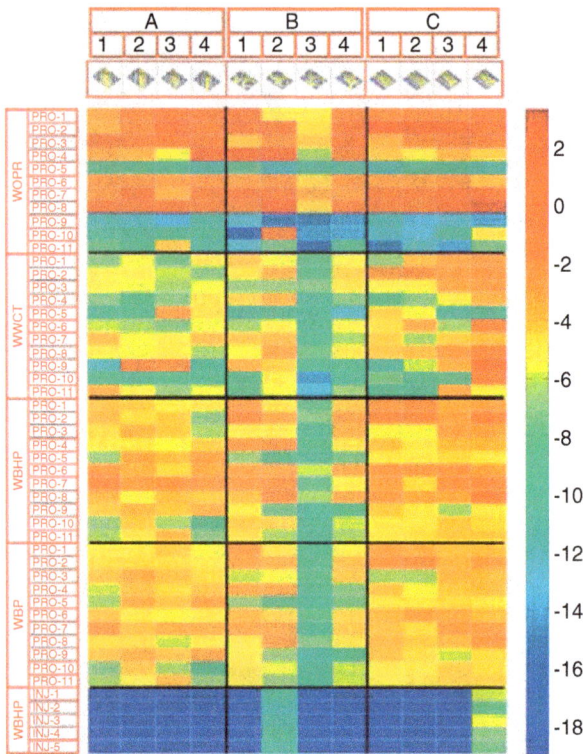

Figure 31

OF heat map of the forecast phase.

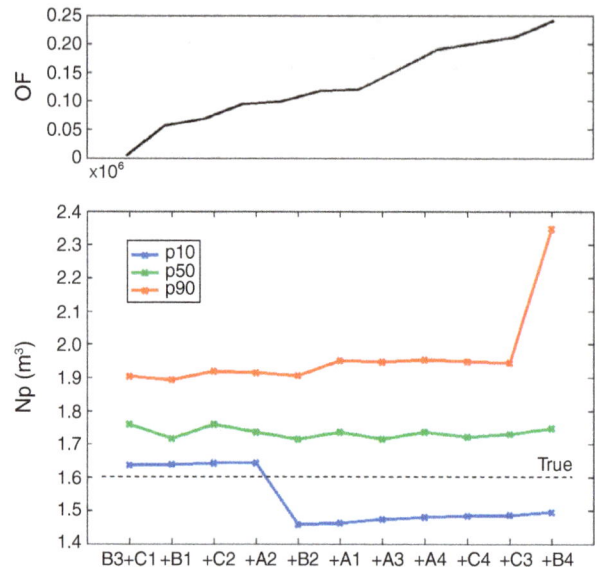

Figure 32

Cumulative oil production at the end of the forecast phase grouped OF ascending.

realizations which had a relatively low ranking after history matching can reliably predict the reservoir behavior.

The field cumulative oil production (FOPT) and the field cumulative water production (FWPT) distribution obtained from the prior and posterior ensembles are presented in Figure 34 to Figure 37. The colored regions identify the [P10, P90] intervals and the dotted colored line represents the median (P50). The continuous line represents the true value. A remarkable increase is observed both in the precision and accuracy of the estimation, *i.e.* a reduction of the distribution spread from the initial to the final models and a shift of the mean of the distribution towards the true value. Thus, the uncertainty associated to the cumulative oil and water production is efficiently reduced. It is worth noting that although case B and case C generally show better OF values they fail to match the true FOPT value within the uncertainty range P10-P90 (*Fig. 34*).

2.13 Facies and Petrophysical Properties

The evolutions of the mean and standard deviation of the ensemble permeability and porosity cases B3 and

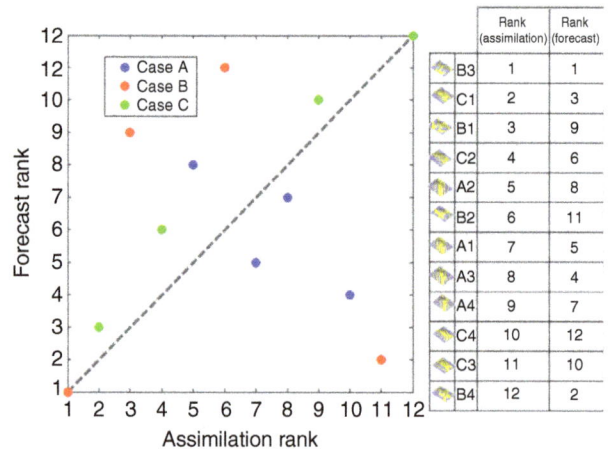

Figure 33

Assimilation *vs* forecast OF ranking.

C1 are presented in Figure 38 to Figure 41. Case B3 and case C1 were selected because they were the models that showed the lowest objective function. In this case as well, the petrophysical distribution of model B3 seems to reproduce in a more realistic way the petrophysical properties of the true cases. Furthermore, the standard deviation for the B3 in both porosity and permeability is substantially lower than in the C1 case.

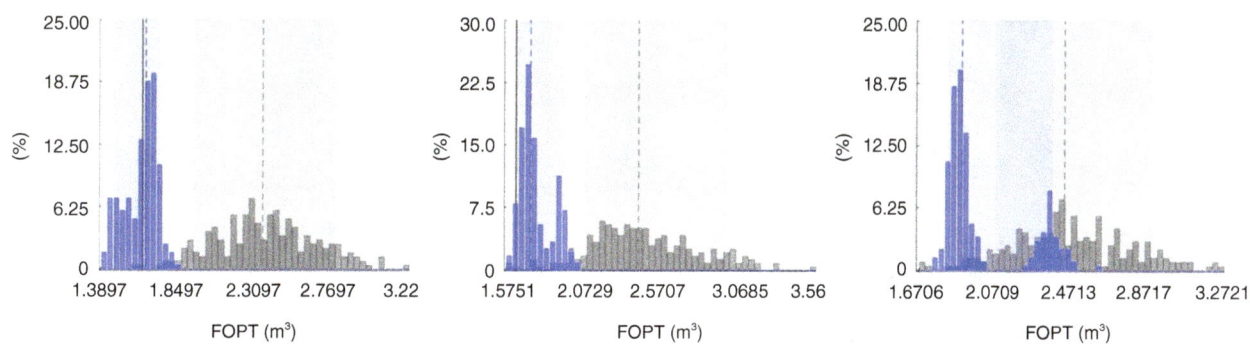

Figure 34

Prior *vs* posterior cumulative oil production at the end of the forecast phase for each facies group.

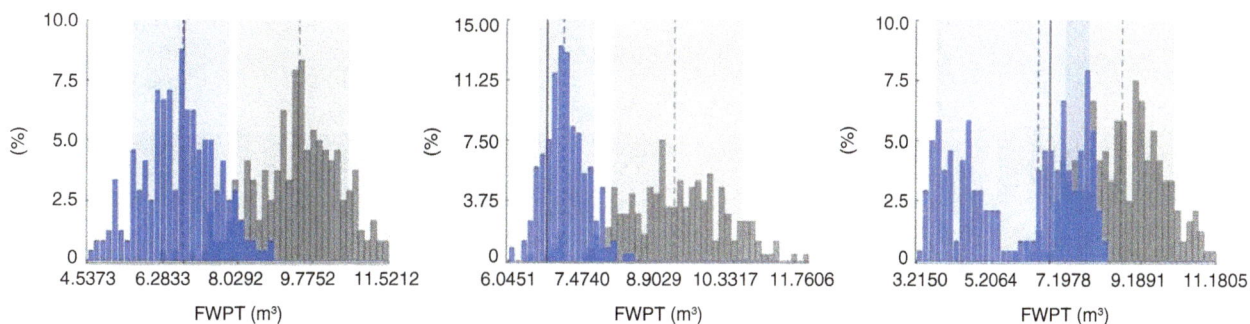

Figure 35

Prior *vs* posterior cumulative oil production at the end of the forecast phase including all facies group.

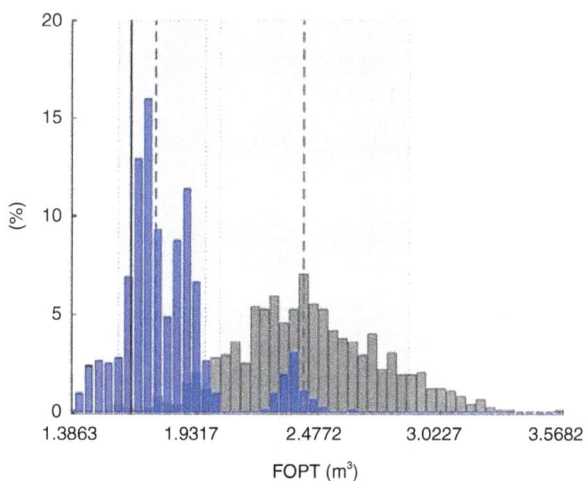

Figure 36

Prior *vs* posterior cumulative water production at the end of the forecast phase for each facies group.

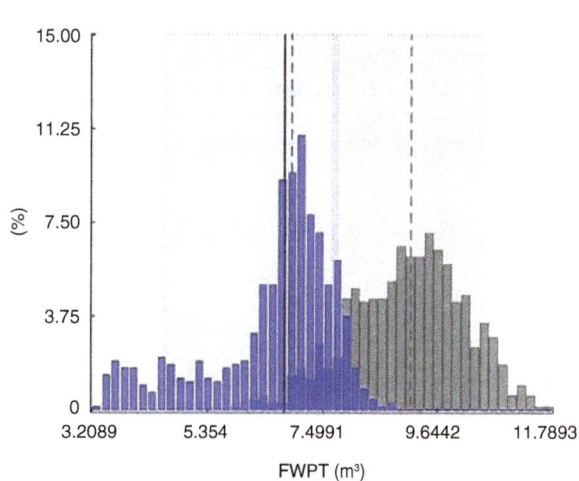

Figure 37

Prior *vs* posterior cumulative water production at the end of the forecast phase including all facies group.

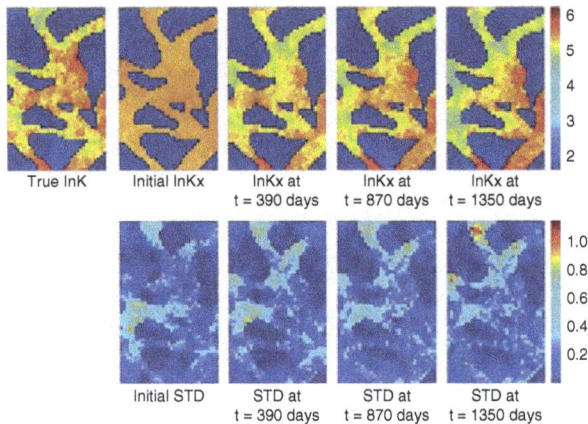

Figure 38

Comparison at different assimilation time (initial, at day 390, at day 870, at day 1 350) between the true ln(PERMx) field for the layer 1 and the ensemble mean in case B3.

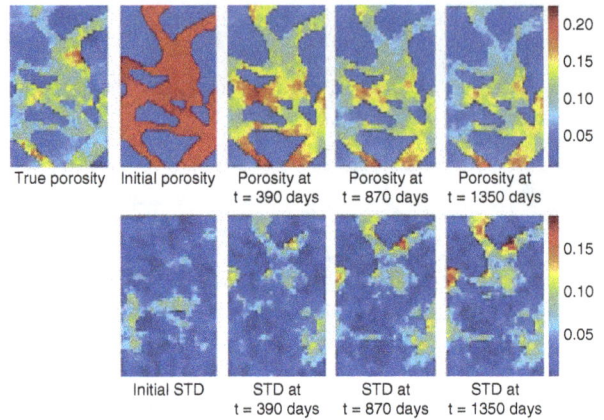

Figure 40

Comparison at different assimilation time (initial, at day 390, at day 870, at day 1350) between the true porosity field for the layer 1 and the ensemble mean in case B3.

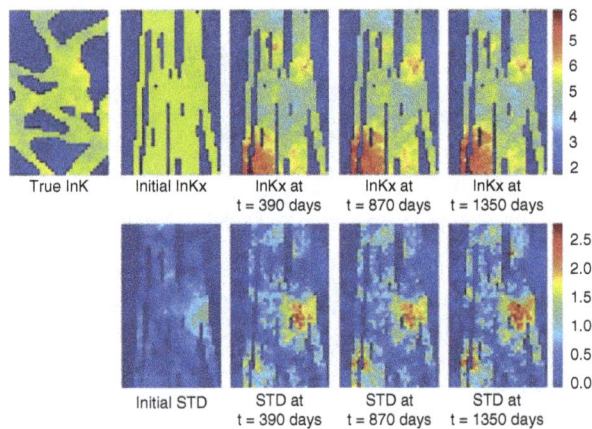

Figure 39

Comparison at different assimilation time (initial, at day 390, at day 870, at day 1350) between the true ln(PERMx) field for the layer 1 and the ensemble mean in case C1.

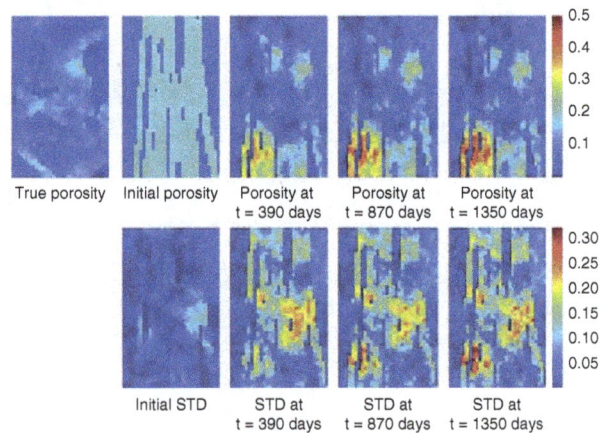

Figure 41

Comparison at different assimilation time (initial, at day 390, at day 870, at day 1 350) between the true porosity field for layer 1 and the ensemble mean in case C1.

CONCLUSIONS

The proposed workflow proved very efficient to handle the uncertainty associated to facies modeling. In the conventional approach only one static model, in which sometimes strong assumptions have to be made due to lack of information or due to the complexity of the system, is used during the history matching process. If geoscientists are not constrained to a single model but can deliver multiple models, a representative description of the model uncertainty can be elaborated and the qualitative judgment is substituted by a quantitative and statistical evaluation in which the dynamic data validates and calibrates the proposed static models. Furthermore, the used techniques for data assimilation showed to work efficiently in providing multiple solutions in a reasonable span of time, in effectively reducing the uncertainty of the model relative to the prior distributions and in significantly shifting the ensemble median towards the true value.

Results also showed that even the adoption of a conceptual model for facies distribution clearly representative of the reservoir internal geometry does not guarantee reliable results in terms of production forecast, thus the statistical approach seems the only way to reduce the uncertainty inherent to reservoir modeling. In addition, the models that showed the best OF values in the assimilation phase were not necessarily able to describe the system behavior in the forecast phase. When the true value of FOPT lays close to the mean FOPT value of the prior distribution the uncertainty of the FOPT remains almost constant, independently of the number of facies realizations considered for the forecast. Conversely, when the prior distribution is biased in respect to the true value, the uncertainty of the model strongly depends on the number of realizations considered for the forecast.

Recently, progresses were made in the automatic update of the facies maps; however, these researches are still under development and not available for application by reservoir engineers in everyday life. Conversely, the proposed approach/workflow is based on well-consolidated techniques and is virtually ready for use by means of available commercial tools, thus it constitutes a step forward for future technologies capable to automatically integrate different facies modeling in the history matching process. To this end the workflow could be extended to use the facies probability maps as soft data for the generation of new facies realizations to further improve the quality of the match in an iterative process. However, it should be mentioned that the computational cost for large models could be relevant.

ACKNOWLEDGMENTS

The authors are very thankful to *Schlumberger* for providing the academic licenses of the software for reservoir modeling used in this research.

REFERENCES

Benetatos C., Viberti D. (2010) Fully Integrated Hydrocarbon Reservoir Studies: Myth or Reality? *American Journal of Applied Sciences* **7**, 11, 1477-1486.

Caers J. (2003) Geostatistical history matching under training-image based geological constraints, *SPE Journal* **8**, 3, 218-226.

Cosentino L. (2001) *Integrated Reservoir Studies*, Editions Technip, Paris, ISBN 2710811170.

Deutsch C.V. (2002) *Geostatistical Reservoir Modeling*, Oxford University Press, ISBN 0195138066.

Evensen G. (1994) Sequential data assimilation with a nonlinear quasi-geostrophic model using Monte Carlo methods to forecast error statistics, *Journal of Geophysical Research: Oceans* **99**, C5, 10143-10162.

Evensen G. (2009) *Data assimilation: The ensemble Kalman filter*, Springer Verlag, Berlin, ISBN 9783642037115.

Hoteit I., Pham D.-T., Triantafyllou G., Korres G. (2008) Particle Kalman Filtering for Data Assimilation in Meteorology and Oceanography, *3rd WCRP International Conference on Reanalysis*, Tokyo, Japan, 28 Jan.-1 Feb.

Hu L.-Y. (2000) Gradual deformation and iterative calibration of Gaussian-related stochastic models, Mathematical. *Geology* **32**, 1, 87-108, doi: 10.1023/A:1007506918588.

Le Ravalec-Dupin M., Da Veiga S. (2011) Cosimulation as a perturbation method for calibrating porosity and permeability fields to dynamic data, *Computers & Geosciences* **37**, 9.

Mattax C.C., Dalton R.L. (1990) *Reservoir Simulation*, SPE Monograph Series Vol. 13, Richardson, Texas, USA, SPE, ISBN:978-1-55563-028-7.

RamaRao B.S., Lavenue A.M., De Marsily G., Marietta M.G. (1995) Pilot point methodology for automated calibration of an ensemble of conditionally simulated transmissivity fields. Part 1. Theory and computational experiments, *Water Resources Research* **31**, 3, 475-493.

Reynolds A.C., He N., Chu L., Oliver D.S. (1996) Reparameterization Techniques for Generating Reservoir Descriptions Conditioned to Variograms and Well-Test Pressure Data, *SPE Journal* **1**, 4, 413-426.

Selberg S., Ludvigsen B.E., Diab A., Harneshaug T. (2006) New Era of History Matching and Probabilistic Forecasting–A Case Study, *SPE Annual Technical Conference and Exhibition*, San Antonio, Texas, 24-27 Sept.

Schulze-Riegert R.K., Axmann J.K., Haase O., Rian D.T., You Y.-L. (2001) Optimization Methods for History Matching of Complex Reservoir, *SPE Reservoir Simulation Symposium*, Houston, Texas, 11-14 Febru.

Schulze-Riegert R.W., Haase O. (2003) Combined Global and Local Optimization Techniques Applied to History Matching, *SPE Reservoir Simulation Symposium*, Houston, Texas, 3-5 Feb., doi: 0.2118/79668-MS.

Stordal A.S., Karlsen H.A., Nævdal G., Skaug H.J., Vallès B. (2011) Bridging the ensemble Kalman filter and particle filters: the adaptive Gaussian mixture filter, *Computational Geosciences* **15**, 2, 293-305.

Stordal A.S., Valestrand R., Karlsen H.A., Nævdal G., Skaug H.J. (2012) Comparing the adaptive Gaussian mixture filter with the ensemble Kalman filter on synthetic reservoir models, *Computational Geosciences* **16**, 2, 467-482.

Tavassoli Z., Carter J.N., King P.R. (2004) Errors in History Matching *SPE Journal* **9**, 3, 352-361.

Valestrand R., Nævdal G., Stordal A.S. (2012) Evaluation of EnKF and Variants on the PUNQ-S3 Case, *Oil & Gas Science and Technology - Rev, IFP Energies nouvelles* **67**, 5, 841-855.

Zitzler E. (1999) Evolutionary Algorithms for Multiobjective Optimization: Methods and Applications, *PhD Thesis,* Swiss Federal Institute of Technology, Zurich.

4

Pre-Spud Mud Loss Flow Rate in Steeply Folded Structures

Zhiyuan Wang[1]*, Baojiang Sun[1] and Ke Ke[2]

[1] School of Petroleum Engineering, China University of Petroleum (East China), Qingdao 266555 - China
[2] SINOPEC Research Institute of Petroleum engineering, Beijing 100101 - China
e-mail: wangzy1209@126.com - sunbj1128@126.com - keke@sripe.cn

* Corresponding author

Résumé — Prédiction du débit de perte des boues de démarrage dans des structures fortement plissées — Dans cet article, nous proposons une nouvelle méthode pour prédire le débit de perte de boues de démarrage avant forage de formations à fracturation tectonique dans des structures fortement plissées. Cette nouvelle méthode est basée sur la stimulation par éléments finis du champ de contraintes paléo-tectoniques et tectoniques, et de la répartition de la fracturation. Les étapes de la méthode sont les suivantes. Tout d'abord, la répartition des contraintes paléo-tectoniques est simulée par l'analyse par éléments finis. La répartition de la fracturation tectonique de la région est obtenue en combinant le critère de rupture des roches avec la distribution des contraintes paléo-tectoniques. Ensuite, les paramètres de rupture tels que la densité, l'ouverture, la porosité et la perméabilité des fractures tectoniques sont calculés en étudiant le processus de reconstruction de contraintes actuelles sur la fracturation tectonique. Enfin, le débit de perte des boues de démarrage est calculé en fonction des paramètres de rupture et des données de base pour un puits donné. La nouvelle méthode permet de prédire le débit de perte de boues de démarrage dans des structures fortement plissées.

Abstract — Pre-Spud Mud Loss Flow Rate in Steeply Folded Structures — In this paper, a new method that predicts the pre-spud mud loss flow rate in formations with tectonic fractures of steeply folded structures is proposed. The new method is based on finite element analysis of the palaeo-tectonic and current tectonic stress field and fracture distribution. The steps of the method are as follows. First, palaeo-tectonic stress distribution is simulated through finite element analysis. The tectonic fracture distribution of the region is obtained by combining rock failure criteria with palaeo-tectonic stress distribution. Afterward, the tectonic fracture density, aperture, porosity and permeability are calculated by studying the rebuilding process of current stress to the fracture parameters. Finally, the mud loss flow rate is calculated according to fracture parameters and the basic data of a given well. The new method enables the prediction of the mud loss flow rate before drilling steeply folded structures.

INTRODUCTION

Steeply folded structures have complex conditions and are prone to develop cracks, thereby leading to serious circulation loss during drilling and danger during the drilling process (Xu et al., 1997; Quintana et al., 2001; Du, 2004). Thus, predicting circulation loss during the pre-spud phase is important. Currently, numerous studies on methods predicting circulation loss concentrate on rock mechanics to describe and evaluate crustal stress (Prensky, 1992; Bell and Aadnøy, 1998; Desroches and Kurkjian, 1999; Nikolaevskiy and Economides, 2000), to analyze and investigate the tectonic fracture distribution, and to obtain pressure loss (Constant David and Bourgoyne, 1988; Rocha and Bourgoyne, 1996; Sadiq and Nashawi, 2000; Rocha et al., 2004). Determining the fracture pressure gradient from well logs is another way to predict circulation loss (Anderson et al., 1973; Holbrook, 1989; Draou and Osisanya, 2000). The methods mentioned above can predict whether circulation loss occurs or not. However, these methods cannot predict the mud loss flow rate before drilling, which is more important during the actual drilling process. Therefore, we propose a new method for predicting the mud loss flow rate in tectonic fractures of steeply folded structures. This new method is based on the numerical simulation of the tectonic stress field and fracture distribution (Dai and Li, 2000; Maerten and Maerten, 2006; Deng et al., 2006) and quantitative description of tectonic fracture parameters related to the mud loss flow rate, such as density, aperture, porosity and permeability (Murray, 1968; Chen and Bai, 1998; Ji et al., 2010a). The main steps of the method are as follows. The geological structure model is established according to the local tectonic fracture distribution and main tectonic movement period. Palaeo-tectonic stress distribution is calculated through finite element analysis, and tectonic fracture distribution of the region is calculated by combining rock failure criteria and palaeo-tectonic stress distribution. Another geological structure model for current stress simulation is established according to local current stress and tectonic features. Tectonic fracture density, aperture, porosity and permeability are calculated by studying the rebuilding process of current stress to the fracture parameters. Finally, the mud loss flow rate is calculated on the basis of fracture parameters and basic data, such as pore pressure profile, mud weight and fracture pressure, among others, in a given well. Taking Block F of a gas field in Sichuan Basin, China, as an example, the tectonic stress field and fracture distribution are analyzed. The predicted value of the mud loss flow rate agrees well with the actual value in Well PG-X1, which means that the method is feasible in predicting the mud loss flow rate before drilling steeply folded structures. This method will be helpful in applying preventive measures during the actual drilling practice in steeply folded structures.

1 GEOLOGICAL STRUCTURE MODEL ESTABLISHMENT

1.1 Characteristic Analysis of Geological Structure

The structure of the site, history of structural evolution, in situ basin structural characteristics, formation sequence and sediment fill were researched and analyzed by using tectonic evolution analysis to comprehend the characteristics of the geological structure (Wei Jia et al., 2008). The structural pattern, formation components, lithology and direction of the structure were determined. The depth of the layers of the structural diagram in the area was also obtained for model establishment.

1.2 Geological Model Establishment

By analyzing the geological structure characteristics, layers with similar lithology were grouped together. Geometry simplification was made by using the depth structural diagram, and the geological model of the area was established.

2 FINITE ELEMENT ANALOGY OF TECTONIC STRESS DISTRIBUTION

The numerical computations of stress distribution can be divided into two sections: palaeo-stress distribution and current stress distribution. As shown in Figure 1, the steps in the evaluation of palaeo-stress distribution are as follows. First, the historical period in which the fracture was formed and the shape of the geological structure pre- and after this period could be determined by using the results of the tectonic evolution analysis. A geological structure model can be established based on the shape of the geological structure before fracture formation. Afterward, the boundary conditions and loads are set by using the trial and error method, which is based on the deformation standard, until the geological structure gained by simulation agrees with the actual structure after fracture formation. Finally, palaeo-stress distribution can be determined by finite element analysis. The current stress distribution can be evaluated in the same way as the palaeo-stress distribution. However, the current geological structure, boundary conditions

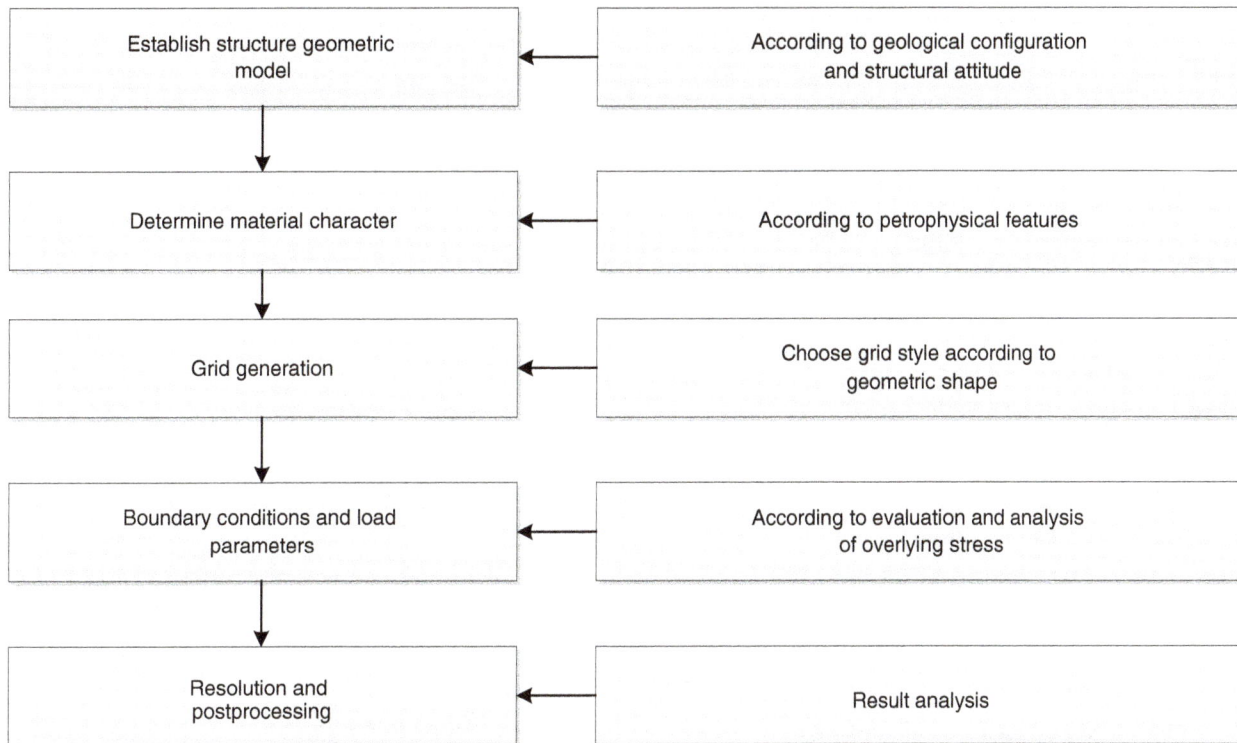

Figure 1

The flow diagram of tectonic fracture finite element analogy.

and loads are used during finite element analysis in the computation of current stress distribution. Current boundary loads can be determined through logging data, data of the indoor core experiment, data of hydraulic fracturing, and other kinds of well data in the area evaluated.

The constitutive law used in the simulation is a linear elastic model, which is commonly used in the fracture simulation process of pay zones (Coblentz and Sandiford, 1994; Kwon and Mitra, 2004).

2.1 Boundary Conditions and Load Parameters of Palaeo-Tectonic Stress Simulation

Accurately setting the factors influencing the control of the structural fracture is difficult because of the complexity of geological structure formation. Therefore, the boundary conditions and the load deformation of palaeo-tectonic stress simulation need repeated trial and error calculations. The deformation standard is applied in this paper. After the boundary conditions and load are loaded, the deformation of the geological body is matched with the practical characteristics. Simulation results are checked with the real structural

features, particularly the shape, length and position of the anticlines, synclines or faults.

2.2 Boundary Conditions and Load Parameters of Current Tectonic Stress Simulation

Calculating the crustal stress in a drilled well is important in determining the boundary conditions and load parameters of the geological model for current tectonic stress simulation. Currently, numerous methods are available in evaluating crustal stress. The equations below are used for the evaluation of crustal stress in this paper (Gazaniol et al., 1995; Helio Santos et al., 1999; Ge et al., 2001). In this study, the fracturing method is used to calculate horizontal principal stress:

$$\sigma_h = p_s \qquad (1)$$

$$\sigma_H = 3\sigma_h - p_f - p_p - S_t \qquad (2)$$

where σ_H and σ_h are the maximum and minimum horizontal principal stress, respectively (MPa); p_p is the pore pressure (MPa); p_s is the instant knock-off pressure (MPa); p_f is fracture pressure (MPa); and S_t is rock tensile strength (MPa).

Hydraulic fracturing points are chosen as the key points of tectonic stress simulation based on the statistical hydraulic fracturing information of the wells drilled in the research area. The horizontal principal stress is obtained at the key points by using Equations (1), (2). Afterward, boundary conditions and load parameters are loaded using the trial and error method until the simulated tectonic stress is nearly consistent with the value calculated from the hydraulic fracturing in the key points. The differential value between simulated tectonic stress and calculated tectonic stress from hydraulic fracturing is less than 5% when few key points exist, whereas the differential value is larger than 5% when many key points exist.

2.3 Procedure of Finite Element Analogy

The procedure of finite element analogy is shown in Figure 1.

3 QUANTITATIVE DESCRIPTIONS FOR TECTONIC FRACTURE PARAMETERS

The parameters of tectonic fracture related to the prediction of the mud loss flow rate include fracture density, aperture, porosity and permeability. The current stress difference is low and cannot create new fractures. Current fracture density has the same value as palaeo-fracture density. Current stress can rebuild palaeo-fractures and can cause fracture aperture, porosity and permeability to change.

3.1 Calculation Methods for Palaeo-Fracture Parameters

3.1.1 Fracture Density Calculation

Rock deformation can accumulate strain energy under stress. Rock failure appears when the release rate of strain energy is equal to the energy required to form the fracture per unit area (surface energy density). Part of the strain energy released is utilized to provide energy for the increasing fracture surface area, and the remaining strain energy is released in the form of an elastic wave. The energy of the elastic wave is minimal and can be neglected.

The Coulomb-Mohr principle was applied in the examination of shear fracture under triaxial compression

stress. The computation for fracture density under triaxial compression stress is as follows (Ji *et al.*, 2010b):

see equation (3) below

$$D_{lf} = \frac{2D_{vf_c}L_1L_3 \sin\theta \cos\theta - L_1 \sin\theta - L_3 \cos\theta}{L_1^2\sin^2\theta + L_3^2\cos^2\theta} \quad (4)$$

$$\theta = 45° - \frac{\varphi}{2} \quad (5)$$

The Griffith-failure principle was applied in the examination of tension fracture under tension stress. When $(\sigma_1 + 3\sigma_3) > 0$, then the formula for fracture density under tensile stress is as follows:

$$D_{Vf} = \frac{(\sigma_1^2 + \sigma_2^2 + \sigma_3^2 - 2\mu(\sigma_1\sigma_2 + \sigma_2\sigma_3 + \sigma_1\sigma_3) - \sigma_t^2)}{2EJ} \quad (6)$$

$$D_{lf} = \frac{2D_{vf}L_1L_3 \sin\theta \cos\theta - L_1 \sin\theta - L_3 \cos\theta}{L_1^2\sin^2\theta + L_3^2\cos^2\theta} \quad (7)$$

$$\theta = \frac{1}{2}\arccos(\frac{\sigma_1 - \sigma_3}{2\sigma_1 + 2\sigma_3}) \quad (8)$$

Let $(\sigma_1 + 3\sigma_3) \leq 0$, $\sigma_3 = -\sigma_t$ and $\theta = 0$. Thus, when θ is equal to zero, the fracture plane is parallel to the direction of σ_1 and perpendicular to the direction of σ_3, so the fracture linear density equals the total number of fractures in the rock volume. The formula for fracture density is as follows:

$$D_{lf} = D_{vf} \qquad \theta = 0 \quad (9)$$

where D_{Vf_c} is the fracture volume density under triaxial compression stress (m^{-2}); D_{Vf} is the fracture volume density under tensile stress (m^{-2}); D_{lf} is the fracture linear density (m^{-1}); σ_1, σ_2 and σ_3 are the maximum principal stress, the intermediate principal stress and the minimum principal stress (Pa); E is the elastic modulus (Pa); μ is the Poisson ratio; L_1 and L_3, respectively, are the side length along the maximum principal stress direction and the minimum principal stress direction of the represent element volume (m); σ_d is the uniaxial compression stress when macro-fractures are going to appear, which can be confirmed by experiment (Pa); J is the surface energy density of fractures (J/m^2); θ is the rupture angle (the angle between the maximum principal stress direction and the fracture orientation) (°); and φ is the angle of internal friction (°).

$$D_{Vf_c} = \frac{(\sigma_1^2 + \sigma_2^2 + \sigma_3^2 - 2\mu(\sigma_1\sigma_2 + \sigma_2\sigma_3 + \sigma_1\sigma_3) - \sigma_d^2 + 2\mu(\sigma_2 + \sigma_3)\sigma_d)}{2EJ} \quad (3)$$

3.1.2 Fracture Aperture Calculation

The formula of fracture aperture b is as follows (Hicks et al., 1996):

$$b = \frac{\varepsilon_f}{D_{lf}} = \frac{|\varepsilon| - |\varepsilon_0|}{D_{lf}} \qquad (10)$$

where ε is the tension strain of the rock cell cube under current stress conditions and ε_0 is the greatest tension strain of rock elastic deformation.

3.1.3 Fracture Porosity Calculation

Fracture porosity is the ratio of total fracture volume and total rock volume. For a single population fracture, the relationship between fracture porosity and fracture volume density and aperture is as follows (Chen and Bai, 1998):

$$\phi_f = b D_{vf} \qquad (11)$$

where ϕ_f is the fracture porosity (%).

For multi-bank fractures, fracture porosity is calculated as follows:

$$\phi_{ft} = \sum_i^m b_i D_{vfi} \qquad (12)$$

where ϕ_{ft} is the fracture total porosity (%); m is the group number of fractures; and b_i is the aperture of the ith group fracture (m).

3.1.4 Fracture Permeability Calculation

Fluid flow in a unitary fracture is mainly confined to a two-dimensional crack plane, and the penetration perpendicular to the crack plane can be neglected. Hence, a faceplate filtration model can be applied when calculating fracture permeability (Song et al., 1999):

$$\begin{bmatrix} K_{fx} \\ K_{fy} \\ K_{fz} \end{bmatrix} = \sum_i^m \begin{bmatrix} K_{fxi} \\ K_{fyi} \\ K_{fzi} \end{bmatrix} = \sum_i^m \frac{b_i^3 D_{lfi}}{12} \begin{bmatrix} \cos^2\theta_i \\ 1 \\ \sin^2\theta_i \end{bmatrix} \qquad (13)$$

where K_{fx}, K_{fy} and K_{fz}, respectively, are the fracture permeability along σ_1, σ_2 and σ_3 directions when multi-bank fractures exist (10^{-3} μm^2), and θ_i is the rupture angle of the ith group fracture.

3.2 Calculation Methods for Current Fracture Parameters

Hicks et al. (1996) proposed Equation (14) to calculate fracture aperture in consideration of the influence of normal stress and shear stress on fracture aperture:

$$b_m = \frac{b_0}{1 + 9\sigma'/\sigma_{nref}} + \Delta b_s + b_c \qquad (14)$$

where b_m is the current fracture aperture (m); b_0 is the initial fracture aperture (m); σ_{nref} is the effective normal stress (MPa); σ' is the effective normal stress when fracture aperture decreases 90% (MPa); b_0 is the aperture increment due to shear displacement (m); b_s is the fracture aperture which is superficial under maximum normal stress (m).

Minimal shear displacement occurred in the current formation. Thus, b_c usually has a low value and can be neglected; the initial fracture aperture can be considered as palaeo-fracture aperture b; σ' is equal to the pressure difference on fractures between normal stress and formation fluid pressure. Thus, Equation (14) transforms into Equation (15):

$$b_m = \frac{b_0}{1 + 9\sigma_n/\sigma_{nref}} \qquad (15)$$

where σ_n is the normal stress on fractures (MPa).

The calculation methods for current fracture parameters can be obtained by replacing fracture aperture b in Equations (10) and (11) with current fracture aperture.

4 CALCULATION OF MUD LOSS FLOW RATE

As shown in Figure 2, the flow path of mud can be divided into three parts based on the flow behavior of drilling fluid in fissured formation: flow in a fracture, flow in mud cake, and flow in formation.

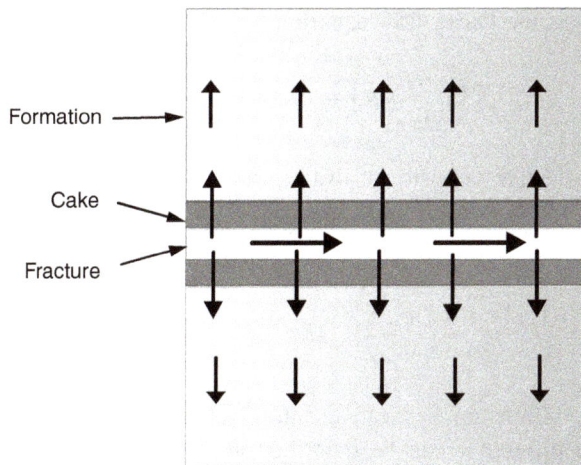

Figure 2

The schematic diagram of mud loss in a fracture, cake and formation.

4.1 Fracture Flow

Equations (16), (17) and (18) can be used to describe the flow behavior of the drilling fluid loss in the fracture if the fractures are supposed to be horizontal. Lavrov and Tronvoll (2004) deduced an equation describing the flow of a power-law fluid from a borehole into a deformable fracture of finite extension:

$$\frac{\partial b}{\partial t} - \frac{nb^{(2n+1)/n}}{(2n+1)2^{(n+1)/n}K_{ci}^{1/n}} \frac{1}{r} \frac{\partial p}{\partial r} \left|\frac{\partial p}{\partial r}\right|^{(1-n)/n}$$

$$- \frac{n}{(2n+1)2^{(n+1)/n}K_{ci}^{1/n}} \frac{\partial}{\partial r}\left[b^{(2n+1)/n} \frac{\partial p}{\partial r} \left|\frac{\partial p}{\partial r}\right|^{(1-n)/n} \right] = 0$$

$$(16)$$

$$\frac{\partial b}{\partial t} = \frac{1}{K_n} \frac{\partial p}{\partial t} \tag{17}$$

The mud loss flow rate is evaluated by the following:

$$u = -\frac{2\pi r_w n b^{(2n+1)/n}}{(2n+1)2^{(n+1)/n}K_{ci}^{1/n}} \frac{\partial p}{\partial r} \left|\frac{\partial p}{\partial r}\right|^{(1-n)/n} \tag{18}$$

where b is fracture aperture (m); t is the time (s); n is the power-law exponent; K_{ci} is the consistency index (Pa · s^n); K_n is the normal fracture stiffness (Pa/m); p is the pressure inside the fracture (Pa); r is the distance from the borehole axis in the fracture plane (m); r_w is the wellbore radius (m); and u is the mud loss flow velocity (m/s).

4.2 Formation Flow

Equations (19-21) can be used to describe the flow behavior of the drilling fluid loss in formation. Equation (19) is the Darcy flow equation:

$$\nabla\left(\rho \frac{K_d}{\mu_e} \nabla p \right) = C_t \phi_d \frac{\partial p}{\partial t} \tag{19}$$

The effective viscosity of drilling fluid changes with the rate of shearing is expressed as follows (Williams, 1990):

$$\mu_e = \frac{1}{\alpha_u} \cdot \frac{3n+1}{8n} \left(\frac{\phi_d}{8K_d} \right)^{\frac{1-n}{2n}} (2K_{ci})^{\frac{1}{n}} \left(-\frac{\partial p}{\partial r} \right)^{\frac{n-1}{n}} \tag{20}$$

The relationship between the flow velocity and the pressure difference can be described as Equation (21) in porous media (Settari, 1988).

$$u = \frac{\phi n}{3n+1} \left(\frac{8K_d}{\phi_d} \right)^{\frac{1-n}{2n}} (2K_{ci})^{\frac{1}{n}} \left(\frac{\partial p}{\partial r} \right)^{\frac{1}{n}} \tag{21}$$

where ϕ is the pressure inside the formation (Pa); ρ is the drilling fluid density (kg/m); ϕ_d is the formation porosity (%); K_d is the formation permeability (µm^2); C_t is the formation compressibility coefficient (Pa^{-1}); α_u is the unit conversion coefficient; and μ_e is the effective viscosity of the drilling fluid.

4.3 Cake Flow

Equations (22-24) can be used to describe the flow behavior of the drilling fluid loss in the mud cake. The mud cake is considered a porous medium with the porosity ϕ_C and the permeability K_C. Equation (22) is the flow equation in mud cake (Warpinski, 1991):

$$\nabla\left(\rho \frac{K_c}{\mu_e} \nabla p \right) = C_c \phi_c \frac{\partial p}{\partial t} \tag{22}$$

The effective viscosity of drilling fluid in the mud cake is expressed as follows:

$$\mu_e = \frac{1}{\alpha_u} \cdot \frac{3n+1}{8n} \left(\frac{\phi_C}{8K_C} \right)^{\frac{1-n}{2n}} (2K_{ci})^{\frac{1}{n}} \left(-\frac{\partial p}{\partial r} \right)^{\frac{n-1}{n}} \tag{23}$$

The relationship between flow velocity and pressure difference can be described as Equation (24) in mud cake (Vitthal and McGowen, 1996).

$$u^n = \frac{\alpha_u^n}{32K_{ci}} \left(\frac{\phi_C n}{3n+1} \right)^n \left(\frac{8K_C}{\phi_C} \right)^{\frac{n+1}{2}} \frac{p_f - p_w}{L} \tag{24}$$

where ϕ_c is the mud cake porosity (%); K_c is the mud cake permeability (µm^2); C_c is the mud cake compressibility coefficient (Pa^{-1}); p_f is the formation pressure (Pa); p_w is the bottom hole flowing pressure (Pa^{-1}); and L is the drilling fluid flow distance (m).

Two main types of drilling fluid loss commonly exist in the fracture formation: loss due to hydraulic fracturing and loss due to pressure difference. When the annular pressure is higher than fracture pressure, the formation will form major fractures and loss due to fracturing will occur. Formation permeability is usually high in steeply folded structures. Thus, the loss due to pressure difference tends to exist when annular pressure is higher than pore pressure. The flow rate of loss due to fracturing can be calculated by using Equations (16-24). The flow rate of loss due to over-pressure can be calculated by using Equations (19-24). However, the mud cake is formed in the borehole wall instead of in the fracture surface. Permeability and porosity are calculated by using the quantitative description method for tectonic fracture parameters mentioned above.

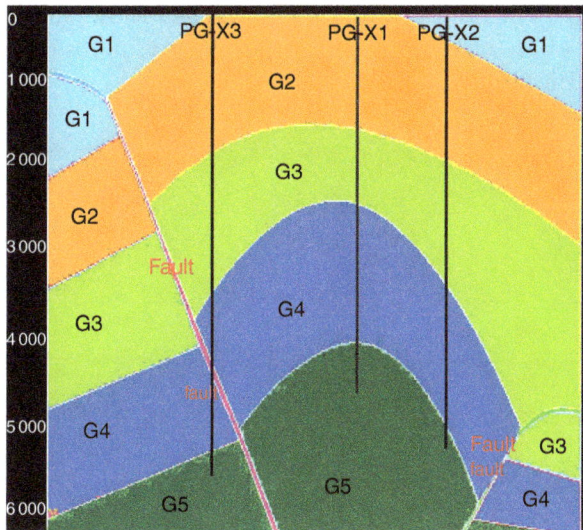

Figure 3

Structural profile of Block F.

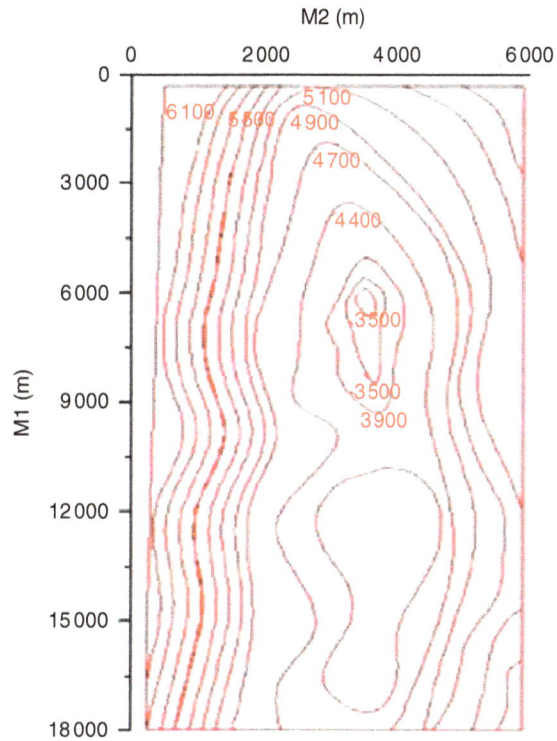

Figure 4

G4 depth structure of Block F.

5 CASE STUDIES ON QUANTITATIVE DESCRIPTION OF TECTONIC FRACTURE AND MUD LOSS FLOW RATE

Figure 3 is a simplified structure profile diagram of Block F of a gas field in Sichuan Basin, China. Structure formations can be divided into five groups, namely, G1, G2, G3, G4 and G5. G4 has had the highest drilling fluid loss in the drilling process; thus, G4 is taken as an example in the finite element analysis of palaeo-stress and current stress and in the quantitative description of tectonic fracture. Figure 4 shows the depth structure of G4. The parameter values of the materials used in the model are listed in Table 1.

The load value and azimuth of palaeo-stress simulation are as follows. The palaeo-maximum principal stress is considered in the EW direction, according to local fracture development. The maximum and minimum principal stress values are 296 MPa and 165 MPa, respectively. These stress values are obtained by trial and error calculation until the deformation agrees with the actual shape characteristic of the geological structure.

The load value and azimuth of current stress simulation are as follows. According to the results of the measured local stress, the current stress azimuth is between NE 76° and NE 88°, and the mean value is 83°. Thus, the angle between the current stress azimuth and the structure alignment direction is small. Moreover, the model profile bears extrusion force, which is mainly caused by the stress in the EW direction and gravity. According to the stress value of the key points and trial and error calculation, the

TABLE 1

Parameters of model materials

Group name	Volume weight (kN/m^3)	Young's modulus (GPa)	Poisson ratio
L1	20	5	0.40
L2	21	25	0.30
L3	22	30	0.28
L4	24	40	0.25
L5	26	47	0.24

maximum and minimum principal stress values are 123 MPa and 90.2 MPa, respectively.

5.1 Results of Tectonic Stress and Fracture Simulation

Tectonic fracture density distribution is calculated by using the results of the palaeo-stress simulation and by using the Coulomb-Mohr criterion and Griffith-failure criterion as criteria of shear failure and tension failure, respectively. Tectonic fracture density, aperture, porosity and permeability are determined by simulating the rebuilding process of current stress to the fracture parameters on the basis of the results of the palaeo-stress simulation. Figures 5-8 show the results of tectonic fracture density, aperture, porosity and permeability.

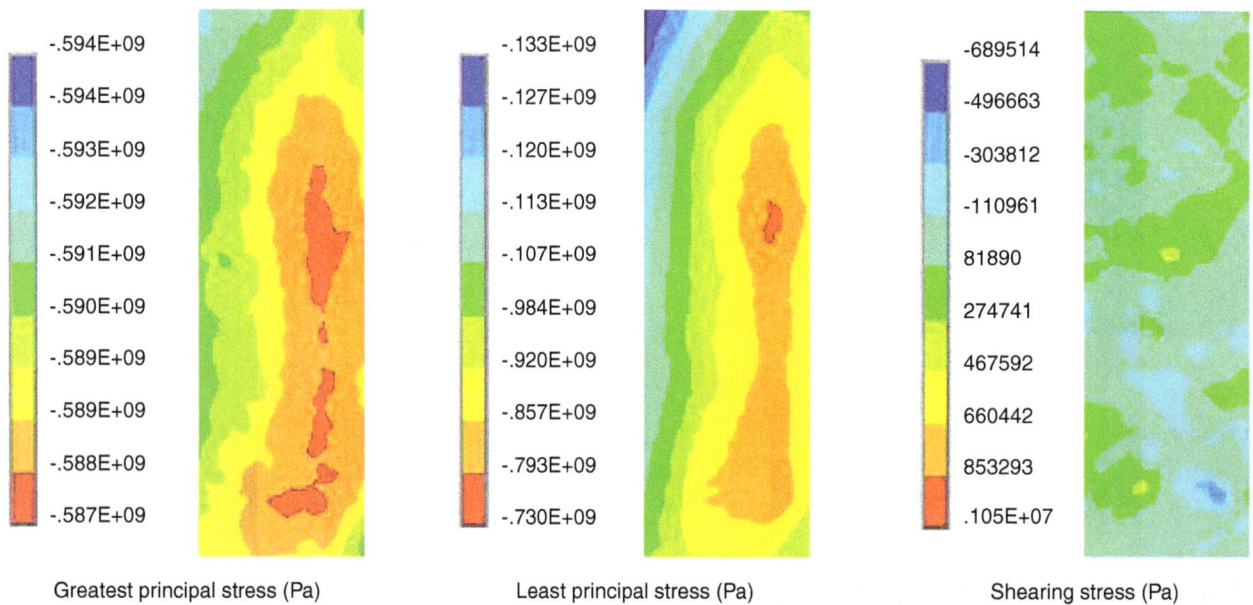

| Greatest principal stress (Pa) | Least principal stress (Pa) | Shearing stress (Pa) |

Figure 5

Tectonic fracture stress distribution of G4 in Block F.

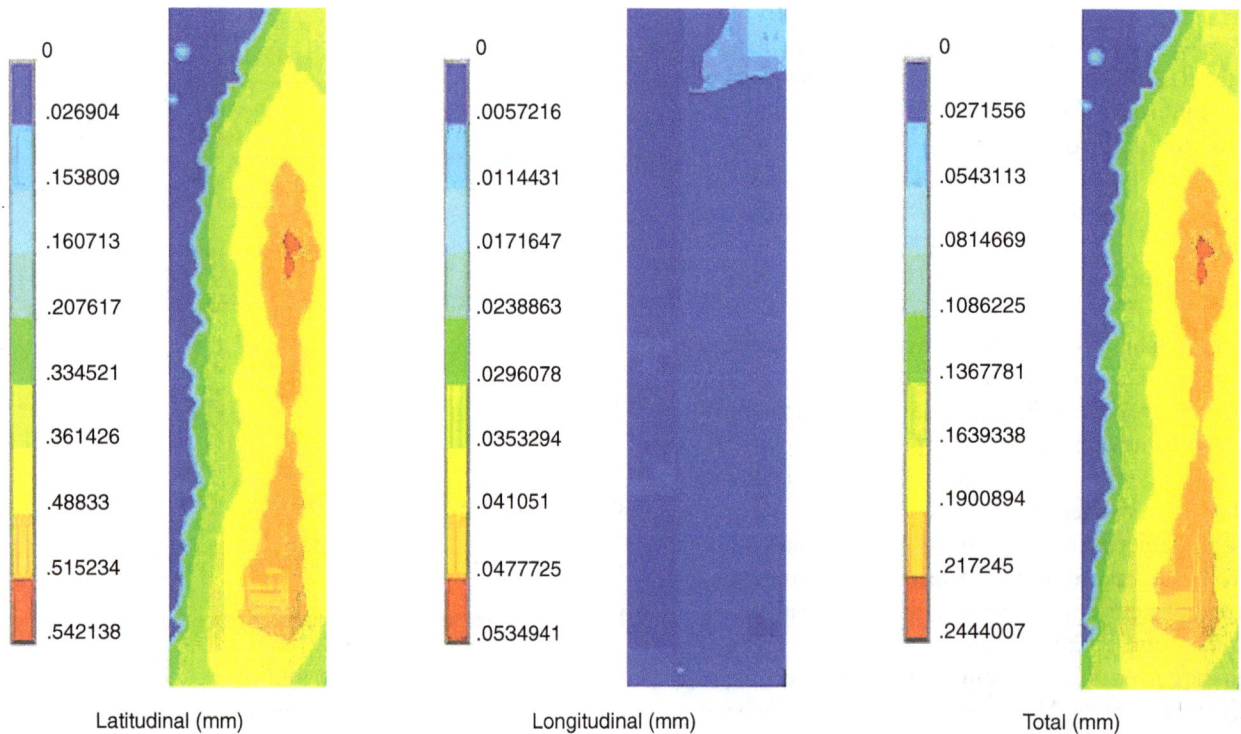

| Latitudinal (mm) | Longitudinal (mm) | Total (mm) |

Figure 6

Tectonic fracture aperture distribution of G4 in Block F.

Figure 7

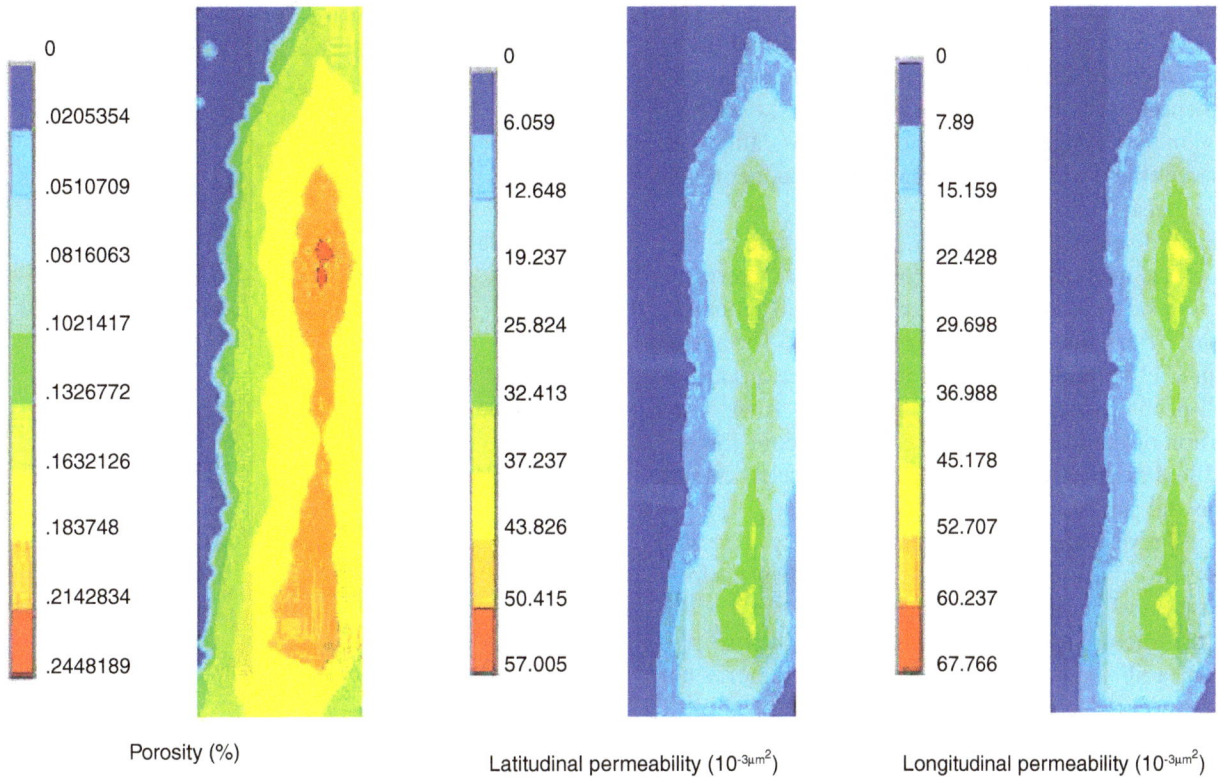

Tectonic fracture porosity and permeability distribution of G4 in Block F.

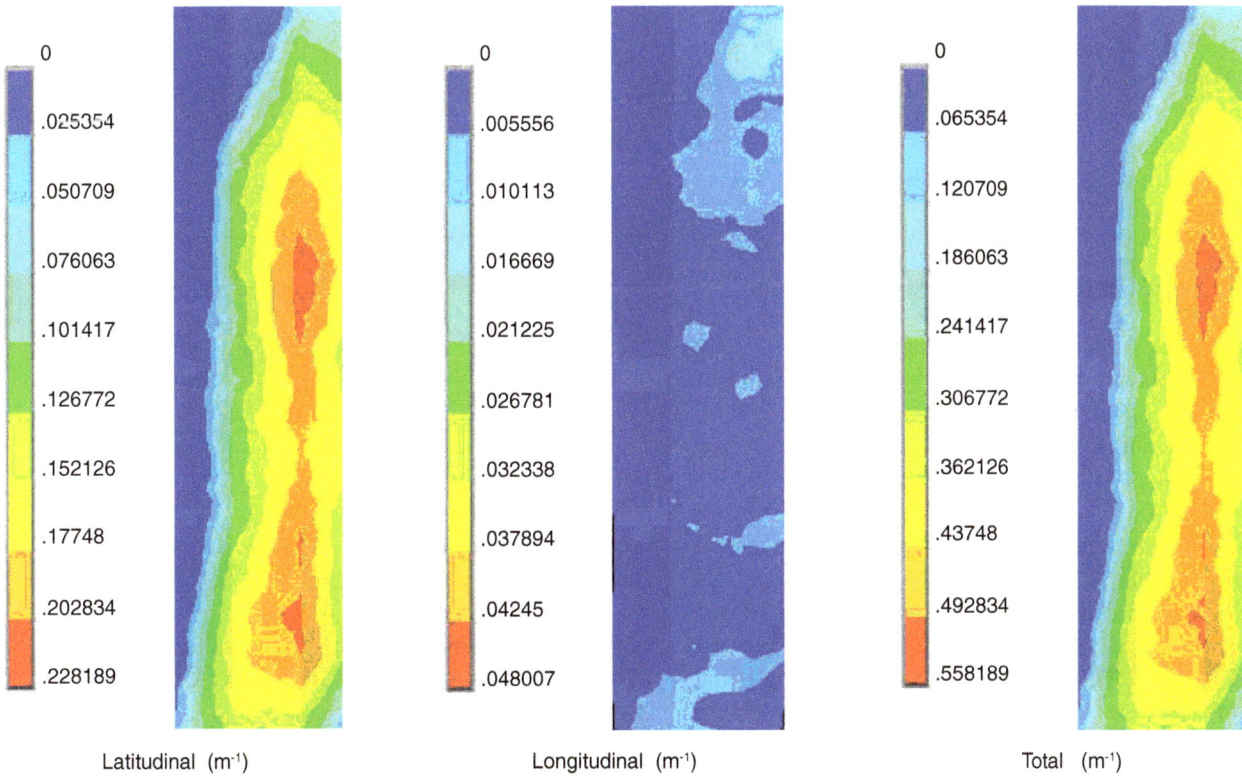

Figure 8

Tectonic fracture linear density distribution of G4 in Block F.

According to the stress distribution shown in Figure 5, the stress at the top of the anticline or fold is greater than the other areas of the anticline. A stress concentration phenomenon exists in the top area. The stress on the top region is stronger than the other areas, and the destructive effects of rock in this region are stronger than the other areas.

Figure 6 displays the tectonic fracture aperture distribution of G4 in Block F. The tectonic fracture aperture in the core of the fold or at the top of the structure is apparently greater compared with other zones. The maximum predictive value is up to 0.5 mm, and serious leakage will probably occur when drilling at this zone. The calculation of the fracture aperture not only contributes to mud loss prediction but also helps in the selection of the type and size of bridging material.

Figure 7 reveals the tectonic fracture porosity and permeability distribution of G4 in Block F. The distribution of porosity and permeability is highly inconsistent with that of the fracture aperture. The area with high porosity and permeability also has a large aperture. The porosity at the top reaches up to 0.25%, and the permeability reaches up to $50 \times 10^{-3} \ \mu m^2$.

Figure 8 shows the fracture linear density distribution. Similarly to Figure 5, the density of the tectonic fracture at the top is greater compared with the other zones in the steeply folded structure. The maximum value is up to $0.55 \ m^{-1}$ or even higher.

The distribution of key fracture parameters of G4 in Block F is determined by using the above analysis. The key fracture parameters at different positions of the whole structure can be obtained by utilizing the same method and finite element analysis applied in G4 in G1, G2, G3 and G5.

5.2 Results of Mud Loss Flow Rate Simulation

Predicting the mud loss flow rate at the different positions of the steeply folded structure can be carried out based on the calculation of fracture aperture, porosity, permeability and density, and by applying fluid flow mechanics. Well PG-X1 had serious mud loss, and its location is marked as "Well PG-X1" in Figure 3. Figure 9 shows the equivalent density profiles of pore pressure, fracture pressure and overburden pressure, as well as the drilling fluid density profile of Well PG-X1. Mud is a power-law fluid, with n equal to 0.75 and K_{ci} equal to $0.03 \ Pa \cdot s^{0.75}$. The main part of the drill string is the 127-mm drill pipe. Table 2 illustrates the well structure of Well PG-X1.

As shown in Figure 10, the mud loss flow rate of Well PG-X1 was predicted by using the proposed method. Severe mud loss occurred when drilling at the bottom

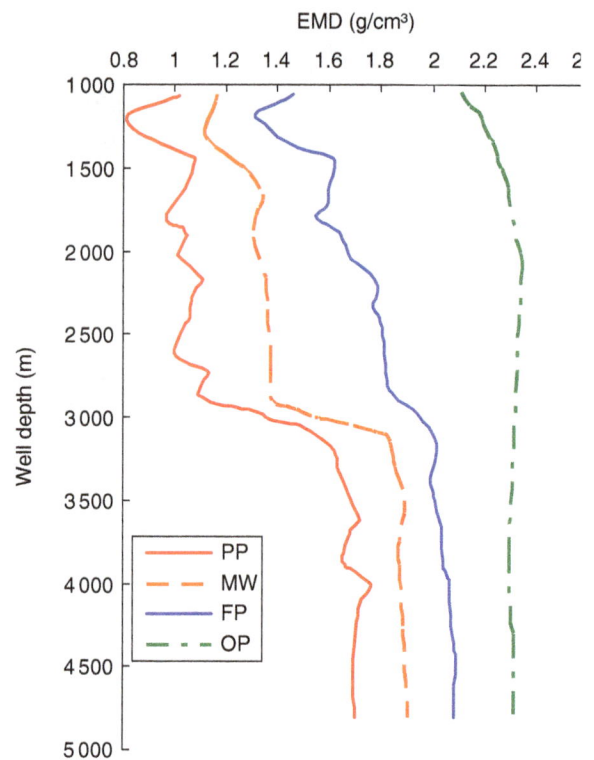

Figure 9

Pore Pressure (PP), Fracture Pressure (FP), Overburden Pressure (OP) and Mud Weight (MW) of Well PG-X1.

TABLE 2
PG-X1 well structure

Well depth (m)	Bit size (mm)	Casing size (mm)	Casing setting depth (m)
540	660.4	508.0	538
2 316	444.5	339.7	2 314
3 932	311.1	244.5	3 930
4 857	215.9	177.8	4 855

of G2 and at most parts of G4. The average mud loss flow rate reached 64 m^3/h, and the total loss volume was approximately 560 m^3 in G4. The predictive value of the mud loss flow rate agrees well with the actual value shown in Figure 10, which verifies the feasibility of the proposed method in predicting the mud loss flow rate.

Two other wells will be drilled in the structure and are marked as "PG-X2" and "PG-X3", as shown in Figure 3. Figure 11 displays the profiles of the mud loss flow rate in Well PG-X2 and Well PG-X3, which are based on the

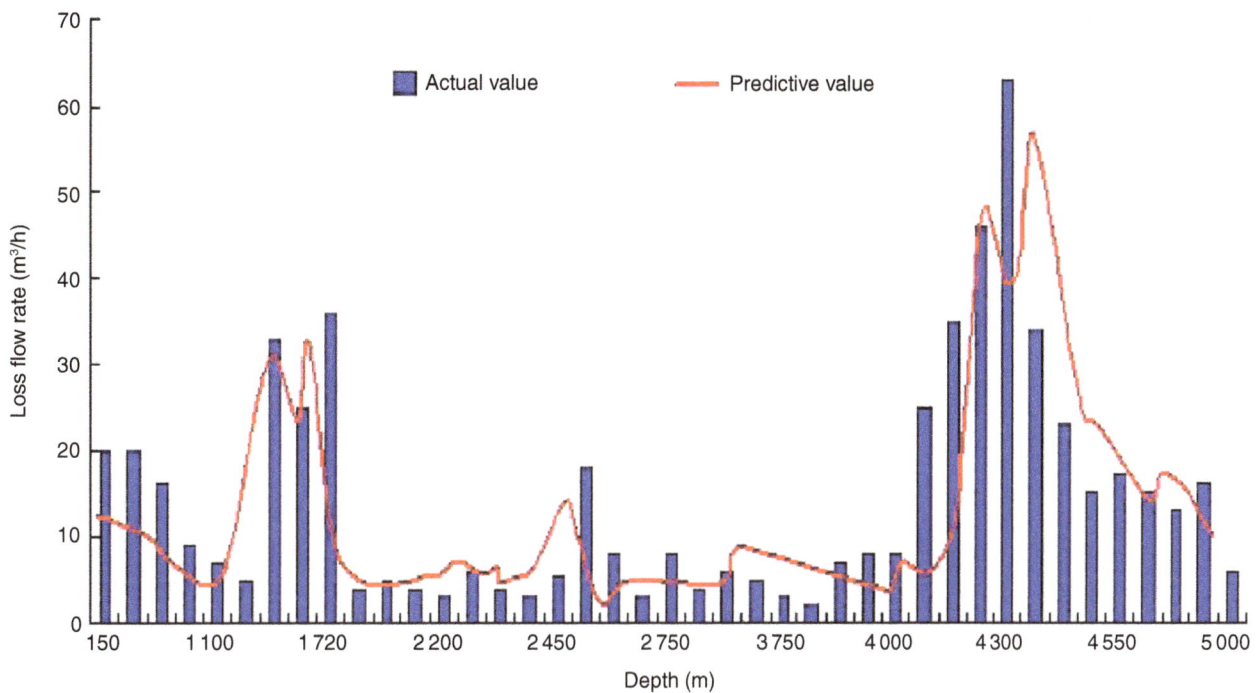

Figure 10

Actual value and predictive value of mud loss flow rate in Well PG-X1.

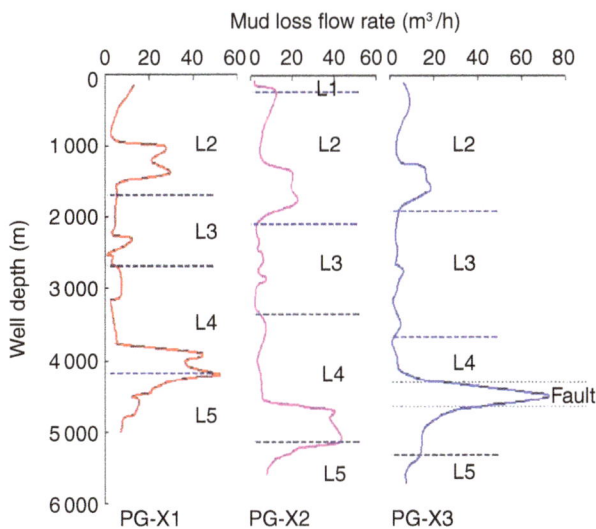

Figure 11

Predictive value of mud loss flow rate in Well PG-X1, Well PG-X2 and Well PG-X3.

results of crustal stress and tectonic fracture distribution simulation. The bottom parts of G2 and G4 have the highest mud loss flow rate in Well PG-X2, which is similar in Well PG-X1. Consequently, the middle part of G4 (4 400 m to 4 600 m), at which the fault crosses the formation, has the highest mud loss flow rate in Well PG-X3. Overall, the mud loss flow rate of Well PG-X1 is higher than those of Well PG-X2 and Well PG-X3. The mud loss flow rate of most parts of Well PG-X3 is low, except for the fault parts where serious cracks have developed. The positions with more serious mud loss lie in the core of the structure, whereas those in the flank of the structure have less mud loss, which is in accordance with the fracture distribution of the structure.

CONCLUSION

By integrating the theory of geological mechanics, rock mechanics and fluid mechanics, we proposed a new method to predict the mud loss flow rate before drilling the tectonic fracture of steeply folded structures. The steps of predicting the mud loss flow rate are as follows: geological structure model establishment, finite element analysis, quantitative calculation of tectonic fracture parameters, and mud loss flow rate prediction. Quantitative description of tectonic fracture parameters related to the mud loss flow rate, such as density, aperture, porosity and permeability, plays an important role in predicting the mud loss flow rate. The density, aperture, porosity and permeability of tectonic fractures at the top of the structure are greater than those of the other zones. The predicted value of the mud loss flow rate agrees well with the actual value in Well PG-X1, which means that the method is feasible in predicting the mud loss flow rate before drilling. The wells have more serious mud loss in the core of the structure, and the wells in the flank of the structure have less mud loss, which is in accordance with the fracture distribution of the structure. This new method makes it possible to predict the mud loss flow rate before drilling, which is helpful in applying preventive measures during the drilling practice in steeply folded structures.

ACKNOWLEDGMENTS

This work is supported by the National Natural Science Foundation of China (No. 51104172, No. 51034007), the National Science and Technology Major Project of China (No. 2011ZX05026-001), "PCSIRT" Project (No. IRT1086) and the Natural Science Foundation of Shandong Province of China (No. ZR2010EL101).

REFERENCES

Anderson R.A., Ingram D.S., Zanier A.M. (1973) Determining Fracture Pressure Gradients from Well Logs, *J. Petrol. Technol.* **25**, 11, 1259-1268, SPE Paper 4135, 10.2118/4135-PA.

Bell J.S., Aadnøy B.S. (1998) Classification of drilling induced fractures and their relationship to *in-situ* stress direction, *Log Analyst* **39**, 6, 216-223.

Chen M., Bai M. (1998) Modeling stress-dependent permeability for anisotropic fractured porous rocks, *Int. J. Rock Mech. Min. Sci.* **351**, 8, 1113-1119.

Coblentz D.D., Sandiford M. (1994) Tectonic stresses in the African Plate: constraints on the ambient lithospheric stress state, *Geology* **22**, 9, 831-834.

Constant David W., Bourgoyne Jr A.T. (1988) Fracture-Gradient Prediction for Offshore Wells, *SPE Drilling Eng.* **3**, 2, 136-140, SPE Paper 15105, 10.2118/15105-PA.

Dai J.S., Li L. (2000) Numerical simulation of tectonic stress field and fracture distribution of Mesozoic and Paleozoic Earthem in Chengdao area, *J. University Petroleum* **24**, 1, 6-9.

Deng Pan, Wei Guo-qi, Yang Yong (2006) To establish and research three-dimentional geological and mathematical model for quantitative prediction of structure fracture, *Natural Gas Geosci.* **17**, 4, 480-481.

Desroches J., Kurkjian A.L. (1999) Application of wireline stress measurement, SPE Paper 58086, 10.2118/58086-PA.

Draou A., Osisanya S.O. (2000) New Methods for Estimating of Formation Pressures and Fracture Gradients from Well Logs, *SPE Annual Technical Conference and Exhibition*, Dallas, Texas, 1-4 Oct., SPE paper 63263, 10.2118/63263-MS.

Du Q.C. (2004) *Geological mechanics analysis and down hole complexity mechanics of the South Edge of Z hunger Basin*, Southwest Petroleum University, Chengdu.

Gazaniol D., Forsans T., Boisson M.J.F., Piau J.-M. (1995) Wellbore failure mechanism in shales: prediction and prevention, *J. Petrol. Technol.* **47**, 7, 589-595.

Ge H.K., Lin Y.S., Ma S.Z. (2001) Modification of Holbrook's Fracture Pressure Prediction Model, *Petroleum Drilling Techniques* **29**, 3, 20-22.

Helio Santos, Placido J.C.R., Wolter C. (1999) Consequences and relevance of drilling vibration on wellbore stability, *SPE/IADC Drilling Conference*, 9-11 March 1999, Amsterdam, Netherlands, SPE Paper 52820, 10.2118/52820-MS.

Hicks T.W., Pine R.J., Willis-Richards J., Xu S., Jupe A.J., Rodrigues N.E.V. (1996) A hydro-thermo-mechanical numerical model for HDR geothermal reservoir evaluation, *Int. J. Rock Mech. Min. Sci. Geomech. Abstr.* **33**, 5, 499-511.

Holbrook P.W. (1989) A New Method for Predicting Fracture Propagation Pressure From MWD or Wireline Log Data, *SPE Annual Technical Conference and Exhibition*, San Antonio, Texas, 8-11 Oct., SPE Paper 19566, 10.2118/19566-MS.

Ji Z.Z., Dai J.S., Wang B.F. (2010a) Quantitative relationship between crustal stress and parameters of tectonic fracture, *Acta Petrolei Sinica* **31**, 1, 68-72.

Ji Z.Z., Dai J.S., Wang B.F. (2010b) Multi-parameter quantitative calculation model for tectonic fracture, *J. China University of Petroleum* **34**, 1, 24-28.

Kwon S., Mitra G. (2004) Three-dimensional finite-element modeling of a thin-skinned fold-thrust belt wedge: Provo salient, Sevier belt, Utah, *Geology* **32**, 7, 561-564.

Lavrov A., Tronvoll J. (2004) Modeling mud loss in fractured formations, SPE Paper 88700.

Maerten L., Maerten F. (2006) Chronologic modeling of faulted and fractured reservoirs using geomechanically based restoration: Technique and industry applications, *AAPG Bull.* **90**, 8, 1202-1226.

Murray G.H. Jr. (1968) Quantitative fracture Study – Sanish pool, McKenzie Co., North Dakota, *AAPG Bull.* **52**, 1, 57-65.

Nikolaevskiy V.N., Economides M.J. (2000) The near-state of stress and induced rock damage, *SPE International Symposium on Formation Damage Control*, Lafayette, Louisiana, 23-24 Feb., SPE Paper 58716, 10.2118/58716-MS.

Prensky S. (1992) Borehole breakouts and *in-situ* rock stress-a review, *The log analyst* **33**, 3, 304-312.

Quintana J.L., Ivan C.D., Blake L.D. (2001) Aphron-Base drilling Fluid: evolving technologies for lost circulation control, *SPE Annual Technology Conference and Exhibition*, New Orleans, Louisiana, USA, 30 Sept.-3 Oct., SPE Paper 71377.

Rocha L.A., Bourgoyne A.T. (1996) A New Simple Method To Estimate Fracture Pressure Gradient (includes associated paper 37685), *SPE Drilling Completion* **11**, 3, 153-159, SPE Paper 28710, 10.2118/28710-PA.

Rocha L.A.S., Falcão J.L., Gonçalves C.J.C., Toledo C., Lobato K., Leal S., Lobato H. (2004) Fracture Pressure Gradient in Deepwater, *IADC/SPE Asia Pacific Drilling Technology Conference and Exhibition*, Kuala Lumpur, Malaysia, 13-15 Sept., SPE Paper 88011, 10.2118/88011-MS.

Sadiq T., Nashawi I.S. (2000) Using Neural Networks for Prediction of Formation Fracture Gradient, *SPE/CIM International Conference on Horizontal Well Technology*, Calgary,

Alberta, Canada, 6-8 Nov., SPE Paper 65463, 10.2118/65463-MS.

Settari A. (1988) A New General Model of Fluid Loss in Hydraulic Fracturing, *SPE J.* **25**, 4, 491-501.

Song Huizhen, Zeng Hairong, Sun Junxiu, Lan Yingang, Huang Fuqiong (1999) Methods of reservoir tectonic fracture prediction and its application, *Seismol. Geol.* **21**, 3, 205-213.

Vitthal S., McGowen J.M. (1996) Fracturing Fluid Leakoff Under Dynamic Conditions Part 2: Effect of Shear Rate, Permeability, and Pressure. *SPE Annual Technical Conference and Exhibition*, Denver, Colorado, 6-9 Oct., SPE Paper 36493, 10.2118/36493-MS.

Warpinski N.R. (1991) Hydraulic fracturing in tight, fissured media, *J. Petroleum Technol.* **43**, 2, 146-151, 208-209.

Wei Jia, Tang Jie, Yue Chengqi, Wu Gangshan (2008) Study of 3-D geological structure model building, *Geophys. Prospect. Petrol.* **47**, 4, 319-327.

Williams B.B. (1990) Fluid loss from hydraulically induced fractures, *J. Pet. Technol.* **22**, 7, 882-888.

Xu T.T., Liu Y.J., Shen W. (1997) *Leak protection and sealing technology*, Petroleum Industry Press, Beijing, pp. 37-43.

5

A Technical and Economical Evaluation of CO_2 Capture from Fluidized Catalytic Cracking (FCC) Flue Gas

Romina Digne*, Frédéric Feugnet and Adrien Gomez

IFP Energies nouvelles, Rond-point de l'échangeur de Solaize, BP 3, 69360 Solaize - France
e-mail: romina.digne@ifpen.fr - frederic.feugnet@ifpen.fr - adrien.gomez@ifpen.fr

* Corresponding author

Résumé — Évaluation technico-économique du captage du CO_2 présent dans les fumées d'une unité FCC (*Fluidized Catalytic Cracking*) — Les contraintes environnementales actuelles relatives aux gaz à effet de serre et parmi eux le CO_2 vont devenir des challenges à relever à court terme. La pression sur l'industrie et par conséquence sur le raffinage afin de limiter et de gérer les émissions de CO_2 va vraisemblablement se renforcer dans les prochaines années.

L'industrie du raffinage contribue pour 2,7 % aux émissions totales de CO_2. Le craquage catalytique en lit fluidisé (FCC) qui est l'un des procédés principaux du raffinage, représente à lui seul 20 % des émissions de CO_2 de la raffinerie. Sachant que ce type d'unité est présente dans une raffinerie sur deux, on comprend bien le défi à trouver des technologies afin d'en gérer les émissions.

Sur la base d'un cas industriel, les objectifs de cette étude sont de déterminer si la technologie HiCapt$^+$, développée pour les centrales électriques, constitue une solution pertinente pour le domaine du raffinage et particulièrement pour le procédé FCC ainsi que d'évaluer le coût additionnel associé qui devra être supporté par les raffineurs.

*Abstract — **A Technical and Economical Evaluation of CO_2 Capture from Fluidized Catalytic Cracking (FCC) Flue Gas** — Environmental issues, related to greenhouse gas and among them CO_2, are becoming short term challenges. Pressure on industries and therefore on refining to limit and manage CO_2 emissions will be reinforced in next few years.*

Refining industry is responsible for about 2.7% of global CO_2 emissions. Fluidized Catalytic Cracking unit (FCC), one of the main process in refining, represents by itself 20% of the refinery CO_2 emissions. As FCC unit is present in half of the refining schemes, it is challenging to find technologies to manage its emissions.

Based on an industrial case, the aims of the presented work are to determine if amine technology HiCapt$^+$, developed for power plant, might be a relevant solution to manage FCC CO_2 emissions and to evaluate the additional cost to be supported by refiners.

INTRODUCTION

Environmental issues and global warming effect are going to strengthen, in short and mid term, greenhouse gas limitations and among them CO_2 emissions. In Europe, pressure on industries have been reinforced, in the last few years by legislation and the two first stages of European Union Emission Trading System (EU ETS). Currently, the cost of a ton of CO_2 emission is quite low (6.5 euros/ton) and this is due to an excess of CO_2 quotas estimated at around 13%.

Stage three of EU ETS is going to be spread for 2013-2020 period. The main differences of this new stage compared to the previous ones will be a decrease of free quotas by 1.75% per year and an increase of no free CO_2 proportion with a final target in 2020 fixed at 70% of total quotas. It means that in 2020, CO_2 quotas will be limited by 21% compared to 2005 situation. In this context, refiners will have to reduce by around 10% their current CO_2 emissions or will have to buy quotas on CO_2 market.

As a reminder, refining industry is responsible for about 2.7% of global CO_2 emissions. The top five of most CO_2 emitters are the vacuum distillation unit (10%), the topping tower (15%), the utility production unit (17%), the steam methane reforming (in the range from 10 to 50% depending on refining scheme) and the Fluidized Catalytic Cracking Unit (FCCU) (20%).

In 2013, refining margins in Europe are low (less than 2 dollars per barrel). This new financial constraint related to CO_2 will be therefore negatively impact for refiners. This implies that there is a need for technological solutions in order to minimize CO_2 emissions for refining processes.

This is especially true for FCCU in the extent that it is one of the main CO_2 contributors and that coke production and combustion during regeneration step are required to run the unit. FCC process converts heavy oil fractions to lighter products such as Liquefied Petroleum Gas (LPG) and gasoline by means of a cracking catalyst. During the reaction step, coke is formed and deposited on the surface of the catalyst, which is then deactivated. To recover catalyst activity, coke is burnt in a regenerator with air and CO_2 is formed. FCC flue gas contain about 10-20% mol of CO_2 for example in full combustion mode. Heat produced in regeneration section is transferred through the catalyst in the reaction section to vaporize the feedstock and to reach the desired riser outlet temperature. The heat balance between regeneration section and reaction section is one of the key points of this process.

The capture of CO_2 from FCC flue gas is therefore a good way to reduce GHG emissions in refineries.

Post-combustion technologies such as CO_2 absorption may be used. The objective of this study is to evaluate the CO_2 capture from flue gas of an industrial FCC unit with available amine technology HiCapt^{+TM} [1] developed by *IFP Energies nouvelles* and *PROSERNAT*. Feasibility and costs have been evaluated.

This work has been carried out within the FCC Alliance program developed by *Total*, *Technip*, *Axens* and *IFP Energies nouvelles*.

1 DESCRIPTION OF HICAPT⁺ AMINE PROCESS FOR CO₂ CAPTURE

Simplified process flow diagrams of HiCapt^{+TM} process for CO_2 post-combustion capture and CO_2 compression are presented in Figure 1 and Figure 2.

Flue gas temperature at inlet of HiCapt^{+TM} unit is around 50°C. In conventional schemes, a water quench tower is mandatory to cool down flue gas to 50°C. Flue gas enters the amine capture unit at atmospheric pressure. A blower is used to increase the pressure in order to compensate for the pressure drop through the absorber and to allow the evacuation of treated gas towards the stack.

Flue gas specifications at inlet of HiCapt$^+$ unit are provided in Table 1.

After the blower, the flue gas is introduced at the bottom of the absorber. This column uses a random or structured packing. The lean solvent is introduced at the top of the absorption section. The solvent is an aqueous solution of MonoEthanolAmine (MEA) at 40 wt%. The solvent and the flue gas flows circulate in counter-current manner through the packing. CO_2 present in flue gas diffuses through the solvent and reacts with MEA. The internal packing enhances mass transfer between the gas and liquid to ensure an optimum efficiency of the CO_2 capture. The absorption zone and the lean MEA flow rate are designed to reach a 90% CO_2 capture.

The decarbonised gas continues its rise through the washing section zone of the absorber, also equipped with packing. This section recovers the MEA and other organic compounds in the vapour (thermodynamic and mechanical entrainments) thanks to water washing at the top of the column. This zone reduces the amounts of degradation products and MEA contained in flue gas (volatile organic compounds, mainly NH_3, MEA, etc.). Part of water extracted from washing zone is sent to the main solvent loop in order to keep a neutral water balance.

The decarbonised flue gas from the absorber is released to the atmosphere. The content of residual MEA is negligible.

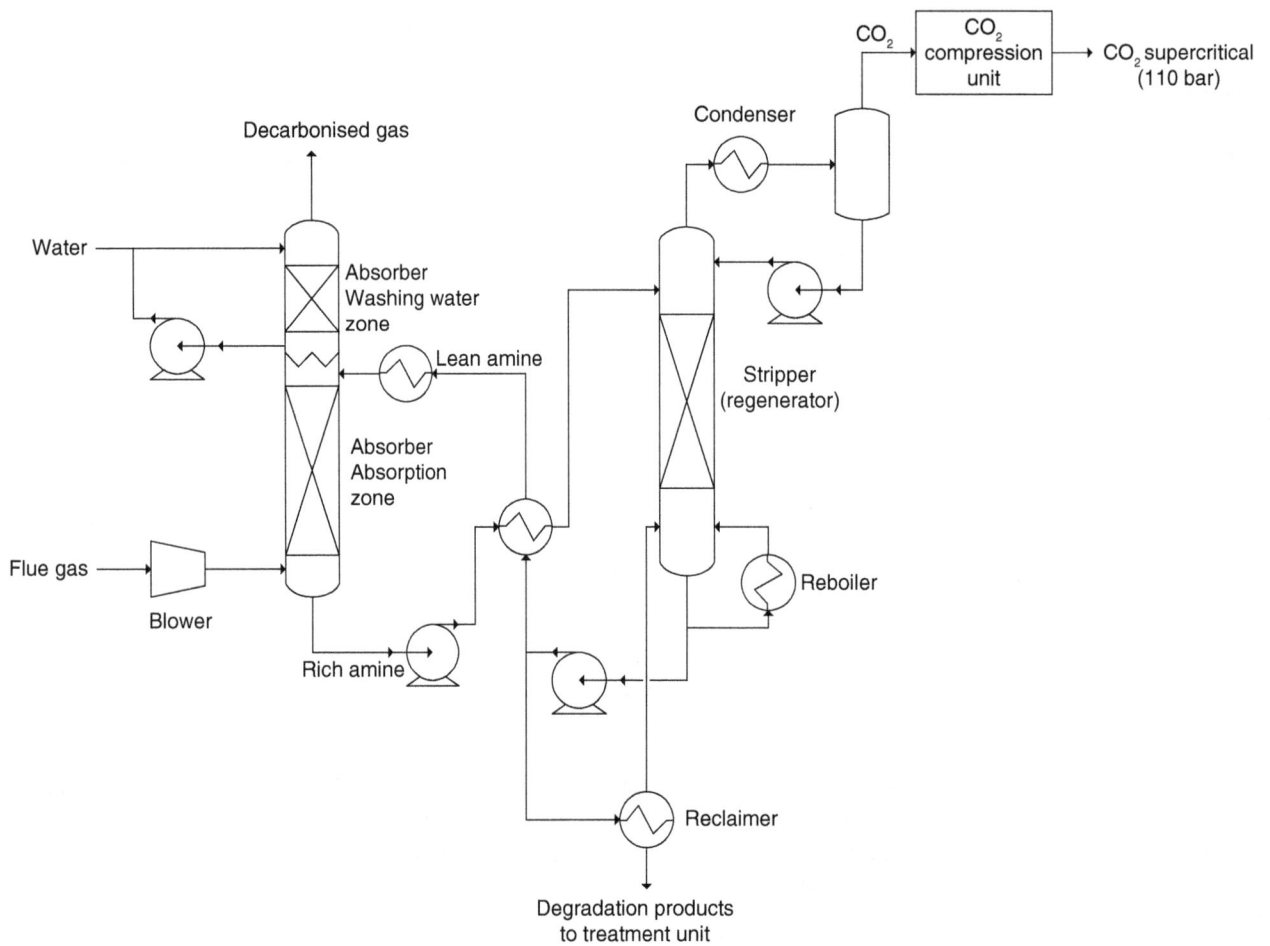

Figure 1

Simplified process flow diagram of HiCapt^{+TM} [1] process.

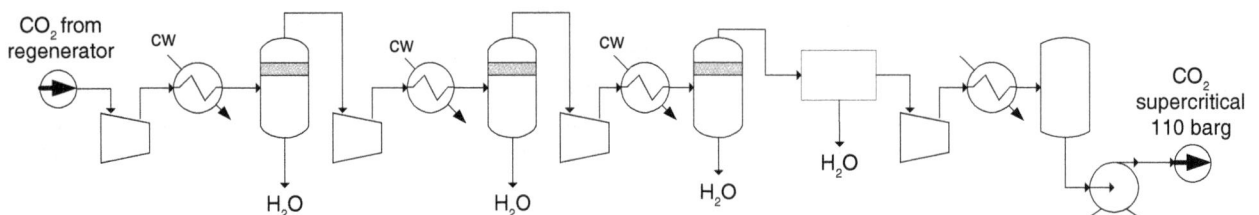

Figure 2

CO_2 compression unit.

The solvent outlet at the bottom of absorber, highly loaded in CO_2, is pumped and then preheated in the heat recovery exchanger, from 60°C to 100°C approximately, by the regenerated solvent coming from the bottom of the stripper (also called regenerator). The preheated solvent is introduced into the stripper, at a pressure between 1 and 2 bar approximately. The rich solvent circulates through the packed column. In the bottom of a regenerator, a reboiler vaporizes a part of the solvent to provide the thermal energy necessary for regeneration.

The regenerated lean solvent is pumped towards the heat recovery exchanger for the preheating of the rich solvent. The temperature of the lean solvent decreases from 120°C to 70°C approximately. A second heat

TABLE 1

Flue gas specifications for HiCapt^{+TM}

Component	Specification
SO_2	< 10-20 mg/Nm^3
NO_2	< 15 mg/Nm^3
NO_x	< 200 mg/Nm^3
Particulates	< 10 mg/Nm^3

exchanger makes it possible to cool the lean solvent to around 50°C before storage and injection in the absorber.

The CO_2 recovered at the top of stripper, containing steam, is sent towards the condenser. The condensate returns towards the regeneration column and the CO_2 flow, with a purity greater than 99.9 mol% (except water content), is then conditioned for its transportation and injection. The conditioning phase needs several stages of compression/condensation and pumping to change CO_2 into a supercritical state at 110 barg, this pressure being able to vary according to specificities of each case.

Moreover, part of the regenerated solvent resulting from the regeneration column is sent towards a batch boiler. This equipment named "reclaimer" enables the vaporization of the solvent (H$_2$O + MEA) to concentrate in an aqueous phase the degradation products (heat stable salt). These by-products are treated by an additional water treatment unit.

The degradation of the solvent is highly limited thanks to the addition of an inhibitor [2] of the reactions due to oxygen. The additive makes it possible to strongly decrease the ammonia emission and to reach the emissions specification at lower costs.

2 INDUSTRIAL FCC UNIT

2.1 FCC Unit Description

To evaluate the feasibility of CO_2 capture from FCC flue gas, an industrial case has been considered. The FCC unit evaluated in this study is located in Europe with processing capacity of 60 000 BPSD (Barrels Per Stream Day). For this case, FCC feed is an hydrotreated vacuum gas oil and the unit operates in maximum gasoline mode (gasoline yield around 51% of fresh feed).

An expander is installed on flue gas from the regenerator. This expander is coupled with the Main Air Blower (MAB). The flue gas at the expander outlet is sent to a waste heat boiler that cools the flue gas and generates a high pressure steam.

The flue gas at waste heat boiler outlet is sent to an ElectroStatic Precipitator (ESP) to reduce particulates content and then vented to the atmosphere trough the stack.

The temperature of the flue gas at the outlet of ESP is around 250°C.

Wet gas or vapors from the main fractionator overhead reflux drum are compressed by the Wet Gas Compressor (WGC). The WGC is a two stage inter-cooled centrifugal machine generally driven by a steam turbine or by an electrical motor. The wet gas compressor of the industrial FCC unit considered for this study is driven by an electrical motor.

2.2 FCC Flue Gas Characteristics

The properties of the FCC flue gas are given after heat recovery in a waste heat boiler and after the electrostatic precipitator. This composition is comparable with values provided by [3].

Flue gas properties are presented in Table 2.

Despite the presence of an electrostatic precipitator, the content of particulates in flue gas is too high for HiCapt^{+TM} process (Tab. 1, 2). Deeper SO_2 and NO_x removals from FCC flue gas are also required for HiCapt^{+TM} process (Tab. 1, 2).

2.3 CO$_2$ Balance of FCC Unit

A simplified CO_2 balance of FCC unit (Tab. 3) has been estimated considering the three main contributors which are:
– coke combustion;

TABLE 2

FCC flue gas characteristics

Pressure	0.04	barg
Temperature	250	°C
Composition		
N_2	77.5	mol%
CO_2	17.7	mol%
H_2O	3.6	mol%
O_2	1.2	mol%
SO_2	134	mg/Nm^3
NO_x	118	mg/Nm^3
CO	15	mg/Nm^3
Particulates	30	mg/Nm^3

- electricity consumption for wet gas compressor and main air blower;
- steam balance (consumption – production).

3 CO$_2$ CAPTURE FROM FCC FLUE GAS WITH HICAPT^{+TM} AMINE PROCESS

3.1 FCC Flue Gas Pre-Treatment before CO$_2$ Capture

SO$_2$ and NO$_x$ removal technologies were investigated. Two different technologies for SO$_2$ and NO$_x$ removals, wet and dry scrubbers, were proposed. In both cases, the temperature of FCC flue gas is 250°C.

3.1.1 Wet Scrubber Technology (*Fig. 3*)

In this case, the gas treatment consists of dust capture by electrostatic precipitator at 250°C, then a catalytic DeNO$_x$ scrubber at 250°C with injection of ammonia solution (NH$_4$OH) 25 wt%. The DeSO$_x$ washing section (with caustic soda) reduces the SO$_2$ concentration at 10 mg/Nm^3. The DeSO$_x$ unit works at low temperature (about 50°C). This technology is mature. The electrofilter is mandatory according to the wet scrubber supplier.

3.1.2 Dry Scrubber Technology (*Fig. 4*)

In case dry scrubber is used, an inlet temperature of 200°C is recommended based on the reactant used

TABLE 3

Simplified CO$_2$ balance of FCC unit

	Value	CO$_2$ emission factor	CO$_2$eq (t/h)
Coke combustion	18.4 t/h	3.4 t CO$_2$eq/t of coke	62.9
Electricity consumption for MAB and WGC	6.1 MW	148 g CO$_2$eq/MJ of electricity[1]	3.3
Steam balance (consumption – production)	−47.8 MW	72 g CO$_2$eq/MJ of steam[2]	−12.4
Total			53.8

[1] Corresponding to world average (2004),
[2] Corresponding to steam production in a boiler with typical refinery fuels.

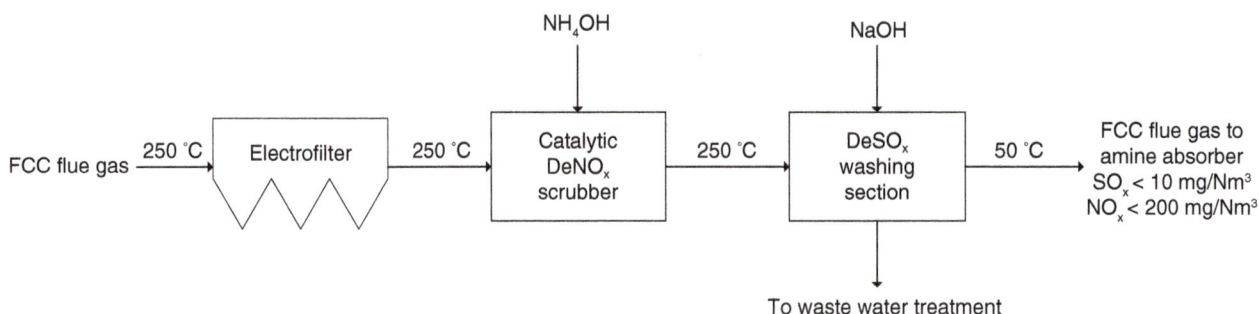

Figure 3

Wet scrubber technology.

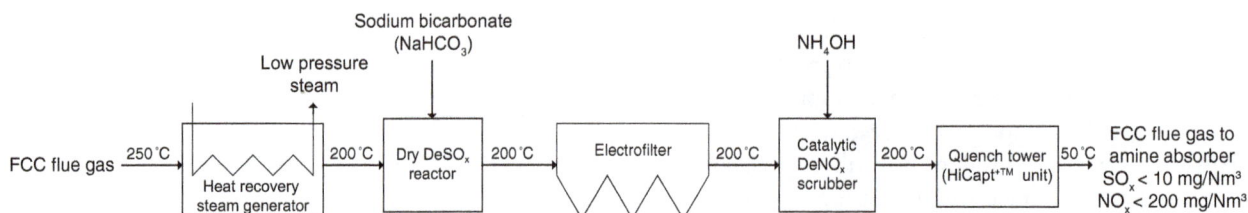

Figure 4

Dry scrubber technology.

(sodium bicarbonate $NaHCO_3$) and the $DeSO_x$ scrubber dry reactor. The gas treatment requires a cyclone to protect the electrofilter. After the electrofilter, the $DeNO_x$ scrubber works at 200°C adding NH_4OH. This technology works at 200°C and it is necessary to reduce the temperature from 250°C to 200°C before the dry $DeSO_x$ reactor. Heat recovery on FCC flue gas can be implemented to recover heat from 250°C to 200°C. 6.3 t/h of Low Pressure (LP) steam can be generated. In the case of dry scrubber, the quench tower at the amine capture section is required to cool down the flue gas from 200°C to 50°C. This technology is considered more complex than the wet scrubber solution.

3.2 Impact of Pre-Treatment on CO₂ Capture

The design of the amine unit is independent of the $DeSO_x/DeNO_x$ technology. The gas composition before $DeSO_x/DeNO_x$ unit implies that it is undersaturated at 50°C and requires make-up water to reach a neutral water balance in the process battery limits.

For a wet scrubber technology, the make-up water comes from the $DeSO_x$ washing tower. In this case, the flue gas could be injected directly at the bottom of the absorber.

For a dry scrubber technology, a quench tower is required before the absorber, to decrease the temperature from 200°C to 50°C. The water saturation occurs in this equipment and the make-up is added to the water cooling loop directly.

3.3 Utilities for Flue Gas Pre-Treatment and CO₂ Capture

The list of utilities used for pre-treatment and CO_2 capture and compression is indicated in Table 4.

TABLE 4

Chemicals and utilities balance for HiCapt^{+TM} process

Caustic soda
Sodium bicarbonate (for dry scrubber case only)
Ammonia aqueous
MEA
Anti-oxidation additive
Electricity
LP steam
Cooling water

The main contributors of HiCapt^{+TM} operating cost are low pressure steam (3 GJ/t of captured CO_2), electricity and cooling water.

3.4 CO₂ Balance of FCC Unit with HiCapt⁺ᵀᴹ Process

As for the reference case, a simplified CO_2 balance has been performed for FCC unit but at that time a more severe flue gas post treatment and the CO_2 capture on flue gas were considered.

This CO_2 balance has been estimated to 14.1 t CO_2eq/h that is to say a reduction of 74% of CO_2 emissions of the reference case.

As an FCC unit represents around 20% of refinery total CO_2 emissions, HiCapt^{+TM} association with an FCC unit will remove more than 14% the refinery emissions and therefore much more than the required target of 10% as presented in the introduction.

Nevertheless, the cost associated with this significant GHG emission reduction has to be evaluated.

3.5 Economical Evaluation of CO₂ Capture with HiCapt⁺ᵀᴹ Process

The following section presents investment and operating cost estimates considering the wet scrubber technology for $DeSO_x/DeNO_x$.

3.5.1 Investment Cost Estimation

ISBL (InSide Battery Limit) cost of FCC unit has been estimated without and with CO_2 capture on flue gas. Results are presented in Table 5.

3.5.2 Operating Cost Estimation

HiCapt^{+TM} shows energy consumption between 3.1 to 3.3 GJ to reduce CO_2 emissions by one ton. This places HiCapt^{+TM} among the most currently effective process technologies. A techno-economic evaluation of HiCapt^{+TM} compared to 30 wt% MEA process shows a reduction of around 15% in the cost of CO_2 captured [1].

Operating cost of an FCC unit has been estimated with and without the CO_2 capture system on flue gas. Operating cost includes utilities, chemicals and catalyst

TABLE 5

ISBL cost

	FCC unit without CO₂ capture	FCC unit with CO₂ capture
ISBL (M$)	Base	Base × 1.25

TABLE 6

Operating cost

	FCC unit without CO_2 capture	FCC unit with CO_2 capture
Operating cost	Base + CO_2 penalty of 75 $/t CO_2 avoided	Base

costs. Results are presented in Table 6. A penalty at 75 €/t of CO_2 avoided must be compare with the cost of CO_2 avoided in a refinery. In a recent study [4], the cost of CO_2 capture lies in the range of 40-263 €/t CO_2-refinery-avoided. Process integration between the capture process and the refinery is a key point to reduce costs, for example using excess heat or combined with heat pumps.

As presented in CO_2 balance, one of the main contributors to CO_2 in FCCU is the compressors and especially the wet gas compressor. In order to go further in FCC CO_2 reduction, it is therefore interesting to investigate and evaluate solutions to reduce utilities required for this compressor.

3.6 Impact of Wet Gas Compressor Driver

Generally, wet gas compressors are driven by condensing steam turbines or electric motors as in the reference case considered previously. In condensing steam turbines, exhaust steam is in a partially condensed state (vapor fraction near 90%) and at a pressure well below atmospheric. Exhaust steam is then condensed with water (*Fig. 5*).

When steam is preferred to drive the wet gas compressor and when HiCapt^{+TM} process is used for CO_2 capture, it is interesting to use a back-pressure steam turbine instead of a condensing steam turbine. The back-pressure steam turbine will consume more high pressure steam but low pressure steam at turbine outlet can be used directly in HiCapt^{+TM} process for amine regeneration (*Fig. 6*).

The back-pressure steam turbine has the advantage to reduce the consumption of the cooling water of the process. A water condenser for vacuum steam condensation is no more needed. The flow rate of cooling water to condense vacuum steam is always very high.

The variation of operating cost and GHG emissions for the system "WGC + amine regeneration" is indicated in Table 7 compared to a total condensing steam turbine.

The back-pressure turbine is therefore a relevant solution to limit CO_2 emissions if there is a specific need of LP steam as it is when amine capture is implemented.

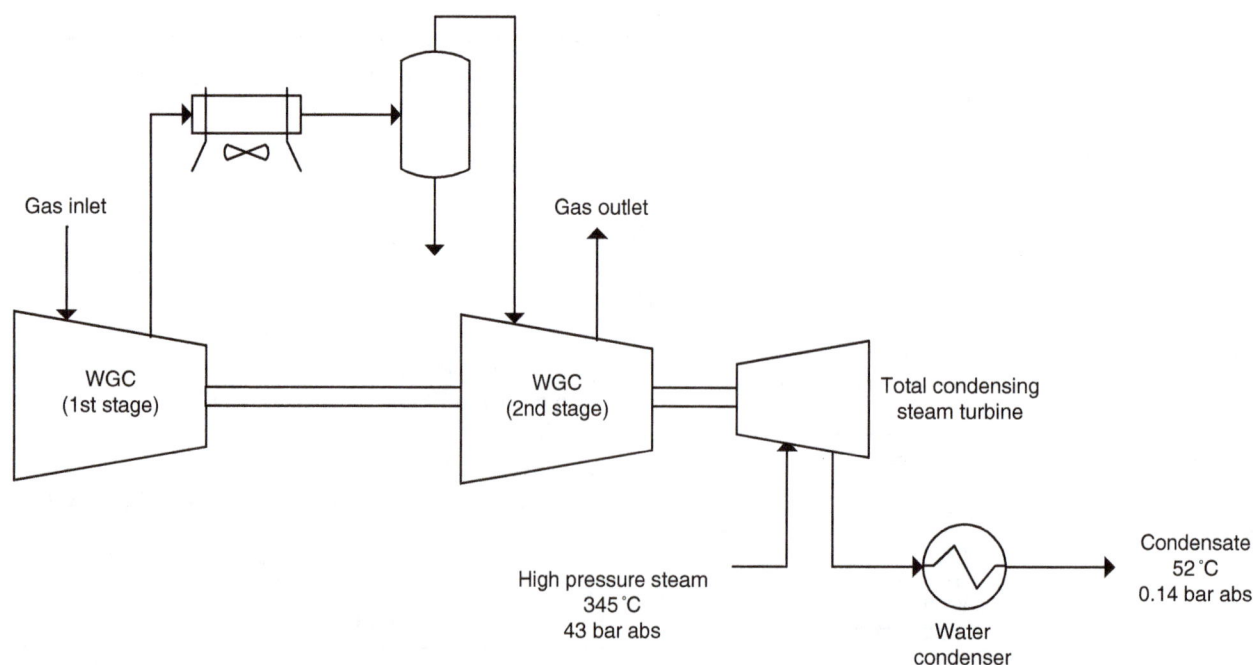

Figure 5

Wet gas compressor driven by a condensing steam turbine.

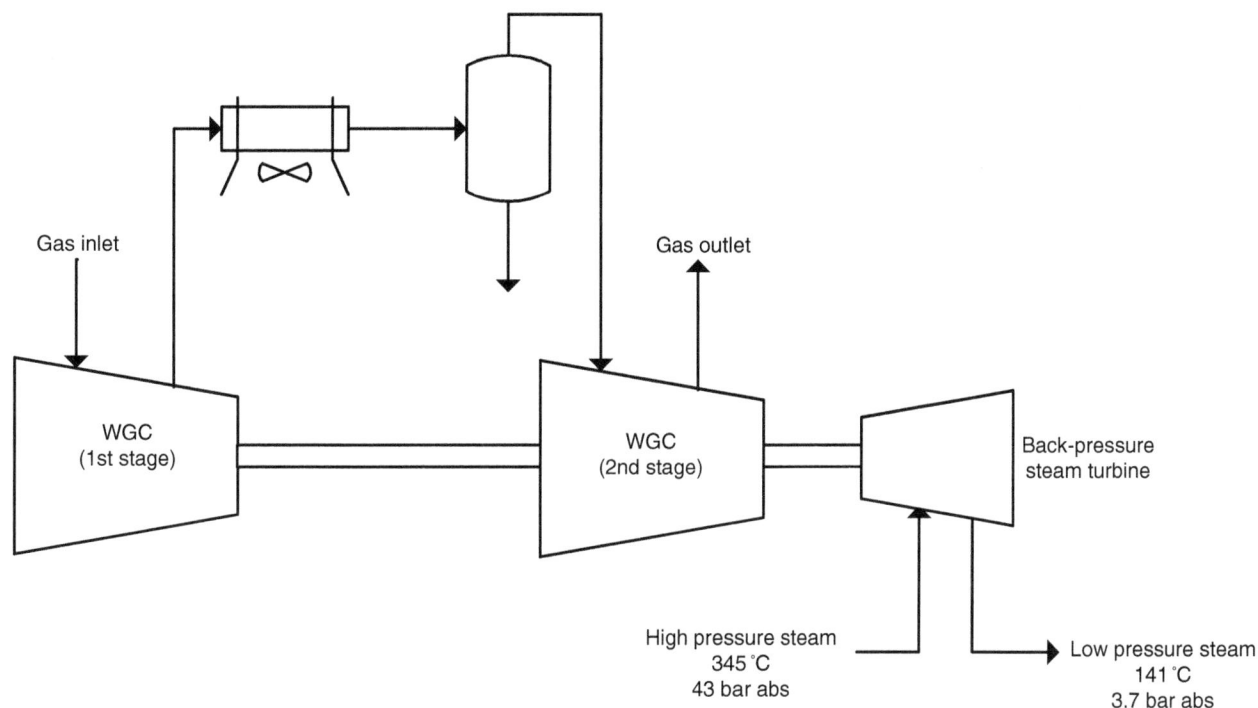

Figure 6

Wet gas compressor driven by a back-pressure steam turbine (HP → LP steam).

TABLE 7

Comparison of operating cost and GHG emissions for total condensing and back-pressure steam turbine

WGC driver type	Total condensing steam turbine	Back-pressure steam turbine (HP → LP steam)
HP steam for WGC driver	Base	Base + 41 t/h
Cooling water for WGC driver	Base	0
LP steam for amine regeneration	Base	Base − 87 t/h
Operating cost[1]	Base	Base − 4.8 M$/year
GHG emissions[2]	Base	Base − 7.3 t CO_2eq/h

[1] Considering following costs: HP steam = 22 $/t, LP steam = 15 $/t, Cooling water = 0.08 $/m³;
[2] Considering following GHG emission factors: HP steam = 221 kg CO_2eq/t, LP steam = 183 kg CO_2eq/t, Cooling water = 0.188 kg CO_2eq/m³.

CONCLUSIONS

The presented work enables to conclude that HiCapt^{+TM} process is a relevant technology to manage CO_2 in FCC flue gas. In a technical point of view, FCC Flue gas can be treated in HiCapt^{+TM} process because HiCapt^{+TM} inlet specifications can be easily reached. Based on a representative industrial case, it was evaluated that 74% of CO_2 emitted in FCC can be captured and this corresponds to a reduction of more than 14% of the total CO_2 emitted in the refinery.

In an economical point of view, an amine capture unit leads to an additional cost estimated at around 25% which is significant but relatively limited. The impact on operating cost is fully in accordance with the one for power plant for which HiCapt^{+TM} process was developed. As amine capture requires LP steam, back pressure turbine for wet gas compressor is an effective option which leads to additional CO_2 gains. In conclusion, HiCapt^{+TM} process is therefore a possible solution to reduce CO_2 emissions for refining processes and especially for FCC.

REFERENCES

1   Lemaire E., Bouillon P.A., Gomez A., Kittel J., Gonzalez S., Carrette P.L., Delfort B., Mougin P., Alix P., Normand L. (2011) New IFP optimized first generation process for post-combustion carbon capture: HiCapt+^{TM}, *Energy Procedia* **4**, 1361-1368.

2   Delfort B., Carrette P.L., Bonnard L. (2009) Additives for inhibiting MEA oxydation in a post-combustion capture process, *IEA Greenhouse Gas R&D's 12th Int. Post Comb. Network Meeting*, Univ. Regine, 29 Sept.-1 Oct.

3   de Mello L.F., Pimenta R.D.M., Moure G.T., Pravia O.R. C., Gearhart L., Milios P.B., Melien T. (2009) A technical and economical evaluation of CO_2 capture from FCC units, *Energy Procedia* **1**, 117-124.

4   Johansson D., Franck P.-A., Berntsson T. (2013) CO_2 capture in oil refineries: Assessment of the capture avoidance cost associated with different heat supply options in a future energy market, *Energy Conversion and Management* **66**, 127-142.

6

Corrosion in CO_2 Post-Combustion Capture with Alkanolamines – A Review

J. Kittel* and S. Gonzalez

IFP Energies nouvelles, Rond-point de l'échangeur de Solaize, BP 3, 69360 Solaize - France
e-mail: jean.kittel@ifpen.fr

* Corresponding author

Résumé — Corrosion dans les procédés utilisant des alcanolamines pour le captage du CO_2 en post-combustion — Les procédés de captage et de stockage du CO_2 occupent une place importante dans les stratégies visant à limiter les émissions industrielles de gaz à effet de serre. Les procédés de captage en post-combustion par solvant chimique de type alcanolamine sont bien adaptés au traitement d'émissions ponctuelles massives issues de combustibles fossiles, rencontrées notamment dans les centrales thermiques au charbon et au gaz, ou les industries sidérurgiques ou de production de ciment. La technologie utilisant le principe d'absorption – désorption par les alcanolamines est une des voies les plus matures à ce jour : elle est en effet déjà mise en œuvre, par exemple pour la désacidification du gaz naturel, bien qu'à une échelle sensiblement plus petite. L'opération de telles unités pour le traitement des fumées de combustion présente toutefois de nombreux challenges, parmi lesquels la corrosion des équipements tient une place importante. Le but de cet article est de présenter une revue des connaissances sur cet aspect particulier. Dans une première partie, l'expérience issue de plusieurs décennies d'utilisation de procédés aux alcanolamines dans le domaine de la production d'huile et de gaz est présentée. Dans une seconde partie, les risques spécifiques associés aux particularités des procédés de captage du CO_2 en post-combustion sont identifiés et discutés. Différentes stratégies de maîtrise de la corrosion sont décrites, et certains axes prioritaires en matière de recherche et développement sont proposés. Enfin, certaines difficultés en vue du transport du CO_2 issu du captage puis de son injection à fin de stockage géologique sont mises en avant, avec des recommandations strictes sur les teneurs maximales en impuretés pour disposer en toute sureté du CO_2 issu de ce procédé.

Abstract — Corrosion in CO_2 Post-Combustion Capture with Alkanolamines – A Review — CO_2 capture and storage plays an important part in industrial strategies for the mitigation of greenhouse gas emissions. CO_2 post-combustion capture with alkanolamines is well adapted for the treatment of large industrial point sources using combustion of fossil fuels for power generation, like coal or gas fired power plants, or the steel and cement industries. It is also one of the most mature technologies to date, since similar applications are already found in other types of industries like acid gas separation, although not at the same scale. Operation of alkanolamine units for CO_2 capture in combustion fumes presents several challenges, among which corrosion control plays a great part. It is the aim of this paper to present a review of current knowledge on this specific aspect. In a first part, lessons learnt from several decades of use of alkanolamines for natural gas separation in the oil and gas industry are discussed. Then, the specificities of CO_2 post-combustion capture are presented, and *their consequences on corrosion risks are discussed. Corrosion mitigation strategies, and research and development efforts to find new and more efficient solvents are also highlighted. In a last part, concerns about CO_2 transport and geological storage are discussed, with recommendations on CO_2 quality and concentration of impurities.*

INTRODUCTION

CO_2 capture roughly consists in separating different gases initially mixed in the combustion fumes. The main goal is to extract CO_2 before it is released in the atmosphere.

The most widely used process uses alkanolamine-based chemical solvents capable of reacting preferentially with CO_2. It is based on reversible chemical reactions between CO_2 and the aqueous amine, leading to the formation of bicarbonate and protonated amine (Reaction 1) and/or to the formation of amine carbamate (Reaction 2) [1].

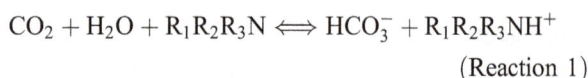

$$CO_2 + H_2O + R_1R_2R_3N \Longleftrightarrow HCO_3^- + R_1R_2R_3NH^+$$
(Reaction 1)

$$CO_2 + 2R_1R_2NH \Longleftrightarrow R_1R_2NCOO^- + R_1R_2NH_2^+$$
(Reaction 2)

In these reactions, R_1, R_2 and R_3 represent alkyl groups or a hydrogen atom.

While the mechanism of Reaction 1 may occur for all types of amine, it proceeds with relatively slow kinetics since it is limited by dissociation of carbonic acid into bicarbonate. It is thus not well adapted to CO_2 post-combustion capture which requires fast reactions.

On the contrary, Reaction 2 is much faster. However, since it leads to the formation of amine carbamate, it is only possible with primary or secondary amines which have hydrogen bond to the nitrogen. Therefore, tertiary amines are usually discarded for CO_2 capture applications, unless an activator is used to increase the reaction rates.

The industrial process of CO_2 capture with alkanolamines is described in Figure 1. It is based on the fact that chemical equilibria of Reaction 1 and Reaction 2 are shifted to the left at high temperature. This property is put into advantage in the industrial process, which consists of successive absorption – desorption in a loop system.

The flue gas which enters in the treatment plant at the bottom of the absorber is typically composed of nitrogen, with 10 to 20% of CO_2 and 5 to 10% O_2, with contaminants such as SO_x and NO_x at trace levels. The gas pressure is typically between 1 and 2 bar. Lean amine is introduced at the top of the column, and chemical reaction between the amine and CO_2 takes place.

At the liquid outlet at the bottom of the absorber, the solvent is enriched in acid gas: one speaks of rich amine. At the top of the absorber, the flue gas has been stripped of its CO_2.

The rich amine is then pre-heated to 90-110°C by a heat exchanger then fed into the top of a regeneration column (stripper). In this part of the unit, the solvent is raised to higher temperature by steam, typically between 120-130°C, which releases the dissolved CO_2. At the liquid outlet of the regenerator, the solvent is hot and contains less acid gas: one speaks of lean amine. The solvent is then cooled by the heat exchanger and sent back to the top of the absorber to start a new cycle. Pure CO_2 is collected at the top of the regenerator. When CO_2 is collected for geological sequestration, it has to be compressed to more than 100 bar for transportation. This compression step also represents an important penalty in terms of energy consumption.

This process using amines for acid gas separation has long been used for natural gas treatment, and it is well known that corrosion represents a major operational issue. A recent evaluation of cost of corrosion in gas sweetening plants concluded that 25% of the maintenance budget was committed to corrosion control [3]. It was also found that approximately half of the maintenance work orders were due to corrosion.

It is thus important to pay great attention to corrosion in the research and development work in progress for CO_2 capture with amines.

The present paper aims at presenting the current knowledge on this topic.

In the first section, experience from several decades in natural gas sweetening is presented. The main types of corrosion are described, and the impact of operational parameters on corrosion is discussed. Corrosion risks associated with the main equipments of the gas separation plant are then discussed individually, with a thorough analysis of industrial corrosion failures reported in the literature.

The second section deals with more recent work on amine process for CO_2 post-combustion capture. MonoEthanolAmine (MEA) represents the benchmark solvent, and a detailed analysis of laboratory and pilot plant data obtained with this amine is proposed. A short paragraph is also dedicated to work in progress for the development of new solvents, aimed at being more efficient, less costly, and sometimes less corrosive than MEA.

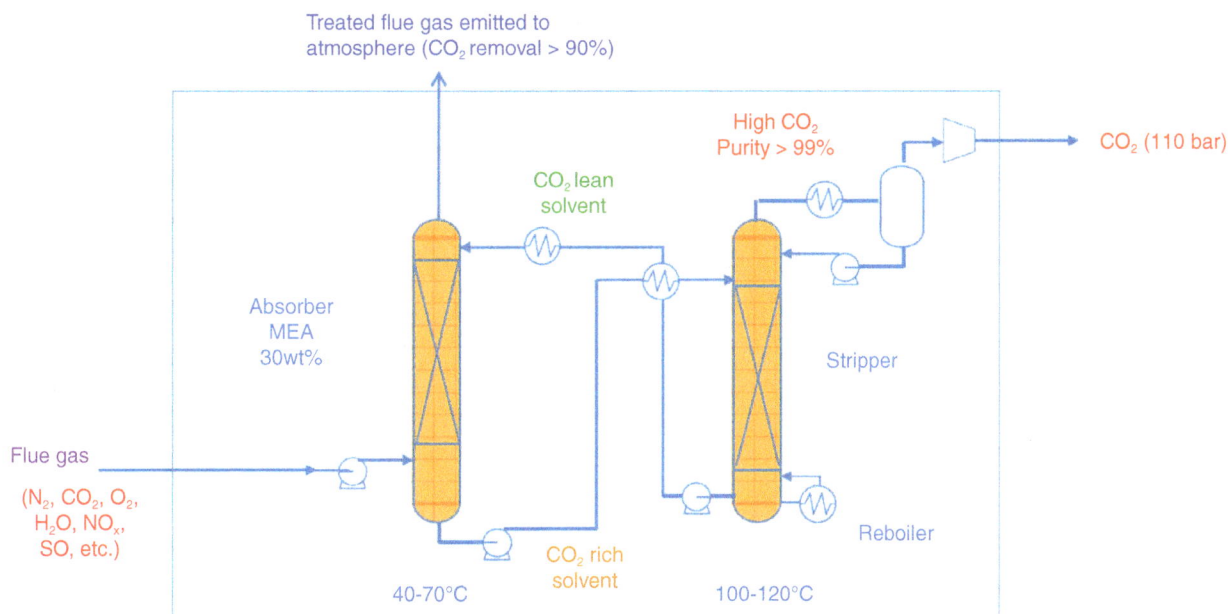

Figure 1

Simplified diagram of a MEA CO_2 capture unit [2].

Finally, a brief look at corrosion issues in the CO_2 leaving the capture plant for transport and storage concludes the paper.

1 EXPERIENCE FROM NATURAL GAS SWEETENING

Using amines for the removal of acid gases is not a new process. It has been used for natural gas treatment or in refineries for several decades, and corrosion has always been considered as one of the major operational problems [4-11]. In the eighties, several industrial failures in gas treating plants were reported, the most important one causing the death of 17 employees [12-14]. Since then, lots of efforts were done to improve the understanding of corrosion processes in amine units. An extensive literature survey is proposed in the next paragraphs.

In such complex units, numerous pieces of equipment are exposed to equally numerous types of corrosion.

An interesting classification of the types of corrosion occurring in gas treatment plants was proposed by Nielsen [15], who identifies:
– wet acid gas corrosion,
– amine solution corrosion.

1.1 Acid Gas Corrosion

Wet acid gas corrosion is encountered in all parts of the unit in contact with an aqueous phase with a high concentration of dissolved acid gases CO_2, H_2S, as well as NH_3 and HCN for refinery units. This type of corrosion is found primarily in zones where the gaseous phases have high concentrations of acid gases and where water may condense, mainly at the bottom of the absorber and the top of the regenerator [15, 16].

For gas containing mostly CO_2, parts of the installation made from carbon steel may suffer fast uniform corrosion, up to several mm/year. In the presence of H_2S, this uniform corrosion is generally delayed by the formation of a protective iron sulfide layer. A minimum H_2S/CO_2 ratio of 1/20 is often considered as sufficient to avoid risks of uniform CO_2 corrosion [17-19]. In the presence of H_2S however, specific cracking phenomena may also be encountered (hydrogen embrittlement, hydrogen induced cracking HIC, sulfide stress cracking SSC, etc.). In the presence of HCN and/or NH_3, the risks of cracking are also increased [16].

1.2 Corrosion by Amine Solution

1.2.1 Mechanisms and Influent Parameters

The second type of corrosive media found in acid gas removal units consists of amine solution. Generally, amines are not intrinsically corrosive, since they associate both high pH and low conductivity. They may nevertheless become corrosive when they absorb CO_2 or H_2S. Furthermore, since the treatment units operate in semi-closed circuit, the solvent may become enriched with possibly corrosive degradation products [15, 18, 19].

No consensus has yet been reached concerning the mechanisms of corrosion by amine solutions. The models proposed vary depending on the type of amine (in particular, primary, secondary and tertiary), the H_2S/CO_2 ratio in the gas to be treated, the possible presence of oxygen either as contaminant in the circuit or as component of the input gas (*e.g.* CO_2 capture in fumes) [8, 10, 18, 20-23].

One may nevertheless identify some systematic trends governing the corrosivity of acid gas chemical solvents.

Acid gas loading (α) and temperature are usually considered as the most important factors. The acid gas loading is defined as the quantity of acid gas absorbed by a defined quantity of solvent and is often expressed in moles of acid gas per mole of amine. Increasing the acid gas loading increases the corrosivity of amine solutions [18, 19, 24, 25].

Temperature generally has an extremely important effect on corrosion phenomena since most electrochemical reactions involved are thermally activated. It is common practice in industry to consider that the corrosion rate is doubled when the operating temperature increases by 10 K to 20 K. For gas treatment units, the effect of temperature is relatively difficult to asses on an individual basis. Temperatures vary widely in the installation, with extreme values ranging from 40°C in the absorber up to 130°C in the reboiler. However, these temperature variations have a significant effect on the chemistry of the solution, in particular the acid gas loading. Taking into account both the loading and the temperature, it is usually considered that the main corrosion risks are encountered in areas with high loading and high temperatures [26]. These conditions are generally found in the rich amine line after the heat exchanger and up to the regenerator input.

The type of amine is also an important factor. Usually, primary amines (*e.g.* MEA) are the most corrosive, secondary amines (*e.g.* DiEthanolAmine, DEA) slightly less and tertiary amines (*e.g.* Methyl DiEthanolAmine, MDEA) exhibit the lowest risks of corrosion [18, 19, 25, 27-30]. Amine concentration also has an influence on corrosion. Excessively high amine concentrations should generally be avoided. Nevertheless, the results obtained from the few laboratory studies conducted on the effect of amine concentration on corrosivity vary widely, between a marked effect [18, 25] and a moderate or null effect [31, 32].

The concentration in degradation products and contaminants can significantly influence corrosion reactions. A distinction must be made between basic and acidic degradation products. Basic amine degradation products mainly result from chain reactions between amine and CO_2, for example the following compounds: HEOD (3-(2-hydroxyethyl)-2-oxazolidone), BHEP (N,N'-bis(2-hydroxyethyl)piperazine), THEED (N,N,N'-tris(2-hydroxyethyl)ethylenediamine). The studies on corrosion by these degradation products date back a number of years, the general conclusion being an absence of specific corrosivity [27, 28]. Most acidic degradation products result from reactions with oxygen. The main products include salts of oxalic, glycolic, formic and acetic acids, which are stronger than carbonic acid. As a result these salts are not thermally regenerated in the process, hence their name: Heat Stable Salts (HSS). The effect of these products on corrosion has been well documented through laboratory tests; they increase corrosion of carbon steel [33-35].

Finally, the solvent flow rate and conditions favourable to turbulence (gas flash, gas injection zones, etc.) may cause risks of erosion-corrosion. This type of corrosion is specific to carbon steels, since stainless steel grades are far more resistant. This type of corrosion is probably aggravated when the content of degradation products becomes too high: some of these products have a chelating effect on iron and may favour more efficient and faster dissolution of the protective deposits exposed to erosion [16, 17, 24, 36].

1.3 Equipments Concerned by Corrosion

This section describes the specific corrosion risks for the main equipment in gas treatment units. A summary is proposed in Table 1.

1.3.1 Absorber

The absorber may suffer several types of corrosion.

Among the parameters which affect corrosion, temperature and acid gas loading may vary on a wide range in the absorber. Temperature might typically evolve between ambient and 80°C, due to the exothermicity of absorption reactions. Acid gas loading may vary from lean (*i.e.* typically below 0.1 mol_{CO_2}/mol_{amine}) at the top of the absorber to rich (*i.e.* typically above 0.4 mol_{CO_2}/mol_{amine}) at the bottom. Therefore, the highest risks of corrosion by amine solution are found in the hot rich section at the bottom of the absorber. Erosion – corrosion represents an aggravating factor. In particular, the high flow rate at gas inlet may lead to turbulence and impingement of solution against the walls, creating conditions favourable to this type of corrosion. The same phenomenon is observed on the plates and in case of excessive flow rates.

TABLE 1

Summary of feedback on corrosion in amine units (CS = Carbon Steel)

Material and type of corrosion	Causes	References
Absorber		
CS – Cracking and mechanical failure	Hydrogen embrittlement or ASCC arising from non PWHT welds	[12, 14, 40, 42, 43]
CS – Uniform corrosion at bottom of absorber	Galvanic coupling with copper deposits from the corrosion inhibitor	[44]
CS – Erosion-corrosion at bottom of absorber	Turbulence at the raw gas inlet	[19, 45]
AISI 410 – Uniform corrosion of the plates	Higher corrosivity of sweet services units in high temperature and high loading zones	[26, 46]
Rich amine lines		
CS – Stress corrosion (ASCC)	No PWHT	[12, 14, 40, 42, 43]
CS – Erosion-corrosion	Excessive flow rates and acid gas flash	[19, 26, 45, 46]
CS – Localised corrosion with perforation	Large quantities of oxygen in the raw gas	[47]
Exchanger		
CS – Amine Stress Corrosion Cracking (ASCC)	No PWHT	[12, 14, 40, 42, 43]
CS – Erosion-corrosion	High temperatures and turbulence	[26, 46]
CS – Erosion-corrosion and pitting	Turbulence and acid gas flash from too high lean loading	[19, 40, 45]
316L and 254SMO – Failure of the exchanger trays	Possible case of stress corrosion	[26, 46]
Regenerator		
CS – Amine Stress Corrosion Cracking (ASCC)	No PWHT	[40]
CS – Erosion-corrosion of the internal parts	High corrosivity of the rich amine	[19, 26, 45, 46]
CS – Serious uniform corrosion	High HSS contents – Acid water condensation zones	[44, 48]
AISI 410 – Uniform corrosion of the trays	Corrosive conditions specific to sweet units, due to high loading and high temperatures	[26, 46]
AISI 304L – Uniform corrosion of internal parts and shells	Specific case of a DiGlycolAmine (DGA) unit	[49]
Reboiler		
CS – Uniform corrosion and erosion-corrosion	Turbulence, high concentration of degradation products	[26, 46, 50]
Lean amine lines		
CS – Amine Stress Corrosion Cracking (ASCC)	No PWHT	[12-14, 42]
CS – Erosion-corrosion	Too high lean loading	[40]
CS – Erosion-corrosion	Significant amine degradation due to the presence of oxygen in the raw gas	[41]

Wet acid gas corrosion may also develop at the bottom of the absorber and on the first plates, if the walls are not wetted sufficiently by the solvent: in this case, water may condense and become loaded with acid gases. Sour service units are prone to risks of hydrogen embrittlement.

For H_2S treatment units, specific risks of Amine Stress Corrosion Cracking (ASCC) are also possible, especially in the lower part of the absorber where the loading is highest. Post Weld Heat Treatment (PWHT) is then essential to reduce these risks [17].

1.3.2 Rich Amine Lines

Corrosion risks are especially high in this section where the amine is loaded with acid gas. Depending on the location before or after the heath exchanger, the temperature varies between 60°C to 110°C. Corrosion-erosion is the most frequent risk encountered with carbon steel lines subjected to high flow rate or flow disturbance. In particular, up to the flash drum, the solvent is pressurised and highly loaded with acid gas, and there is a high risk of degassing, which may aggravate the turbulence effects. Similarly, between the rich/lean amine exchanger and the regenerator, the risks of degassing remain high and are combined with a higher solvent temperature.

In the rich amine lines, it is commonly admitted that the solvent flow rate should not exceed 1.8 m/s [17].

1.3.3 Rich/Lean Amine Exchanger

This equipment is exposed to a wide range of highly specific corrosion risks.

On the rich amine side, the risks of erosion-corrosion mentioned in the previous paragraph still remain, especially if the rich amine inlet has been badly designed [4, 15].

When stainless steel plate exchangers are used, the main risks are stress corrosion cracking (especially at welds or in case of repairs) and crevice corrosion.

Sour service units are prone to risks of hydrogen embrittlement on the rich amine side of the exchanger, if it is made of carbon steel.

1.3.4 Regenerator and Acid Gas Outlet (Condenser, Reflux Drum)

The solvent at the top of the regenerator is still rich and already at high temperature. The intrinsic corrosivity is therefore very high and there is a serious risk of erosion-corrosion in case of turbulence. If there is no significant turbulence in the medium, extensive uniform corrosion of carbon steel is frequently observed. For this type of corrosion, it would also appear that the risks are greater in sweet service units (only CO_2), where fast uniform corrosion has been observed [26, 37, 38].

Wet acid gas corrosion is another major risk, especially at the top of the regenerator and in the acid gas outlet lines. The condenser is also highly sensitive. Experience has shown that keeping the gas flow rates above 8 m/s limits these risks considerably by preventing accumulation of condensates [26].

For sour service units, hydrogen embrittlement may occur at the top of the regenerator and in the acid gas outlet lines, in case of inappropriate choice of metal (non sour service carbon steel) or heat treatment (failure to carry out PWHT) [17].

1.3.5 Reboiler

Due to the high temperatures, the reboiler is relatively sensitive to corrosion and fouling [39]. The risks are elevated by solutions containing high concentrations of degradation products. If the regenerator fails to operate correctly (insufficient stripping), solvent still loaded with acid gas may be brought up to the reboiler, significantly increasing the risk of corrosion. Excessive temperatures also represent a recognised risk factor.

1.3.6 Lean Amine Lines

An extensive survey launched following the explosion of an absorber in 1984 detected cases of amine stress corrosion cracking in the lean amine lines, due mainly to failure to carry out PWHT [13, 14].

A case of erosion-corrosion has been reported for a MDEA unit in Indonesia [40], possibly due to an excessive lean loading ($\alpha > 0.02$ mol_{CO_2}/mol_{amine}).

Another case is reported for a MDEA unit, where very severe corrosion of the lean amine parts developed just a few months after starting the unit [41]. This case would seem to have been caused by very fast degradation of MDEA to form bicine, due to the presence of oxygen in the gas to be treated, at a concentration of 90-100 ppmv.

1.4 From Natural Gas Sweetening to CO_2 Capture

If both natural gas sweetening and CO_2 capture can be described by amine absorption – desorption process of Figure 1, three major differences have to be mentioned as concerns corrosion risks evaluation, as illustrated in Table 2:
- gas composition and partial pressures,
- nature of the amine used,
- lean loading level.

In natural gas processing, the gas to be treated usually has high pressure up to 100 bar, and might contain a significant proportion of CO_2 and/or H_2S, up to several tens of percent. Oxygen contamination is not supposed

TABLE 2

Main differences between amine processes for natural gas treatment and CO_2 post-combustion capture

Parameter	Natural gas treatment	CO_2 post-combustion capture
Gas composition	High P_{CO_2} (1-100 bar) No O_2	Low P_{CO_2} (< 0.5 bar) 5-10% O_2
Type of amine	Secondary or tertiary	Primary (MEA)
Lean loading level	< 0.1 mol_{CO_2}/mol_{amine}	0.25 mol_{CO_2}/mol_{amine}

to be present. The main goal of the process is to recover natural gas with a minimum amount of acid gas contaminants. Concerning H_2S, complete removal is usually expected, while CO_2 removal efficiency depends on the application (between 2-3% for conventional applications, but down to less than 50-100 ppmv for Liquefied Natural Gas, LNG). Optimisation of the process then allows using secondary or tertiary amines, and requires complete regeneration of the solvent, *i.e.* hardly no acid gas is present in the solvent at the outlet of the regenerator column, with a lean loading typically below 0.1 mol_{CO_2}/mol_{amine} [16].

On the other hand, CO_2 capture from combustion fumes responds to different constraints, and has slightly different objectives. The first important factor is in the composition and pressure of the gas to be treated. Usually, it contains up 10-15% CO_2, for a total pressure close to 1 bar. The CO_2 partial pressure is then extremely low, while the emitted fumes flow-rate is extremely high. It is therefore required to have a solvent capable of a very fast absorption reaction with CO_2, which is generally the case of primary amines, but not secondary or tertiary amines. Additionally, the presence of up to 5% oxygen in the flue gas is also an important factor, since it might react with the amine to form corrosive degradation products. Lastly, operating conditions are aimed at finding a compromise between a good CO_2 removal, without penalising too much the power plant efficiency. For this reason, CO_2 regeneration is not complete in CO_2 capture processes: then, the lean amine loading is generally not zero, but preferably around 0.25 mol_{CO_2}/mol_{amine} in the case of MEA.

2 RECENT INVESTIGATIONS ON CORROSION IN CO$_2$ POST-COMBUSTION CAPTURE PROCESSES

2.1 Laboratory Studies on MEA

Several papers presenting laboratory corrosion measurements in MEA solutions have been published in the last years [21, 23, 25, 51, 52]. Most of these studies were

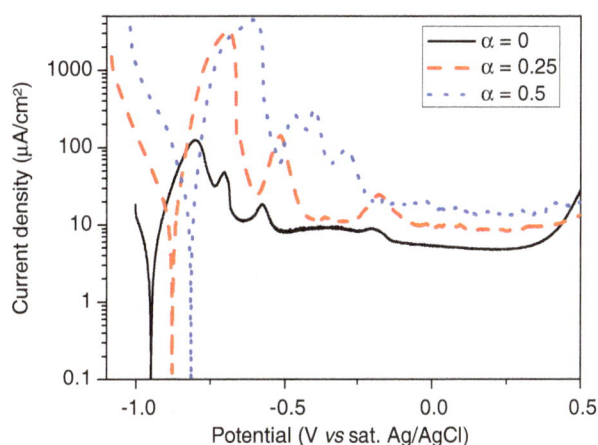

Figure 2

Typical polarization curves of carbon steel in MEA at different CO_2 loading (results taken in reference [52]).

performed in glass cells at moderate temperature (max. 80°C) and at ambient pressure, and used electrochemical measurements to examine the impact of several parameters on corrosion. Most of the time, corrosion rate was evaluated by extrapolating the cathodic region to the corrosion potential in order to determine the corrosion current density.

The electrochemical behaviour of carbon steel in 30% MEA solution at 80°C for different levels of CO_2 loading is illustrated in Figure 2. These results were taken in [52], where a detailed description of experimental conditions can be found. These curves are typical of an active corrosion behaviour. Both the cathodic and the anodic reaction rates are increased with an increase of CO_2 loading. A passive region is also observed at high overpotential, with a plateau current value increasing with CO_2 loading.

The impact of amine concentration was studied in [21, 25, 52]. At constant CO_2 loading, a weak increase of corrosion rate is observed when the MEA concentration is raised from 6 to 30% [25]. At higher concentration, between 30% and 55% MEA, polarisation curves show very similar corrosion rates [21, 52].

This weak impact of MEA concentration on corrosion might sound contradictory with past experience of natural gas treatment, where a classical rule-of-thumb indicates that MEA should not exceed 20-30% [15, 53]. However, it is also well admitted that concentrated amine solutions are more prone to degradation, forming corrosive by-products [53, 54]. Therefore, the increased corrosivity with amine concentration is more a consequence of increased degradation than an intrinsic property of concentrated solutions.

The impact of temperature on corrosion in MEA was mainly evaluated between ambient temperature and 80°C in laboratory conditions. As expected, the rate of electrochemical reactions increases with temperature. Both the cathodic and anodic reactions are affected [21, 25, 52, 54]. This result is a typical consequence of thermally activated electrochemical reactions.

The impact of CO_2 loading is usually considered as the most influencing parameter. As illustrated in Figure 2, the evolution of acid gas loading from 0 to 0.5 induces a tenfold increase of the corrosion current density. CO_2 loading affects both electrochemical reactions, but the impact is more pronounced on the cathodic side [21, 23, 25, 51, 52].

A few other studies focussed on the impact of impurities, $i.e.$ oxygen [21, 25, 51, 52], HSS [35, 51, 55], SO_2 and NO_2 [56, 57]. All these impurities were found to be detrimental to corrosion, by an increase of the rates of electrochemical reactions. HSS is also known to decrease the protectivity of corrosion scale by a chelating effect. Concerning SO_2 and NO_2, the main impact is probably on amine degradation rather than on corrosion.

For all these impurities, it can be concluded that the amplitude of corrosion rate increase remains of second order in comparison with the impact of CO_2 loading and temperature.

A synthetic view of the evolution of carbon steel corrosion rate with CO_2 loading and temperature is presented in Figure 3 [52]. At a given CO_2 loading, it appears quite clearly that the corrosion rate follows a linear evolution with the reciprocal of temperature, confirming thermal activation. From this diagram, corrosion rates can be estimated at any temperature and CO_2 loading encountered in the unit. It appears that even in the presumably least severe conditions, $i.e.$ moderate temperature (50°C) and lean loading (0.25 mol_{CO_2}/mol_{amine}), carbon steel corrosion still exceeds 100 μm/year. On the other hand, in the hot rich amine, corrosion rate above several mm/year is predicted, and this level was confirmed by pilot plant experiments [58].

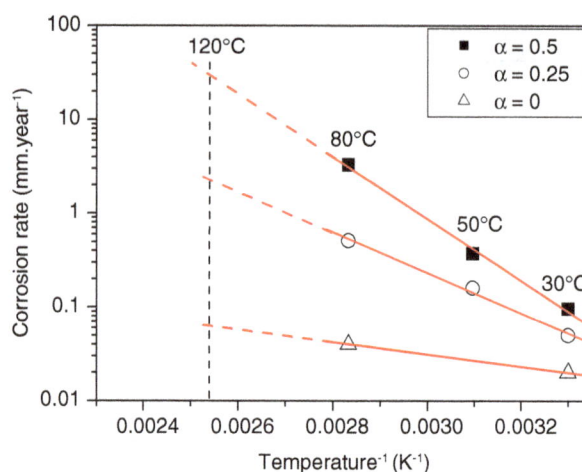

Figure 3

Impact of temperature and CO_2 loading on the corrosion rate of carbon steel in 30% MEA [52].

From all laboratory studies, it appears that carbon steel grades cannot be used safely in all operating conditions encountered in a MEA capture unit.

It also appears from most laboratory studies that austenitic stainless steel grade 316L seems to perform well [35, 52].

2.2 Corrosion Prevention Strategies

Two alternatives are usually considered as corrosion protection measures. The first one consists in the addition of corrosion inhibitors. Use of corrosion inhibitors is often recommended when the operator wants to minimise investment costs and make most components from carbon steel. The second alternative consists in using stainless steel grades in those parts of the plant exposed to extremely corrosive conditions. Although investment costs are usually increased, it allows reducing operational costs, and permits more versatility. advantages and drawbacks of these two options for CO_2 capture units are discussed below.

Corrosion inhibition is commonly used in acid gas treatment plants, where it often allows using carbon steel grades rather than more expensive alloys. Two types of inhibitors are found.

Oxidizing passivators react electrochemically with the steel surface to promote the formation of a stable protective passive layer. Most of the time, these inhibitors consist of inorganic molecules. For gas sweetening applications, they are usually presented as the most efficient [20, 59]. Sodium metavanadate is probably the most cited molecule, and the first patents for gas treatment applications date 1936, in US patent No. 2.031.632. Positive feedback from field experience was reported in [11], about a MEA unit in a hydrogen refining plant. Other salts of heavy metals can be used, like antimony, cobalt, bismuth or nickel. However, these types of inhibitors are largely becoming obsolete, mainly because of high toxicity and the cost of waste disposal. As a more environmental friendly alternative, copper salts are also known to inhibit corrosion in amine units. The first reference to this property was found in US patent No. 2.377.966 issued in 1945. However, inhibiting corrosion with copper salts presents some difficulty for application in CO_2 capture plants due to the high concentration of oxygen in the gas. Indeed, a catalysis of amine oxidative degradation can be caused by such salts of heavy metals [60, 61]. Furthermore, in the stripper section where dissolved oxygen vanishes, precipitation of metallic copper might happen, causing severe bimetallic corrosion and consumption of the inhibitor [62]. In spite of these problems, copper carbonate and sodium metavanadate are often used as reference inhibitors in recent studies for CO_2 capture with MEA. The excellent performance of both products was confirmed in [62-65]. It was confirmed, however, that in the absence of oxygen, copper carbonate was less efficient and a pitting tendency could be observed, associated with the precipitation of metallic copper on steel surfaces [62].

Film forming inhibitors represent the other family of molecules commonly used for gas treatment applications. This type of molecules adsorbs on the steel surface to form a thin layer which impedes access of the corrosive solution to the surface. However, for this specific application, they are considered to be less efficient than inorganic compounds [20, 59]. For CO_2 capture application, investigation on low-toxic organic inhibitors with promising results was proposed in [65]. However, it was concluded that none of the tested molecules was as efficient as sodium metavanadate.

Although a few companies propose processes including corrosion inhibitors [66], the most followed strategy is to use corrosion resistant alloys in those parts of the plant exposed to extremely corrosive conditions. For acid gas treatment units, it is therefore recommended to use stainless steel at least in the rich amine parts of the unit. However, some authors recommend stainless steel even in lean solvent sections [26, 46, 67-69], in order to operate at higher flow-rates, and also to provide versatility to the unit for easier solvent swapping.

In the case of CO_2 capture with MEA, carbon steel presents high levels of corrosion in the entire parts of the plant, either rich or lean solvent sections as illustrated in Figure 3. Therefore, stainless steel represents an interesting alternative when one wants to avoid using proprietary and often ecotoxic inhibitors. Furthermore, most industrial applications to date consist in demonstration pilot plants, for which versatility is a premium requirement. To serve this demand, using corrosion resistant grades allows easier solvent swap for benchmark studies with new solvents.

2.3 Corrosion Aspects in the Development of New Solvents

CO_2 capture by 30% MEA represents the reference technology for post-combustion applications. However, it is well admitted that the cost of CO_2 removal with this process is too high, and needs to be reduced.

Thus, a lot of research programs aims at developing new solvents, with a particular focus on the energy consumption needed at regeneration step. Minimizing solvent degradation is also often considered as the second immediate priority. Even if this research is usually not driven by corrosion aspects, testing the corrosivity of new molecules is often considered at the early stages of research and development efforts, and a few papers were found in the recent literature.

An extensive study on more than 20 amines including degradation and corrosion evaluations was proposed by Martin et al. [70]. Molecules with a greater stability and less corrosivity than MEA were identified, and a Quantitative Structure Property Relationship (QSPR) model was built that can be applied to predict degradation.

Solutions composed of MEA and piperazine were tested in [71]. Such solutions showed more corrosion than in pure MEA.

A great deal of attention is also given to activated solutions of tertiary amines. MDEA using piperazine as activator is one of the favourite subject of investigation in the recent years [72-74]. Corrosion in this type of solution was studied in [75], showing the same detrimental impact of CO_2 loading and temperature as already observed in MEA or in other amine solutions.

Sterically hindered amines have also received a great deal of attention due to good performances in terms of absorption and desorption capacity. It is also thought that sterical-hindrance might limit the interactions between the amine and the steel surface, thus reducing the corrosivity. Investigation of AMP (2-Amino-2-Methyl-1-Propanol) was proposed in [65]. In comparison with MEA, AMP shows less corrosivity at elevated temperature and in lean loading conditions.

Other good corrosion performances were mentioned in several papers presenting new proprietary solvents, without a precise description of the amine nature.

This is the case of the DMX^{tm} amine developed by IFP Energies nouvelles, showing carbon steel corrosion below 10 μm/year at temperature up to 160°C in CO_2 loaded conditions [76].

2.4 Corrosion Monitoring at Pilot Plant Scale

Corrosion evaluation at the pilot scale was performed at several locations worldwide. MEA 30% is usually used as benchmark solvent, to which newly developed solvents with promising features are compared.

The CASTOR pilot plant, located in Denmark at a coal-fired power station was used for process evaluations with MEA and other proprietary solvents [77]. Corrosion monitoring was performed with weight loss coupons at different locations in the pilot plant. Carbon steel corrosion at a rate exceeding millimeters' per years was found at the outlet of the stripper with 30% MEA. During the same study, stainless steel grades (304L and 316L) performed extremely well, without significant weight loss corrosion [58]. In the same paper, results obtained on the ITC pilot plant from the University of Regina (Canada) were also presented, confirming that corrosion in 30% MEA was the highest in the rich section from the heat exchanger to the stripper. This pilot plant also revealed corrosion in the stripper overhead, where CO_2 saturated water is subject to condensation.

Other corrosion results in pilot plants operated with 30% MEA were recently published. Different types of stainless steels and polymeric materials as well as concrete were tested in the lignite-fired Niederaussem power station (Germany) [78, 79]. Carbon steel corrosion was evaluated in the Tarong coal-fired power station (Australia), confirming once more the high corrosivity of the rich MEA [80]. Similar results were also obtained in the Brindisi pilot plant (Italy) [81].

Another study presented pilot plant experiments on a proprietary solvent specified by *Toshiba*. It was found that the corrosion rate in this solvent was the highest in the rich amine section, similarly as in MEA or other amines [82]. It was shown also that corrosion and amine degradation was increased when SO_2 was present in the flue gas [83].

3 CONSEQUENCES FOR CO_2 TRANSPORT AND STORAGE

Another important aspect of the CO_2 Capture and Storage (CCS) chain is the transport of CO_2 from the post-combustion capture unit to the storage site. Most of the time, dense phase CO_2 at pressure above 100 bar will circulate in a pipeline network. Transportation of high pressure dry CO_2 through carbon steel pipelines is commonly used for enhanced oil recovery for more than several decades, with extremely positive feedback. As long as the water content is maintained below saturation, corrosion will be insignificant. On the other hand, as soon as free water can condensate, it is highly concentrated in CO_2 and extremely rapid and localised corrosion can occur in range above 10 mm/year. Based on the good experience of Enhanced Oil Recovery (EOR), it has long been considered that transportation of post-combustion CO_2 was not a major issue, and this area received much less attention and funding for R&D programs. However, recent publications showed that it was not as straightforward, and that corrosion risks had to be examined carefully [84-97]. Indeed, it appears that water solubility in dense phase CO_2 is strongly modified in presence of impurities like SO_x and NO_x, or O_2. Furthermore, when SO_2 is present with water and oxygen, extremely corrosive sulphuric acid might form.

Understanding the impact of all impurities present in the CO_2 after post-combustion capture on water solubility is therefore a major task in order to develop CCS safely. At the present time, there are no commonly accepted specifications for the maximum levels of impurities required for pipeline transport. Some recommendations were proposed after two European projects within the sixth framework programme (FP6), Dynamis[1] and Encap[2], and often referred as Dynamis CO_2 quality recommendations (*Tab. 3*) [98, 99]. These recommendations were established from a set of criteria, including health and safety, operation and design, but it was acknowledged that "the mechanisms of the water related corrosion for CO_2 pipelines of carbon steel are not fully understood". As an illustration, several experimental studies have been carried out below or close to the the Dynamis recommendations, and showed some risks of corrosion.

According to Dugstad *et al.* [91] water solubility in pure CO_2 at 100 bar and in the temperature range 4-25°C exceeds 1 900 ppmv. This was confirmed by corrosion experiments in pure CO_2 with 500 ppm to 1 200 ppm H_2O, giving no corrosion of carbon steel coupons. However, when 500 ppmv H_2O and 100 ppmv NO_2 is present, the corrosion rate was high (> 0.1 mm/year). Considering several impurities at the same time, these authors also showed that corrosion could take place with as low as 200 ppmv H_2O when SO_2 and O_2 were present [94].

---

[1] Project website www.dynamis-hypogen.com.

[2] Project website www.encapCO2.org.

TABLE 3

Dynamis CO_2 quality recommendations (concentrations in ppmv) [98, 99]

Component	Concentration	Limitation
H_2O	500 ppm	Design and operational considerations
H_2S	200 ppm	Health and safety considerations
CO	2 000 ppm	Health and safety considerations
CH_4	Storage in aquifer: 4%EOR: 100-1 000 ppm	As proposed in ENCAP project
N_2	4%	As proposed in ENCAP project
Ar	4%	As proposed in ENCAP project
H_2	4%	Further reduction of H_2 is recommended because of its energy content
SO_x	100 ppm	Health and safety considerations
NO_x	100 ppm	Health and safety considerations
CO_2	> 95.5%	

It seems therefore essential to continue to work on the impact of impurities on high pressure CO_2 corrosion, to gain more confidence in CO_2 quality recommendations.

Lastly, after CO_2 transport through a pipeline network, underground injection might inevitably put free water in contact with high pressure CO_2 and impurities. In this situation, corrosion rate of carbon steel easily reaches several tens of mm/year, and even some corrosion resistant alloys might suffer corrosion above hundreds of μm/year [84, 96].

CONCLUSIONS

Corrosion issues are encountered in the whole CCS chain.

In the CO_2 post-combustion capture process, the corrosive environment mainly consists of aqueous amine. Identification of corrosion risks and mitigation strategies benefit from several decades experience of natural gas treatment with amines. Nevertheless, important differences have to be kept in mind. Currently, the most widespread process uses primary amines which are the more prone to degradation and corrosion, in comparison with secondary or tertiary amines. For this process using MEA, current challenges consist in limiting the degradation by oxygen, and to mitigate corrosion of the installation.

It was shown also that corrosion is often taken into account at the early stages of research programmes dedicated to new molecules. Laboratory tools and pilot plants equipped with corrosion monitoring devices are helpful for this purpose.

Impact on corrosion of impurities in the post-combustion CO_2 was also emphasized. Commonly accepted recommendations for the composition of CO_2 might not provide complete assurance of the absence of corrosion of steel pipelines used for transport. More work has to be done to predict the impact of impurities on water solubility.

At the very end of the chain, it was shown also that formation water and supercritical CO_2 might combine to form a very corrosive environment for the injection facilities.

For both transport and storage steps, experience from EOR and oil and gas production should be put to good advantage to find appropriate mitigation strategies. Nevertheless, it seems important also to address the specificities of high pressure CO_2 coming from the capture process, and especially the impact of impurities.

REFERENCES

1 Kohl A.L., Nielsen R.B. (1997) *Gas Purification*, fifth edition, Gulf Publishing Company, Houston, Texas.

2 Bouillon P.A., Carrette P.L., Faraj A., Lemaire E., Raynal L. (2011) IFPEN solutions for lowering the cost of post-combustion CO_2 capture, *ICCDU International Conference on Carbon Dioxide Utilization*, 27-30 June, Dijon, France.

3 Tems R., Al-Zahrani A. (2006) Cost of corrosion in gas sweetening and fractionation plants, *Corrosion/2006*, Paper No. 444, NACE International.

4 Dingman J.C., Allen D.L., Moore T.F. (1966) Minimize corrosion in MEA units, *Hydrocarbon Processing* **45**, (9), 570-575.

5 Lang F.S., Mason J.F. (1958) Corrosion in amine gas treating solutions, *Corrosion* **14**, (2), 65-68.

6   Montrone E.D., Long W.P. (1971) Choosing materials for CO_2 absorption systems, *Chemical Engineering*, Jan. 25.

7   Polderman L.D., Dillon C.P., Steele A.B. (1955) Degradation of monoethanolamine in natural gas treating service, *Oil Gas Journal* **16**, (5), 180-183.

8   Riesenfeld F.C., Blohm C.L. (1950) Corrosion problems in gas purification units employing MEA solutions, *Petroleum Refiner* **20**, (4), 141-150.

9   Riesenfeld F.C., Hugues C.L. (1951) Corrosion in amine gas treating plants, *Petroleum Refiner* **30**, (2), 97-106.

10 Riesenfeld F.C., Blohm C.L. (1951) Corrosion resistance of alloys in amine gas treating systems, *Petroleum Refiner* **30**, (10), 107-115.

11 Williams E., Leckie H.P. (1968) Corrosion and its prevention in a monoethanolamine gas treating plant, *Materials Protection* **7**, (7), 321-326.

12 McHenry H.I., Read D.T., Shives T.R. (1987) Failure analysis of an amine-absorber pressure vessel, *Materials Performance* **26**, (8), 18-24.

13 Richert J.P., Bagdasarian A.J., Shargay C.A. (1987) Stress corrosion cracking of carbon steel in amine systems, *Corrosion/87*, Paper No. 187, NACE International.

14 Richert J.P., Bagdasarian A.J., Shargay C.A. (1989) Extent of stress corrosion cracking in amine plants revealed by survey, *Oil Gas Journal* **5**, 45-52.

15 Nielsen R.B., Lewis K.R., McCullough J.G., Hansen D.A. (1995) Controlling corrosion in amine treating plants, Proceedings of the *Laurance Reid Gas Conditioning Conference*, Norman, Oklahoma.

16 Nielsen R.B., Lewis K.R., McCullough J.G., Hansen D.A. (1995) Corrosion in refinery amine systems, *Corrosion/95*, Paper No. 571, NACE International.

17 EFC (2003) Publication No. 46, Avoiding environmental cracking in amine units, The European Federation of Corrosion, *Woodhead Publishing Ltd*, Cambridge, England.

18 Dupart M.S., Bacon T.R., Edwards D.J. (1993) Understanding corrosion in alkanolamine gas treating plants. 1. proper mechanism diagnosis optimizes amine operations, *Hydrocarbon Processing* **72**, 75-79.

19 Dupart M.S., Bacon T.R., Edwards D.J. (1993) Understanding corrosion in alkanolamine gas treating plants. 2. Case histories show actual plant problems and their solutions, *Hydrocarbon Processing* **72**, 89-94.

20 Kosseim A.J., McCullough J.G., Butwell K.F. (1984) Corrosion-Inhibited Amine Guard St Process, *Chemical Engineering Progress* **80**, 64-71.

21 Soosaiprakasam I.R., Veawab A. (2008) Corrosion and polarization behavior of carbon steel in MEA-based CO_2 capture process, *International Journal of Greenhouse Gas Control* **2**, (4), 553-562.

22 Tomoe Y., Shimizu M., Kaneta H. (1996) Active dissolution and natural passivation of carbon steel in carbon dioxide loaded alkanolamine solutions, *Corrosion/96*, Paper No. 395, NACE International.

23 Veawab A., Aroonwilas A. (2002) Identification of oxidizing agents in aqueous amine-CO_2 systems using a mechanistic corrosion model, *Corrosion Science* **44**, (5), 967-987.

24 Bich N.N., Vacha F., Schubert R. (1996) Corrosion in MDEA sour gas treating plants: correlation between laboratory testing and field experience, *Corrosion/96*, Paper No. 392, NACE International.

25 Veawab A., Tontiwachwuthikul P., Chakma A. (1999) Corrosion behavior of carbon steel in the CO_2 absorption process using aqueous amine solutions, *Industrial Engineering Chemistry Research* **38**, (10), 3917-3924.

26 Bonis M.R., Ballaguet J.P., Rigaill C. (2004) A critical look at amines: a practical review of corrosion experience over four decades, *83rd annual GPA convention*, 14-17 March, New-Orleans, LO.

27 Blanc C., Grall M., Demarais G. (1982) The part played by degradation products in the corrosion of gas sweetening plants using DEA and MDEA, Proceedings of the *Laurance Reid Gas Conditioning Conference*, 8-10 March, Norman, OK.

28 DuPart M.S., Rooney P.C., Bacon T.R. (1999) Comparing laboratory and plant data for MDEA/DEA blends, *Hydrocarbon Processing* **78**, (4), 81-86.

29 DuPart M.S., Rooney P.C., Bacon T.R. (1999) Comparison of laboratory and operating plant data of MDEA/DEA blends, Proceedings of the *49th Laurance Reid Gas Conditioning Conference*, 21-24 Feb., Norman, OK.

30 Veldman R.R. (2000) Alkalonamine solution corrosion mechanisms and inhibition from heat stable salts and CO_2, *Corrosion/2000*, Paper No. 496, NACE International.

31 Guo X.P., Tomoe Y. (1999) The effect of corrosion product layers on the anodic and cathodic reactions of carbon steel in CO_2-saturated MDEA solutions at 100°C, *Corrosion Science* **41**, (7), 1391-1402.

32 Vazquez R.C., Rios G., Trejo A., Rincon R.E., Uruchurtu J., Malo J.M. (2000) The effect of diethanolamine solution concentration in the corrosion of steel, *Corrosion/2000*, paper No. 696, NACE International.

33 Rooney P.C., Bacon T.R., DuPart M.S. (1996) Effect of heat stable salts on MDEA solution corrosivity, *Hydrocarbon Processing* **75**, (3), 95-103.

34 Rooney P.C., DuPart M.S., Bacon T.R. (1997) Effect of heat stable salts on MDEA solution corrosivity .2, *Hydrocarbon Processing* **76**, (4), 65-71.

35 Tanthapanichakoon W., Veawab A., McGarvey B. (2006) Electrochemical investigation on the effect of heat-stable salts on corrosion in CO_2 capture plants using aqueous solution of MEA, *Industrial Engineering Chemistry Research* **45**, (8), 2586-2593.

36 Al-Zahrani A., Al-Luqmaun S.I. (2006) Methodology of mitigating corrosion mechanisms in amine gas treating units, *Corrosion/2006*, Paper No. 441, NACE International.

37 Raut N., Chaudhari R.M., Naik V.S. (2009) Failure of amine regenerating column of amine treatment unit, *Corrosion/2009*, Paper No. 334, NACE International.

38 Xie J., Simanzhenkov V., Santos B., Ikeda K., Davies L. (2010) Corrosion of UNS S30403 stainless steel trays in an amine unit, *Corrosion/2010*, Paper No. 187, NACE International.

39 Moore M.A., Qarni M.M., Lobley G.R. (2008) Corrosion problems in gas treating systems, *Corrosion/2008*, Paper No. 08419, NACE International.

40 Safruddin S., Safruddin R. (2000) Twenty years experience in controlling corrosion in amine unit, Badak LNG plant, *Corrosion/2000*, Paper No. 497, NACE International.

41 Howard M., Sargent A. (2001) Operating experiences at Duke energy field services Wilcox plant with oxygen contamination and amine degradation, Proceedings of the *51st Laurance Reid Gas Conditioning Conference*, 25-28 Feb., Norman, OK.

42 Kane R.D., Wilhelm S.M., Oldfield J.W. (1989) Review of hydrogen induced cracking of steels in wet H_2S refinery service, *Materials Property Council*, 28 March, Paris, France.

43 Teevens P.J. (1990) Toward a better understanding of of the cracking behavior of carbon steel in alkanolamine sour gas sweetening units: its detection, monitoring and how to avoid it, *Corrosion/90*, Paper No. 198, NACE International.

44 DeHart T.R., Hansen D.A., Mariz C.L., McCullough J.G. (1999) Solving corrosion problems at the NEA Bellingham Massachusetts carbon dioxide recovery plant, *Corrosion/99*, Paper No. 264, NACE International.

45 Dupart M.S., Bacon T.R., Edwards D.J. (1991) Understanding and preventing corrosion in alkanolamine gas treating plants, Proceedings of the *41st Laurance Reid Gas Conditioning Conference*, 4-6 March, Norman, OK.

46 Kittel J., Bonis M.R., Perdu G. (2008) Corrosion control on amine plants: new compact unit design for high acid gas loadings, *Sour Oil & Gas Advanced Technology Conference*, 27 April-1 May, Abu Dhabi, UAE.

47 Pearson H., Shao J., Norton D., Dandekar S. (2005) Case study of effects of bicine in CO_2 only amine treater service, Proceedings of the *55th Laurance Reid Gas Conditioning Conference*, Norman, OK.

48 Fan D., Kolp L.E., Huett D.S., Sargent M.A. (2000) Role of impurities and H_2S in refinery lean DEA system corrosion, *Corrosion/2000*, Paper No. 495, NACE International.

49 Tomoe Y., Miyata K., Ihara M., Masuda K., Efird K.D. (2002) Evaluation of corrosion resistance of metallic materials for DGA regenerators in dynamic conditions, *Corrosion/2002*, Paper No. 350, NACE International.

50 Jordan T.J., Nozal P.J., Azodi A. (2006) Handling trace oxygen at the saunders gas processing facility, Proceedings of the *56th Laurance Reid Gas Conditioning Conference*, Norman, OK.

51 Fleury E., Kittel J., Vuillemin B., Oltra R., Ropital F. (2008) Corrosion in amine solvents used for the removal of acid gases, *Eurocorr 2008*, The European Federation of Corrosion, Edinburgh, UK, 7-11 Sept.

52 Kittel J., Fleury E., Vuillemin B., Gonzalez S., Ropital F., Oltra R. (2012) Corrosion in alkanolamine used for acid gas removal: From natural gas processing to CO_2 capture, *Materials and Corrosion* **63**, (3), 223-230.

53 Wagner R., Judd B. (2006) Fundamentals - Gas sweetening, Proceedings of the *56th Laurance Reid Gas Conditioning Conference*, Norman, OK.

54 Lawal A.O., Idem R.O. (2006) Kinetics of the oxidative degradation of CO_2 loaded and concentrated aqueous MEA-MDEA blends during CO_2 absorption from flue gas streams, *Industrial Engineering Chemistry Research* **45**, (8), 2601-2607.

55 Duan D., Choi Y.S., Nesic S., Vitse F., Bedell S.A., Worley C. (2010) Effect of oxygen and heat stable salts on the corrosion of carbon steel in MDEA-based CO_2 capture process, *Corrosion/2010*, Paper No. 191, NACE International.

56 Kladkaew N., Idem R., Tontiwachwuthikul P., Saiwan C. (2009) Corrosion Behavior of Carbon Steel in the Monoethanolamine-H_2O-CO_2-O_2-SO_2 System: Products, Reaction Pathways, and Kinetics, *Industrial Engineering Chemistry Research* **48**, (23), 10169-10179.

57 Kladkaew N., Idem R., Tontiwachwuthikul P., Saiwan C. (2009) Corrosion Behavior of Carbon Steel in the Monoethanolamine-H_2O-CO_2-O_2-SO_2 System, *Industrial & Engineering Chemistry Research* **48**, (19), 8913-8919.

58 Kittel J., Idem R., Gelowitz D., Tontiwachwuthikul P., Parrain G., Bonneau A. (2009) Corrosion in MEA units for CO_2 capture: Pilot plant studies, *Energy Procedia* **1**, (1), 791-797.

59 Pearce B., DuPart M.S. (1987) Corrosion in gas conditioning plants - An overview, *Corrosion/87*, Paper No. 39, NACE International.

60 Bello A., Idem R.O. (2006) Comprehensive study of the kinetics of the oxidative degradation of CO_2 loaded and concentrated aqueous monoethanolamine (MEA) with and without sodium metavanadate during CO_2 absorption from flue gases, *Industrial Engineering Chemistry Research* **45**, (8), 2569-2579.

61 Goff G.S., Rochelle G.T. (2006) Oxidation inhibitors for copper and iron catalyzed degradation of monoethanolamine in CO_2 capture processes, *Industrial Engineering Chemistry Research* **45**, (8), 2513-2521.

62 Soosaiprakasam I.R., Veawab A. (2009) Corrosion inhibition performance of copper carbonate in MEA- CO_2 capture unit, *Energy Procedia* **1**, (1), 225-229.

63 Soosaiprakasam I.R., Veawab A. (2007) Inhibition performance of copper carbonate in CO_2 absorption process using aqueous MEA, *Corrosion/2007*, Paper No. 396, NACE International.

64 Tanthapanichakoon W., Veawab A. (2005) Polarization behavior and performance of inorganic corrosion inhibitors in monoethanolamine solution containing carbon dioxide and heat-stable salts, *Corrosion* **61**, (4), 371-380.

65 Veawab A., Tontiwachwuthikul P. (2001) Investigation of low-toxic organic corrosion inhibitors for CO_2 separation process using aqueous MEA solvent, *Industrial Engineering Chemistry Research* **40**, (22), 4771-4777.

66 Reddy S., Johnson D., Gilmartin J. (2008) Fluor's econamine FG plus technology for CO_2 capture at coal-fired power plants, *Power Plant Air Pollutant Control Mega Symposium*, Baltimore, MD, 25-28 Aug.

67 Rennie S. (2006) Corrosion and materials selection for amine service, *Materials and Testing Conference*, Fremantle, Australia, 30 Oct.-2 Nov.

68 Rooney P.C., DuPart M.S. (2000) Corrosion in alkanolamine plants: causes and minimization, *Corrosion/2000*, Paper No. 494, NACE International.

69 Titz J.T., Asprion N., Katz T., Wagner R. (2003) Corrosion in amine solutions used for acid gas removal, Proceedings of the *53rd Laurance Reid Gas Conditioning Conference*, Norman, OK.

70 Martin S., Lepaumier H., Picq D., Kittel J., de Bruin T., Faraj A., Carrette P.L. (2012) New amines for CO_2 capture. IV. Degradation, corrosion, and quantitative structure property relationship model, *Industrial Engineering Chemistry Research* **51**, (18), 6283-6289.

71 Nainar M., Veawab A. (2009) Corrosion in CO_2 capture process using blended monoethanolamine and piperazine, *Industrial Engineering Chemistry Research* **48**, (20), 9299-9306.

72 Ali B.S., Aroua M.K. (2004) Effect of piperazine on CO_2 loading in aqueous solutions of MDEA at low pressure, *International Journal Thermophysics* **25**, (6), 1863-1870.

73 Bishnoi S., Rochelle G.T. (2002) Absorption of carbon dioxide in aqueous piperazine/methyldiethanolamine, *AIChE Journal* **48**, (12), 2788-2799.

74 Derks P.W.J., Hogendoorn J.A., Versteeg G.F. (2006) Absorption of carbon dioxide into aqueous solutions of MDEA and piperazine, *CHISA 2006 – 17th International Congress of Chemical and Process Engineering*, Prague, Czech Republic, 27-31 August.

75 Zhao B., Sun Y., Yuan Y., Gao J., Wang S., Zhuo Y., Chen C. (2011) Study on corrosion in CO_2 chemical absorption process using amine solution, *Energy Procedia* **4**, 93-100.

76 Kittel J., Gonzalez S., Lemaire E., Raynal L. (2012) Corrosion in post-combustion CO_2 capture plants - comparisons between MEA 30% and new processes, *Eurocorr 2012*, The European Federation of Corrosion, Istanbul.

77 Knudsen J.N., Jensen J.R.N., Vilhelmsen P.J., Biede O. (2009) Experience with CO_2 capture from coal flue gas in pilot-scale: Testing of different amine solvents, *Energy Procedia* **1**, (1), 783-790.

78 Moser P., Schmidt S., Sieder G., Garcia H., Stoffregen T. (2011) Performance of MEA in a long-term test at the post-combustion capture pilot plant in Niederaussem, *International Journal of Greenhouse Gas Control* **5**, (4), 620-627.

79 Moser P., Schmidt S., Uerlings R., Sieder G., Titz J.T., Hahn A., Stoffregen T. (2011) Material testing for future commercial post-combustion capture plants. Results of the testing programme conducted at the Niederaussem pilot plant, *Energy Procedia* **4**, 1317-1322.

80 Pearson P., Cousins A., Cottrell A.J., Duncombe B., Feron P.H.M., Hollenkamp T.F., Huang S., Meuleman E. (2013) Corrosion in amine post combustion capture plants, *Eurocorr 2012*, The European Federation of Corrosion, Istanbul.

81 Lemaire E., Bouillon P.A., Mangiaracina A., Normand L., Laborie G. (2012) Results of the 2.25 t/h post-combustion CO_2 capture pilot plant of ENEL at the Brindisi coal power plant and last R&D developments for Hicapt$^+$ process, *SOGAT - CO_2 forum*, Abu-Dhabi, UAE, 29 March.

82 Gao J., Wang S., Sun C., Zhao B., Chen C. (2012) Corrosion behavior of carbon steel at typical positions of an amine-based CO_2 capture pilot plant, *Industrial Engineering Chemistry Research* **51**, (19), 6714-6721.

83 Gao J., Wang S., Zhou S., Zhao B., Chen C. (2011) Corrosion and degradation performance of novel absorbent for CO_2 capture in pilot-scale, *Energy Procedia* **4**, 1534-1541.

84 Zhang X., Zevenbergen J., Spruijt M.P.N., Benedictus T. (2012) Corrosion of steels in CO_2 transport and storage environments, *Eurocorr 2012*, The European Federation of Corrosion, Istanbul.

85 Xiang Y., Wang Z., Yang X., Li Z., Ni W. (2012) The upper limit of moisture content for supercritical CO_2 pipeline transport, *Journal Supercritical Fluids* **67**, (7), 14-21.

86 Xiang Y., Wang Z., Xu C., Zhou C., Li Z., Ni W. (2011) Impact of SO_2 concentration on the corrosion rate of X70 steel and iron in water-saturated supercritical CO_2 mixed with SO_2, *Journal Supercritical Fluids* **58**, (2), 286-294.

87 Xiang Y., Wang Z., Yang X., Ni W., Li Z. (2011) Corrosion behavior of X70 steel in the supercritical CO_2 mixed with SO_2 and saturated water, Proceedings of the *Twenty-first International Offshore and Polar Engineering Conference*, Maui, Hawaii, USA, 19-24 June.

88 Ruhl A.S., Kranzmann A. (2012) Corrosion behavior of various steels in a continuous flow of carbon dioxide containing impurities, *International Journal Greenhouse Gas Control* **9**, (7), 85-90.

89 Lucci A., Demofonti G., Spinelli C.M. (2011) CO_2 anthropogenic pipeline transportation, Proceedings of the *Twenty-first International Offshore and Polar Engineering Conference*, Maui, Hawaii, USA, 19-24 June.

90 Farelas F., Choi Y.S., Nesic S. (2012) Effects of CO_2 phase change, SO_2 content and flow on the corrosion of CO_2 transmission pipeline steel, *Corrosion*/2012, NACE International, C2012-0001322.

91 Dugstad A., Halseid M., Morland B., Siversten A.O. (2012) Corrosion in dense phase CO_2 with small amounts of SO_2, NO_2 and water, *Eurocorr 2012*, The European Federation of Corrosion, Istanbul.

92 Dugstad A., Halseid M. (2012) Internal corrosion in dense phase CO_2 transport pipelines - State of the art and the need for further R&D, *Corrosion*/2012, Paper No. 1452, NACE International, *Corrosion 2012*, Salt Lake City, Utah, 11-15 March.

93 Dugstad A., Morland B., Clausen S. (2011) Corrosion of transport pipelines for CO_2 - Effect of water ingress, *Energy Procedia* **4**, 3063-3070.

94 Dugstad A., Clausen S., Morland B. (2011) Transport of dense phase CO_2 in C-steel pipelines- When is corrosion an issue? *Corrosion*/2011, Paper No. 70, NACE International.

95 Dugstad A., Halseid M., Morland B. (2011) Corrosion in dense phase CO_2 pipelines - State of the art, *Eurocorr 2011*, The European Federation of Corrosion, Stockholm.

96 Choi Y.S., Nesic S. (2011) Effect of water content on the corrosion behavior of carbon steel in supercritical CO_2 phase with impurities, *Corrosion*/2011, Paper No. 377, NACE International.

97 Chambers B., Kane R., Yunovich M. (2010) Corrosion and selection of alloys for carbon capture and storage (CCS) systems: Current challenges, *SPE International Conference on CO_2 capture, storage and utilization*, The New-Orleans, LO, 10-12 Nov.

98 de Visser E., Hendriks C., Barrio M., Mølnvik M.J., de Koeijer G., Liljemark S., Le Gallo Y. (2008) Dynamis CO_2 quality recommendations, *International Journal Greenhouse Gas Control* **2**, (4), 478-484.

99 de Visser E., Hendriks C., de Koeijer G., Liljemark S., Barrio M., Austegard A., Brown A. (2007) Dynamis CO_2 recommendations, FP6 European Project No. 019672 report No. D3.1.3.

7

Development of Innovating Materials for Distributing Mixtures of Hydrogen and Natural Gas. Study of the Barrier Properties and Durability of Polymer Pipes

Marie-Hélène Klopffer[1]*, Philippe Berne[2] and Éliane Espuche[3]

[1] IFP Energies nouvelles, 1-4 avenue de Bois-Préau, 92852 Rueil-Malmaison Cedex - France
[2] CEA, LITEN, DTNM, LCSN, 38054 Grenoble - France
[3] Université Lyon 1, CNRS, UMR5223, Ingénierie des Matériaux Polymères, 15 Bd A. Latarjet, 69622 Villeurbanne - France
e-mail: marie-helene.klopffer@ifpen.fr - philippe.berne@cea.fr - eliane.espuche@univ-lyon1.fr

* Corresponding author

Abstract — With the growing place taken by hydrogen, a question still remains about its delivery and transport from the production site to the end user by employing the existing extensive natural gas pipelines. Indeed, the key challenge is the significant H_2 permeation through polymer infrastructures (PolyEthylene (PE) pipes, components such as connecting parts). This high flow rate of H_2 through PE has to be taken into account for safety and economic requirements.

A 3-year project was launched, the aim of which was to develop and assess material solutions to cope with present problems for hydrogen gas distribution and to sustain higher pressure compared to classical high density polyethylene pipe. This project investigated pure hydrogen gas and mixtures with natural gas (20% of CH_4 and 80% of H_2) in pipelines with the aim to select engineering polymers which are more innovative than polyethylene and show outstanding properties, in terms of permeation, basic mechanical tests but also more specific characterizations such as long term ageing and behaviour. The adequate benches, equipments and scientific approach for materials testing had been developed and validated.

In this context, the paper will focus on the evaluation of the barrier properties of 3 polymers (PE, PA11 and PAHM). Experiments were performed for pure H_2 and CH_4 and also in the presence of mixtures of hydrogen and natural gas in order to study the possible mixing effects of gases. It will report some round-robin tests that have been carried out. Secondly, by comparing data obtained on film, polymer membrane and on pipe section, the influence of the polymer processing will be studied. Innovative multilayers systems will be proposed and compared on the basis of the results obtained on monolayer systems. Finally, the evolution of the transport properties of the studied polymers with an ageing under representative service conditions will be discussed.

Résumé — **Développement de nouveaux matériaux pour la distribution de mélanges de gaz naturel et d'hydrogène. Étude des propriétés barrière et de la durabilité de tubes polymères** — Avec la place croissante prise par l'hydrogène se pose la question de son transport et de sa distribution par le très vaste réseau existant de conduites de gaz naturel. Le principal verrou à lever est le taux de perméation sensiblement plus fort pour l'hydrogène, à travers les parois des canalisations qui sont essentiellement en PolyÉthylène (PE). Il en résulte des implications potentielles en termes de sécurité mais aussi de pertes économiques.

Un projet d'une durée de trois ans a été lancé avec pour objectifs de proposer et de qualifier des polymères plus performants et innovants que le PE vis-à-vis de cette introduction massive d'hydrogène mais aussi capables de supporter des niveaux de pression plus élevés. Différents polymères ont été caractérisés en termes de propriétés barrière, de comportement mécanique, mais également de comportement en vieillissement vis-à-vis de l'hydrogène pur ou en mélange avec le gaz naturel (20 % de méthane pour 80 % d'hydrogène). Les bancs d'essais, les équipements et les procédures nécessaires pour ces tests de matériaux ont été développés et validés.

Dans ce contexte, l'article se concentrera sur l'évaluation des propriétés barrière de 3 polymères : PE, PA11 et PAHM. Il rendra compte d'essais de comparaison interlaboratoires qui ont été menés. Les expériences en hydrogène et méthane purs ou combinés permettront de quantifier les éventuels effets de mélange. En second lieu, l'influence de la mise en forme sera établie grâce à la comparaison des données obtenues sur des films fins, des membranes ou des tronçons de tube. Des solutions innovantes de systèmes multicouches seront proposées et comparées sur la base des résultats obtenus en monocouches. Enfin, l'article discutera l'évolution des propriétés de transport des polymères étudiés sous l'effet d'un vieillissement en conditions de service représentatives.

INTRODUCTION

With the development of hydrogen as an energy vector, its delivery and transport from the production site to the end user remains an issue. Indeed, new steel pipeline is cost intensive and the use of existing natural gas infrastructure raises the question of its durability in presence of hydrogen for metallic parts and its tightness for polymer parts, respectively. Actually, the key challenge is the high hydrogen permeation rate through existing polymer infrastructures used for natural gas distribution (Poly-Ethylene (PE) pipes, components as connecting parts) due to its small size in comparison with methane [1]. Due to safety and economic reasons, one of the main concerns is to limit leakage of hydrogen by permeation through the pipe.

The project called PolHYtube was relative to the development and study of innovating materials for hydrogen distribution networks and has benefited from a grant by the French National Research Agency (ANR, *Agence Nationale de la Recherche*). It has involved several industrial and academic partners. This 3-year project has investigated pure hydrogen and gas mixtures (20% CH_4 - 80% H_2) in pipelines made of engineering polymers to develop and assess material solutions to cope with current problems for H_2 distribution and to sustain higher pressure compared to classical high density polyethylene pipe.

Test benches and protocols for testing materials in terms of mechanical and barrier properties were first developed and validated on reference materials [2]. Materials such as a High Density PolyEthylene (HDPE) and PolyAmide 11 (PA11) have been studied. HDPE is a

semi-crystalline polymer considered as a reference material as it is used today in natural gas distribution pipes. PA11 should allow a higher operating pressure combined with better gas-barrier performances.

On the other hand, technical polymers and assemblies (other semi-crystalline or amorphous thermoplastics, multi-layers, polymer blends, etc.) have been proposed and studied to improve gas-barrier performances compared to polyethylene [3]. Step by step, permeation and basic mechanical tests have been performed and then more specific characterisations have been done for long-term ageing under various conditions in order to finally determine the materials that could meet all the specifications required by hydrogen distribution. The design of a pipe prototype was also carried out at the end of the project and an economic study was performed for the different potential solutions [2, 4].

One of the objectives of this paper is to evaluate the impact of hydrogen introduction in the existing natural gas infrastructures in terms of modification of the polymer pipes barrier properties in comparison with natural gas. For this purpose, the transport properties of HDPE films of different thicknesses in presence of pure hydrogen but also with mixtures of natural gas and hydrogen have been measured in different conditions of temperature and pressure and moreover, in different laboratories (*IFP Energies nouvelles, IMP, CEA*). Furthermore, a specific device for the evaluation of hydrogen permeability of pipes has been developed at *IFP Energies nouvelles* (IFPEN) to test them under hydrogen in service conditions. This apparatus allowed us to perform experiments on pipes sections in more realistic operating conditions.

By comparing data obtained on film and pipes sections samples exposed to various temperatures, pressures and particularly different mixtures of natural gas and hydrogen, it was possible to evaluate the influence of the polymer processing on its barrier properties.

An important point has also concerned the evolution of the material barrier properties with time and with exposure to hydrogen. Therefore, the properties of the unaltered material and of samples after ageing in presence of hydrogen in various conditions were compared to assess the long term behaviour in service. After a brief review of the equipments specifically developed by *CEA* for this type of study, the results obtained on both PE100 and PA11 will be presented.

The last item of the paper is relative to the evaluation of innovative systems (such as other thermoplastics, multilayers) with enhanced barrier characteristics in comparison with PA11 and HDPE reference materials and the detailed analysis of their respective behaviour at different temperatures and hydrogen pressures.

1 GAS PERMEATION THEORY

In a general way, gas permeation in a polymer can be defined as the susceptibility of this material to be penetrated and crossed by the gas molecules. It is described by a solution-diffusion mechanism. At a given temperature, the transport of a gas molecule through a homogeneous polymer matrix can be described as a three-step process [1, 5]: sorption of the component at the upstream face of the membrane, followed by diffusion/solution through the material cross section under the influence of the applied driving force (pressure gradient which corresponds to a chemical potential gradient) and finally desorption at the downstream face of the film [6]. In a Fickian transport mechanism, the interfacial equilibrium is supposed to be fast compared to the diffusion step, which is then the governing step of the transport mechanism. Both the solubility and diffusion parameters are dependent on the characteristics of the membrane material and the gases, and can be studied separately with various sorption and diffusion models [1, 6-12].

The permeability coefficient, denoted Pe, is, in a Fickian mechanism, the product of the solubility coefficient, S, and the diffusion coefficient, D:

$$Pe = DS \qquad (1)$$

Diffusion is the process by which a small molecule (organic liquids, vapours, gases, etc.) is transferred in the system due to random molecular motions. Therefore, D is a kinetic term that is related to the free volume and the molecular mobility in the polymer phase.

The solubility coefficient has a thermodynamic origin and depends on the molecule-polymer interactions, on the polymer free volume as well as on the ability of the gas to condense. It is related to the local concentration of the gas C dissolved in the polymer and to the gas pressure by the following relation:

$$C = Sp \qquad (2)$$

Generally, the units are the following: D is given in (cm^2/s), S in $cm^3(STP)/cm^3$ polymer.bar, C in $cm^3(STP)/cm^3$ polymer. Consequently, Pe is expressed in $cm^3(STP)/cm.s.bar$.

Barrer [13] was the first one who showed that the diffusion of small size molecules in rubbery polymers is a thermally activated process. A great number of data in literature suggests that the transport coefficients (namely S, D and Pe) depend on the temperature, at a given pressure, *via* Arrhenius's law on a narrow range of temperature [11]:

$$S(T) = S_0 \ \exp\left(-\frac{\Delta H_S}{R T}\right) \qquad (3)$$

$$D(T) = D_0 \ \exp\left(-\frac{E_D}{R T}\right) \qquad (4)$$

$$Pe(T) = Pe_0 \ \exp\left(-\frac{E_P}{R T}\right) \qquad (5)$$

The pre-exponential terms represent the limit values of the various coefficients of transport for an infinite molecular agitation ($T \to \infty$). E_P represents the apparent activation energy for the permeation process and is equal to the sum of E_D, the apparent activation energy of the diffusion process, and ΔH_S, the heat of solution needed for the dissolution of a permeant mole in the matrix. These parameters depend on the chemical structure and on the morphology of the polymer matrix: amorphous or semicrystalline structure, value of the temperature relative to the characteristic temperatures such as the glass transition temperature Tg and the melting temperature Tm.

Gas permeability rates are essential properties for the use of polymers for gas distribution and transport pipes. The improvement in their barrier performance implies that either D or S presents a very low value in the operating conditions. For that reason, semicrystalline polymers are interesting candidates due to their morphology and structure. Indeed gas transport through semicrystalline polymers is generally studied in the context of the two-phase model proposed and developed by Michaels and Parker [14] and Michaels and Bixler [15, 16].

For isotropic HDPE with spherulitic structures, these authors have shown that the sorption and the diffusion took place exclusively in the amorphous regions. The crystalline zones act as excluded volumes for the sorption process and are impermeable barriers for the diffusion process.

The gas barrier properties of polymeric materials can be controlled by different parameters. The degree of crystallinity is an important parameter and as the sorption and diffusion processes take place in the amorphous phase of the polymer, its chemical structure is a key point to take into account [1], explaining why important variations can be observed in the resulting barrier properties.

The aim of this work is to determine the hydrogen and methane permeability in a wide range of temperature (from 10 to 85°C) and pressure (from 5 to 20 bar) for different thermoplastic semicrystalline polymers that could be used as alternative to HDPE in target pipeline. The evolution of the barrier properties after an ageing performed in an hydrogen environment was also studied for some polymers, and multilayer systems based on the most interesting materials were also proposed.

2 EXPERIMENTAL SECTION

2.1 Materials

An important point to be taken into account for the initial choice of the base thermoplastic polymers is their easy processing by extrusion or co-extrusion. As a technical plastics producer, *Arkema* participated in this project and provided some polymer materials. In particular, *Arkema* shared with the other partners its experience in developing polyamide 11 pipes for gas distribution in US. This allows the project team to well define the technical targets and specifications of the project (pipe diameter, pipe design (*i.e.* the thickness to diameter ratio), internal pressure).

Three thermoplastic polymers have been evaluated as base materials:
- a high density polyethylene HDPE (PE100 grade- PE XS10B from *Total Petrochemical*) currently used for natural gas distribution in Europe. It has been taken as the reference material for this study;
- two polymers of the polyamide family:
  - PA11 (Rilsan TL): which is the material proposed by *Arkema* for making gas distribution pipes in US;
  - PAHM: a polyamide with higher mechanical properties, under development by *Arkema*.

Extruded films (thickness around 1 mm) as well as monolayer pipes with lengths up to 1 m and sections of 6/8 and 26/32 (internal/external diameter in mm) have

been provided for each of these base materials to be studied under permeation conditions.

In addition to these base materials, more complex systems consisting of multilayer systems have also been studied. For this second route, Ethylene Vinyl Alcohol (EVOH), which is known for its high gas barrier properties [17] has been combined with the base materials to improve their properties. Their structure was the following Polymer/EVOH/Polymer with an additive to promote adhesion between EVOH and the polymer matrix.

2.2 Morphology Characterization

The characteristic thermal parameters of HDPE, PA11 and PAHM materials have been determined by Differential Scanning Calorimetry (DSC). The analyses were performed at 10°C/min with a DSC 2020 apparatus from *TA Instruments*. Samples of about 10 mg were taken from extruded sheets of 1 mm thickness. To take off the water eventually contained in the polar films, PA11 and PAHM samples were dried 1 night at 60°C under vacuum before analysis. All samples were semi crystalline samples with a melting temperature of 128°C, 189°C and 247°C for HDPE, PA11 and PAHM, respectively. The crystallinity rates have been calculated for HDPE and PA11 from the melting enthalpy (ΔH_m) and ΔH_∞ according to the law:

$$X_c = \frac{100 \Delta H_m}{\Delta H_\infty} \qquad (6)$$

with ΔH_∞ equal to 290 J/g and 226 J/g for HDPE and PA11, respectively. X_c is equal to $58 \pm 2\%$ and $20 \pm 2\%$ for HDPE and PA11, respectively. It is noteworthy that the glass transition temperature of these materials is equal to $-120°C$ and 47°C, respectively.

The thermograms relative to PAHM exhibited a ΔCp change at 80°C assigned to the glass transition temperature. Moreover, the DSC analysis performed on PAHM showed a very small endothermic peak, meaning that the crystallinity rate of this material was very small in comparison with the one of PE and PA11. It was not possible to determine the crystallinity rate of this material due to the unknown value of ΔH_∞ for this material.

2.3 Permeation Measurements

Due to safety requirements, one of the main concerns about polymer pipes is their permeability to hydrogen, which may induce critical leakages of gaseous hydrogen. Permeation measurements were performed by IFPEN, *CEA* and *IMP* for hydrogen but also for pure methane and H_2/CH_4 mixtures in various conditions of pressure, temperature and gas mixtures compositions.

The experimental technique available at IFPEN concerns gas or gas mixtures permeation through a polymer membrane or a pipe section, in a flowing stream of vector gas and detection by gas chromatography. In that kind of measurement, one or several diffusing species cross the polymer to reach an opened cavity swept away by an inert gas stream. This carrier gas fulfils its purpose by transporting the different molecules towards an appropriate detector measuring the present gas proportion. Then, it is possible to determine the intrinsic transport coefficients of each of the gases constituting the initial mixture. It allows determination of the permeability coefficients of pure gas in numerous polymers but also the permeability coefficients of gas mixtures, such as CH_4-H_2 [18, 19]. Practically, many permeability tests (> 70) were carried out during this study in order to identify the influence of four parameters: temperature (at least 3), pressure (5 and 20 bar), compositions of gases (pure CH_4, 80% H_2 – 20% CH_4, pure H_2) and polymer processing (pipe section and membrane). Most experiments have been carried out twice in order to evaluate the repeatability of the tests. For polymer membranes, the experimental device with two permeation cells is showed in Figure 1.

Considering the study of polymer pipe section, which is a geometry more representative of the application, two different permeation cells are available dependent on the geometry of the section pipe that is to say (0.5 m length, 1 mm thickness) or (1 m length, 3 mm thickness) and are represented in Figure 2. These equipments have allowed us to compare data obtained on disc samples and pipes sections faced to various temperatures, pressures and different mixtures of natural gas and hydrogen in order to evaluate the processing influence on the barrier properties. The key point is that this experimental technique is very sensitive and allows the measurement of very small flows of diffusing molecules (up to 0.06 mL/h).

The permeation cell available at *IMP* (*Fig. 3*) consists in two compartments (upstream and downstream compartments) separated by the film to be studied. The cell is thermostated at a constant temperature chosen between 10 and 40°C. A preliminary high vacuum desorption is performed to ensure that static vacuum pressure changes in the downstream compartment are smaller than the pressure changes due to the gas diffusion.

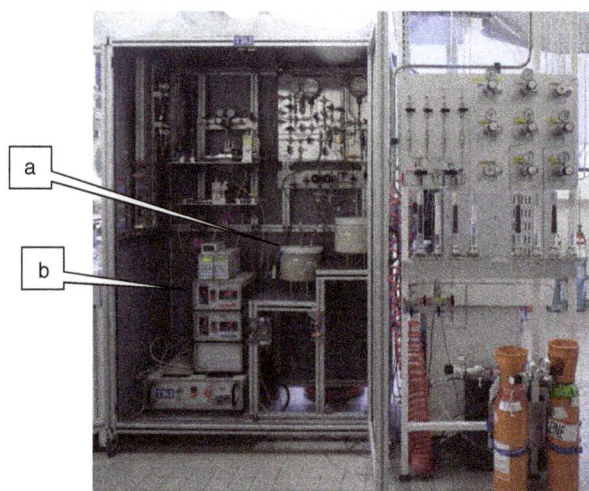

Figure 1

Permeation device dedicated to polymer membranes at IFPEN with a) one permeation cell and b) the temperature regulation.

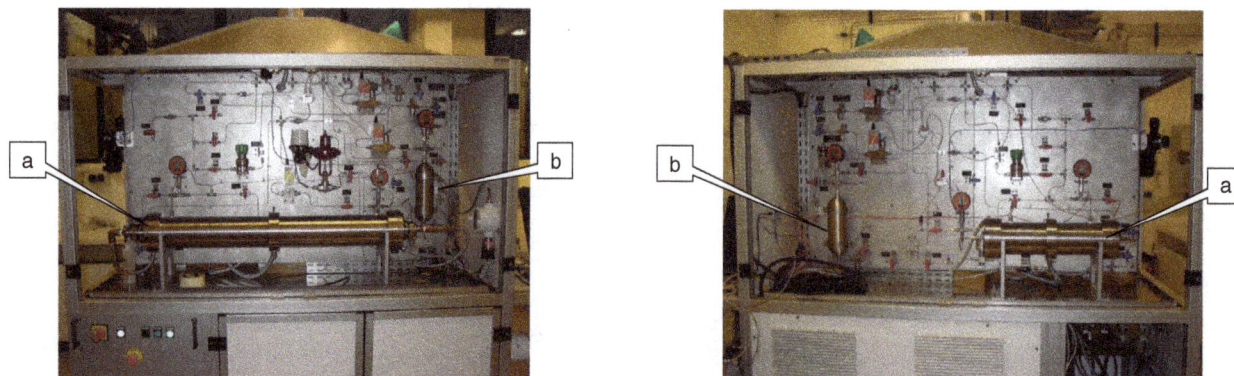

Figure 2

The two permeation cells dedicated to the study of polymer pipes sections of different lengths at IFPEN with a) one permeation cell and b) the gas tank.

Figure 3

The permeation cell available at *IMP* with a) pressure sensor, b) by-pass, c) upstream pressure valve, d) permeation cell and e) downstream pressure valve.

For gas permeation experiments, a 5.0×10^5 Pa gas pressure of hydrogen or methane is introduced in the upstream compartment. The pressure variations in the downstream compartment are measured as a function of time with a 10 torr datametrics pressure sensor. The permeability coefficient Pe is calculated from the slope of the steady-state line.

2.4 Long-Term Ageing Tests

CEA has participated in this project with the ageing of polymers under hydrogen and hydrogen/methane mixture, with the following tasks:
- ageing of polymer membranes and pipe sections by contact with hydrogen or hydrogen-methane (80-20%) mixture under controlled pressure (5, 20 bar) and temperature (20 to 80°C) conditions;
- measurement and monitoring of the permeation coefficient during ageing of the samples;
- study of the variation of transport coefficients with time.

During ageing, the transfer of gas through the polymers is seen as a permeation mechanism governed by diffusion and classical laws can be applied.

The key point is that permeation measurements are made on a same polymer sample during its ageing without pressure relaxation thanks to the design of the ageing cells (*Fig. 4*). They comprise an upstream chamber that is continuously kept under the desired gas pressure thanks to a buffer reservoir, and a downstream chamber that is normally open to the atmosphere – the sample, membrane or tube, being sandwiched between the two chambers. Permeation measurements are made on an average every second month. Whenever they are required, the downstream chamber is flushed with a neutral gas (helium or nitrogen) that transports the permeated gases to a gas chromatograph allowing detection and quantification of the hydrogen or methane flux through the sample. This technique is basically the same as the one used for the permeation measurements (previous section). It allows *in situ* measurement of the permeation coefficients

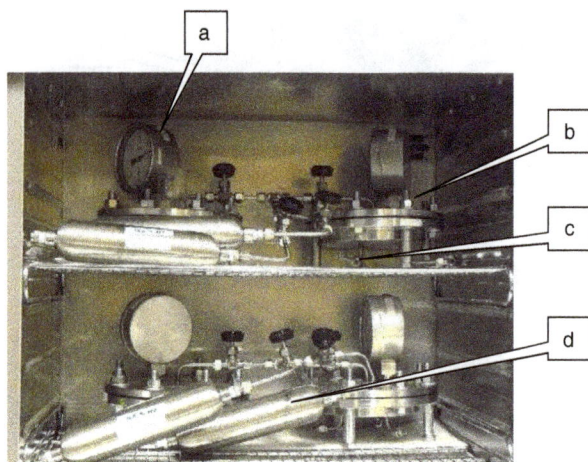

Figure 4

Ageing cells in an oven, connected to buffer H_2 reservoirs
with a) pressure indicator, b) test cell (membrane is between
the 2 flanges), c) connections to measurement device (not
connected on this picture) and d) buffer reservoirs.

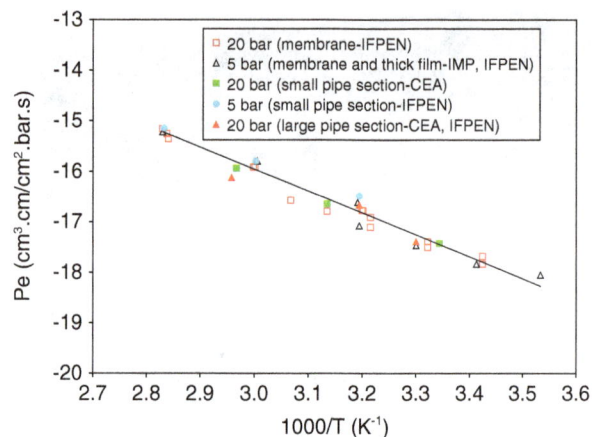

Figure 5

Evolution of the permeability coefficient with temperature
in an Arrhenius representation for pure H_2 through HDPE
at two different pressures, for various geometries. The measurements have been performed on different devices.

without having to remove the samples from the ageing
cells and to disrupt the experimental conditions [20].
One drawback is that uncertainty is rather high ($\pm$ 20
to 30% depending on operating conditions) because
the equipment, especially the pressure indicators, have
not been specifically selected for precise measurement
of the permeation rate.

Temperature of the test cells is controlled by keeping
them in an oven. The upstream chambers are filled with
dry gas from a cylinder while the downstream chambers
are open to the atmosphere and therefore exposed to
ambient humidity.

A total of 17 membrane samples and 8 small-diameter
(8 mm) and 8 large-diameter (32 mm) tubes have been
submitted to ageing and periodic monitoring for 1 year.

3 PERMEATION RESULTS

Permeation properties of the three base materials were first
measured and compared. An important point is that very
few information concerning hydrogen permeation through
these polymers and polymer assemblies can be found in the
literature making this study particularly original.

3.1 Characterization of the Monomaterial Systems

3.1.1 HDPE: Round Robin from H_2 Permeation Experiments

Permeation properties of a HDPE were studied and considered as reference properties, due to the use of this
material in pipe manufacturing. All the measurements

were performed either on HDPE membranes of around
1 mm thickness or on pipe sections of different radius
at IFPEN and *CEA*. At *IMP*, the experimental device,
initially developed for the study of thin films, was used
for these thick samples. In all cases, tests were carried
out twice in order to evaluate the repeatability of the
tests. On the other hand, the influence of the parameters
such as the temperature, the pressure and the geometry
was studied.

All the different data obtained with pure hydrogen
on HDPE membranes, pipes of various radius and in
different conditions of temperature and two pressures
have been reported in Figure 5. The evolution of the
permeability coefficient was represented in a logarithmic scale as a function of the reciprocal of temperature. The first point is that, as expected, the
permeability coefficient increases with temperature
and is well represented by an Arrhenius law, that is
to say that the plot of the logarithm of the permeability coefficient as a function of $1/T$ defined a straight
line. This is quite normal, because, in the temperature
range studied, no transition temperature is associated
either to the polymer or to the gases [1, 11, 21]. Moreover, from this curve and with the determination of the
Arrhenius law relative to hydrogen (*Tab. 1*), it is thus
possible to calculate its permeability value at lower
temperatures by extrapolation. It has to be noticed
that, in the studied range of pressure, no influence of
the pressure was found (from 5 to 20 bar).

The most important feature is that whatever the
equipment used to determine permeation coefficients,

TABLE 1

Activation energies and pre exponential terms for H_2 and CH_4 permeation through HDPE, PA11 and PAHM

Polymer-gas system	HDPE-H_2	HDPE-CH_4	PA11-H_2	PA11-CH_4	PAHM-H_2	PAHM-CH_4
E_P (kJ/mol)	36	38	38	46	25	55
Pe_0 (cm^3.cm^{-1}.s^{-1}.bar^{-1})	5.4×10^{-2}	4.2×10^{-2}	6.3×10^{-2}	9.4×10^{-2}	4.8×10^{-4}	8.6×10^{-1}

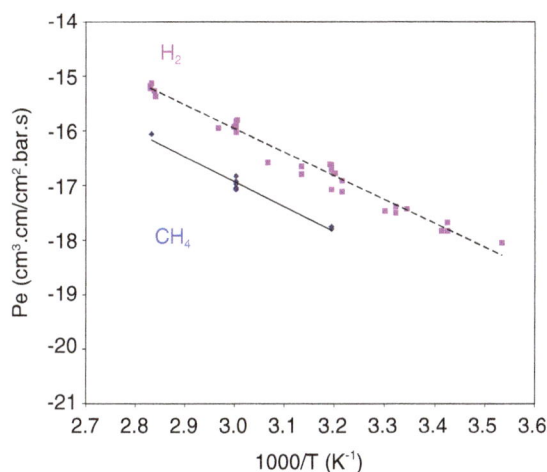

Figure 6

Evolution of the permeability coefficients of H_2 and CH_4 through HDPE for pure and mixtures of gases, various geometries (membranes and pipes of various sections).

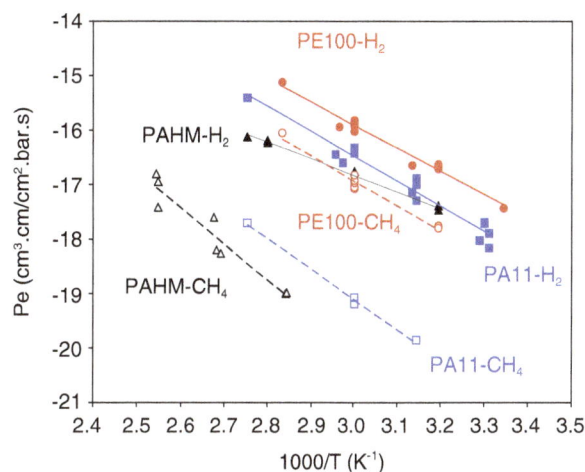

Figure 7

Comparison of the permeability coefficients of H_2 and CH_4 through HDPE, PA11 and PAHM (for different pipe sections, pure gas and mixtures at two pressures). The closed symbols are relative to H_2 and the open symbols to CH_4.

all the experimental data were found to be in good agreement and did not present a significant discrepancy (maximal value of 20%). These round-robin tests have validated that the various methods used by the project's partners, to detect and quantify the diffusing molecules through the polymer are reproducible. Furthermore, it has shown that for HDPE, there is no influence of the processing conditions on its barrier properties.

3.1.2 HDPE: Comparison for H_2, CH_4 and their Gas Mixtures

Figure 6 shows the evolution of the permeability coefficients obtained for H_2 and CH_4 through HDPE in a representation of Arrhenius' plot. It has to be noticed that some data have been obtained with pure gases and some other with mixtures (80% H_2 – 20% CH_4). Results showed that the permeability of hydrogen is higher than that of methane. The activation energy values obtained for H_2 and CH_4 are roughly the same

(36 and 38 kJ/mol, (Tab. 1), the uncertainty on these calculated data can be estimated at least 10%). On that plot, data relative to pure H_2 or H_2 in a mixture are represented with the same symbols as they were really close considering the natural scattering. The same remark applies to methane independently of the sample geometry. This implies that in the case of mixtures of hydrogen and natural gas, no particular interaction could be noticed so no mixture effect. Consequently, for a given temperature, each gas (either H_2 or CH_4) keeps its intrinsic permeability coefficient independently of the other gas.

3.1.3 Comparison of Polyamide Systems with HDPE

The next step has consisted in determining the transport properties of CH_4 and H_2 through PA11 and comparing them with those obtained on PE as illustrated by Figure 7. The experimental values obtained for the two polyamides as for PE in the different laboratories were found to be in good agreement and for each polymer, the permeability coefficient follows an Arrhenius law.

In the case of hydrogen, the permeation values have been determined at temperatures below and above the glass transition temperature in the polyamides. As its size is really small, the hydrogen molecule is not sensitive to differences in polymer chain mobility associated with the transition between glassy and rubbery state [22]. For methane, all the tests have been performed at temperatures above Tg.

Moreover, by comparing the behaviour of the two gases, one can see that the permeability of H_2 is larger than that of CH_4 whatever the considered temperature is and whatever the polymer is. On the other hand, no influence of the applied pressure could be detected in the studied range (5 to 20 bar) and no mixture effect.

The permeability coefficients decreased going from HDPE to polyamide 11. Although the crystallinity was higher for PE than for PA11, the barrier properties were higher for PA11 than for PE. The gas barrier properties were then closely related to the cohesive energy density of the amorphous phase in each polymer. Indeed, the polar structure of polyamide type polymer allows the formation of intra and intermolecular hydrogen bonds leading to a higher cohesion in the amorphous phase and as a consequence to lower gas permeability [1]. The permeability decrease observed between PE and PA11 is higher for methane than for hydrogen which is in accordance with the size difference of the two molecules.

The H_2 permeability coefficients of PA11 and of the new high performance polyamide were in the same range showing the interest of this last material for target pipeline for which combined high mechanical and barrier performances are required. The activation energy corresponding to the permeation of H_2 through PAHM is slightly lower than in PA11. This surprising result can probably be related to a higher cohesive energy density of the polymer amorphous phase associated to a significant rigidity of the polymer chains. Indeed, as aforementioned, PAHM has the lowest crystallinity degree and the highest glass transition temperature. Comparing the permeation level of both polyamides with the two gases, it appears that the new grade PAHM seems to present better barrier properties. It can lastly be noticed that the hydrogen permeability coefficient of PA11 and PAHM is in the same range as the one of methane for PE showing the interest of these two materials for new pipes development. Polyamide 11 pipes are already commercially available which makes them an interesting solution for hydrogen distribution.

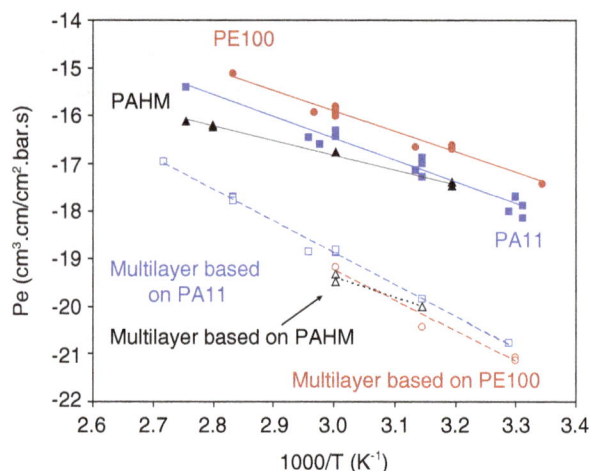

Figure 8

Permeability coefficients of all the tested systems to hydrogen as a function of temperature in an Arrhenius representation.

3.2 Study of Multilayers Systems

The effect of the addition of a gas barrier layer of EVOH to the different polymers was investigated with respect to hydrogen permeation.

Multilayers pipes based on PE, PA11 and PAHM respectively, including a thin gas barrier layer composed of EVOH (thickness between 50 and 160 μm), have been provided with the same geometrical characteristic as for the monolayer pipes. The permeation data relative to hydrogen and the different systems *i.e.* monolayers (PE, PA11, PAHM) or multilayers configurations (based on the homopolymer but comprising an EVOH layer) are illustrated in Figure 8.

The hydrogen permeability coefficient in presence of EVOH is strongly decreased compared to a pipe made of pure PE or PA (PA11 or PAHM). This demonstrates that the layer of EVOH drives the gas barrier properties of these multilayer systems as there is no significant difference between them. By the way, the activation energies in the multilayer systems are higher than in the monolayer references (around 54 kJ/mol compared to 37). It has to be noticed that this energy is lower in the case of the multilayer based on PAHM which agrees with the results obtained on monolayer systems.

Lastly, compared to the permeability coefficient of methane through PE, the H_2 permeability coefficient through multilayer systems is 15 to 30 times smaller. It can be concluded that EVOH is effective as a gas barrier layer but to conclude on the feasibility of this solution, the behaviour in fatigue has to be assessed.

3.3 Ageing Effect on HDPE and PA11

The ageing of materials under hydrogen environment is a crucial point. Indeed, hydrogen embrittlement of steels is considered as a potential risk for transmission and for distribution pipeline explaining why polymers are preferred. Different ageing parameters such as the pressure, the temperature, hydrogen effect have been studied here on the durability of HDPE and PA11.

For PE and PA11, a great number of membranes and small or large diameter pipes have been submitted to ageing and periodic monitoring for 1 year. Exploitation of the measurements on membranes has been made according to the methodology exposed in the experimental part. Since the permeation coefficient has been shown to follow Arrhenius' law, the measurements are all converted to an equivalent at 20°C by using Equation (5) with the value for E_p in Table 1. Besides, the permeation coefficient has been found independent of pressure and gas composition, which makes direct comparison possible in spite of the various experimental conditions (Fig. 9, 10, for HDPE and PA11 respectively).

The most important result is that under our conditions of ageing, the permeation coefficient is constant, equal to 2×10^{-8} cm^3.cm^{-1}.s^{-1}.bar^{-1} for PE sample and 8×10^{-9} cm^3.cm^{-1}.s^{-1}.bar^{-1} for PA11 sample at 20°C under H$_2$, within measurement uncertainty. It is unaffected by one year of ageing, whatever the experimental conditions. Analysis of the available results for tube samples suggests that the same ratio exists between the permeation coefficients of PE and PA11 tubes and that they are also essentially unaltered by ageing.

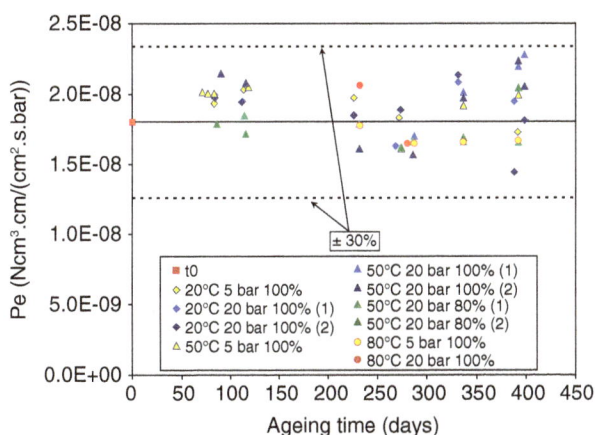

Figure 9

Evolution of the H$_2$ permeability coefficient in HDPE at 20°C as a function of ageing time (various conditions of T, P, gas composition).

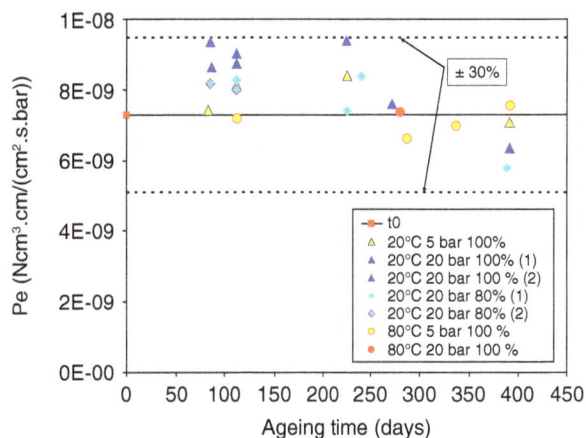

Figure 10

Evolution of the H$_2$ permeability coefficient in PA11 at 20°C as a function of ageing time (various conditions of T, P, gas composition).

CONCLUSION

Permeation measurements were performed on unaged PE100 material samples in various conditions of pressure, hydrogen content and temperature and moreover in different laboratories. First of all, there is no significant discrepancy between the results obtained on various equipments, the reproducibility of data is very good.

Our second objective was to study the possible mixing effects of gases on the transport properties of a polymer membrane. These effects are observed by comparing the membrane performance with respect to the mixed feed gas composition (80% H$_2$ – 20% CH$_4$).

Some results obtained by using specific permeability equipments dedicated to gas mixtures have been presented. The key point is that, in the case of mixture of methane and hydrogen through PE100, no particular gas-gas or gas-polymer interaction has been revealed whatever the composition as well as no pressure effect.

Concerning the evolution of permeability with temperature, it is well represented by an Arrhenius law in the range (10-85°C). It is thus allowed to extrapolate permeability coefficients at lower temperatures if no characteristic transition relative to the polymer occurs.

Two types of polyamide materials were compared to a high density polyethylene. The hydrogen permeation of these two polyamides has been shown to be equivalent to that of methane through PE, which makes them interesting for hydrogen distribution. The effect of the addition of a gas barrier such as EVOH to PE and PA11 was also investigated with respect to hydrogen permeation (before and after an exposure to hydrogen) and found to be beneficial.

Furthermore, new specific devices for the evaluation of gas permeability through pipes have been developed to test pipes sections under hydrogen in service conditions (pressure, temperature, composition of the mixture of natural gas and hydrogen). In the case of PE100 and PA11, a good correlation of the measurements between pipe sections and discs samples was obtained which means that no influence of the polymer processing could be noticed.

But the most important point is that the effect of ageing under an hydrogen environment has also been studied. No evolution of the barrier properties of the PE100 or the PA11 system has been observed after more than one year of ageing in the presence of hydrogen whatever the conditions (in terms of pressure, temperature, gas composition).

Polymer materials tested in PolHYtube project such as multilayers systems based on a EVOH layer seem very promising and interesting solutions for the different identified targets in hydrogen distribution.

ACKNOWLEDGMENTS

Authors would like to gratefully acknowledge funding from the French National Agency of Research (ANR) (PolHYTube project; PAN-H program) and partnerships of the project: *Arkema* and G. Hoschstetter for providing the materials, *Air Liquide*, *Institut P'*.

REFERENCES

1 Crank J., Park G.S. (1968) *Diffusion in polymers*, Academic Press, London and New-York, pp. 1-414.

2 Klopffer M.H., Berne P., Castagnet S., Weber M., Hochstetter G., Espuche E. (2010) Polymer pipes for distributing mixtures of hydrogen and natural gas: evolution of their transport and mechanical properties after an ageing under an hydrogen environment, *18th World Hydrogen Energy Conference 2010, WHEC 2010*, Essen, Germany.

3 Lafitte G., Espuche E., Gerard J.F. (2011) Polyamide 11/ poly(hydroxy amino ether) blends: Influence of the blend composition and morphology on the barrier and mechanical properties, *European Polymer Journal* **47**, 10, 1994-2002.

4 Klopffer M.H., Berne P., Weber M., Castagnet S., Hochstetter G., Espuche E. (2012) New materials for hydrogen distribution networks: materials development & technico-economic benchmark, *Diffusion in materials - DIMAT 2011, Defect and Diffusion Forum*, 407-412.

5 Stannett V. (1978) The transport of gases in synthetic polymer membranes - an historic perspective, *Journal of Membrane Science* **3**, 97-115.

6 Naylor T.V. (1989) Permeation properties, in *Comprehensive Polymer Science*, Pergamon Press.

7 Crank J. (1975) *The Mathematics of Diffusion*, Oxford University Press, Oxford.

8 Koros W.J., Hellums M.W. (1985) Transport properties, *Encyclopedia of Polymer Science and Technology*, John Wiley & Son's.

9 Neogi P. (1996) Transport Phenomena in Polymer Membranes, *Diffusion in Polymers*, Marcel Dekker Inc, New-York.

10 Rogers C.E. (1964) Permeability and Chemical Resistance, in *Engineering Design for Plastics*, Baer E. (ed.), Reinhold, New-York.

11 Rogers C.E. (1985) Permeation of gases and vapours in polymers, in *Polymer Permeability*, Comyn J. (ed.), Elsevier Applied Science.

12 Stern S.A. (1994) Polymers for gas separation, *Journal of Membrane Science* **94**, 1-65.

13 Barrer R.M. (1937) Nature of the diffusion process in rubber, *Nature* **140**, 106-107.

14 Michaels A.S., Parker R.B. (1959) Sorption and flow of gases in polyethylene, *Journal of Polymer Science* **41**, 53-71.

15 Michaels A.S., Bixler H.J. (1961) Solubility of gases through polyethylene, *Journal of Polymer Science* **50**, 393-412.

16 Michaels A.S., Bixler H.J. (1961) Flow of gases through polyethylene, *Journal of Polymer Science* **50**, 413-439.

17 Cai Y., Wang Z., Yi C.H., Bai Y.H., Wang J.X., Wang S.C. (2008) Gas transport property of polyallylamine-poly(vinyl alcohol)/polysulfone composite membranes, *Journal of Membrane Science* **310**, 1-2, 184-196.

18 Klopffer M.H., Flaconnèche B., Esterlé K., Lafontaine M. (2005) Experimental method of permeability measurements of H_2 and H_2-CH_4 mixtures through Polyethylene, *2nd European Hydrogen Energy Conference, EHEC 2005*, Zaragoza, Spain.

19 Klopffer M.H., Flaconnèche B., Odru P. (2007) Transport properties of gas mixtures through polyethylene, *Plastics, Rubber and Composites* **36**, 5, 184-189.

20 Foulc M.P., Nony F., Mazabraud P., Berne P., Klopffer M.H., Flaconnèche B., Ferreira Pimenta G., Müller Syring G., Alliat I. (2006) Durability and transport properties of polyethylene pipes for distributing mixtures of hydrogen and natural gas, *16th World Hydrogen Energy Conference 2006, WHEC 2006*, Lyon.

21 Klopffer M.H., Flaconnèche B. (2001) Transport properties of gases in polymers: Bibliographic review, *Oil & Gas Science and Technology* **56**, 3, 223-244.

22 Costello L.M., Koros W.J. (1994) Effect of Structure on the Temperature-Dependence of Gas-Transport and Sorption in A Series of Polycarbonates, *Journal of Polymer Science Part B-Polymer Physics* **32**, 4, 701-713.

8

New Insights in Polymer-Biofuels Interaction

Emmanuel Richaud[1]*, Fatma Djouani[1], Bruno Fayolle[1], Jacques Verdu[1] and Bruno Flaconneche[2]

[1] Arts et Métiers ParisTech, CNRS, PIMM UMR 8006, 151 bd de l'Hôpital, 75013 Paris - France
[2] IFP Energies nouvelles, 1-4 avenue de Bois-Préau, 92852 Rueil-Malmaison - France
e-mail: emmanuel.richaud@ensam.eu

* Corresponding author

Résumé — **Avancées dans la compréhension des interactions polymères-biocarburants** — Cet article traite des interactions polymères-biocarburants et en particulier des effets des biocarburants sur le polyéthylène (PE) employé pour des applications automobiles. L'objectif est de développer un modèle prédictif pour la durée de vie des réservoirs en polyéthylène vieillissant au contact de carburants contenant de l'éthanol ou du biodiesel. La principale conséquence d'un vieillissement au contact d'éthanol est la diminution de la vitesse d'extraction des antioxydants du PE. La vitesse d'extraction obéit à une loi du premier ordre et sa constante de vitesse obéit à la loi d'Arrhenius. L'interaction entre le PE et les biodiesels a été étudiée au travers de systèmes réels (méthyl ester de soja et de colza) comparés à deux systèmes modèles (méthyl oléate et méthyl linoléate). Il en est principalement ressorti que l'interaction entre biodiesel et polyéthylène se décomposait en deux parties : une première liée au vieillissement physique dû à la pénétration du biodiesel dans le PE et l'autre à un vieillissement chimique au cours duquel polyéthylène et biodiesel s'oxydaient simultanément. L'étude du transport des méthyl esters dans le PE a révélé que la cinétique de diffusion ne dépendait que de la température et de la masse molaire du carburant. L'étude de l'interaction chimique a mis en évidence que les méthyl esters s'oxydent plus rapidement que le PE et contribuent à accélérer son oxydation. Un premier modèle de co-oxydation a été proposé pour rendre compte de ce phénomène.

Abstract — *New Insights in Polymer-Biofuels Interaction* — *This paper deals with polymer-fuel interaction focusing on specific effects of biofuels on polyethylene (PE) in automotive applications. The practical objective is to develop a predictable approach for durability of polyethylene tanks in contact of ethanol based or biofuel based fuels. In the case of ethanol, the main consequence on PE durability is a reduction of the rate of stabilizer extraction; this latter phenomenon can be modeled by first order kinetics with a rate constant that obeys the Arrhenius equation. Concerning biodiesels, the study was focused on soy and rapeseed methyl ester which were compared to methyl oleate and methyl linoleate used as model compounds. Here, PE-fuel interactions can be described as well as physical interaction, linked to the oil penetration into the polymer, as chemical interaction linked to an eventual co-oxidation of PE and oil. Both aspects were investigated. Concerning biofuel transport in PE, it appeared that the oil diffusivity depends only of temperature and oil molar mass. Some aspects of the temperature dependence of the oil solubility in PE are discussed. About chemical interaction between oil and PE, it was put in evidence that unsaturated fatty esters promote and accelerate PE oxidation. A co-oxidation kinetic model was proposed to describe this process.*

1 INTRODUCTION

1.1 Strategy

Fuels from vegetable sources did not constitute a novelty at the beginning of the 21th century, since alcohol from sugar-cane was widely used in Brazil more than 30 years ago. What is new is the trend to generalize worldwide their use and to diversify fuel types (oils for Diesel engines based on fatty esters coexist now with ethanol) as well as vegetable sources, depending on the country of production. Automotive designers and manufacturers have to face many problems linked to the use of these fuels. We will focus here on the problems linked to eventual interactions between these biofuels (fatty acids methyl esters and ethanol) and polymer components (tanks, pipes, joints and other parts) in contact with them. Although leaching effects cannot be totally excluded, it will be considered here that most of the possible interactions result from the fuel penetration in the polymer. They are schematized in Figure 1.

Since fuel penetration in the polymer is the first step of any ageing process, its analysis must be the first step of any quantitative study of such processes. Two key characteristics determine the behavior of the polymer-fuel system in this domain: the fuel solubility S and diffusivity D in the polymer. These quantities are generally determined from sorption or permeation experiments. Generally, when the solubility is low, fuel penetration

doesn't affect significantly polymer properties. It will be considered that a case of high solubility would result from a non-adequate polymer choice; this case will not be examined here.

Behind their apparent simplicity, these phenomena hide some difficulties linked for instance:
- to the fact that fuels from vegetable sources are not pure compounds, they are more or less complex mixtures;
- to the fact that the physical properties of their elementary components are not always well known. An important part of the research will thus consist to choose pure compounds representative of the industrial mixtures (and to demonstrate the validity of this choice) and to use all the theoretical resources of molecular physics to determine the physical properties not reported in literature.

Polymer ageing processes can be divided in two main categories: physical ageing in which there are no chemical modifications of macromolecules, and chemical ageing in which there are changes of the macromolecular structure and thus of physical properties which depend of this structure.

Physical ageing processes can be ranged in two categories:
- processes of polymer plasticization leading to a premature fracture under stress: static stress cracking or fatigue;
- processes of stabilizer extraction by the fuel leading to an acceleration of chemical ageing.

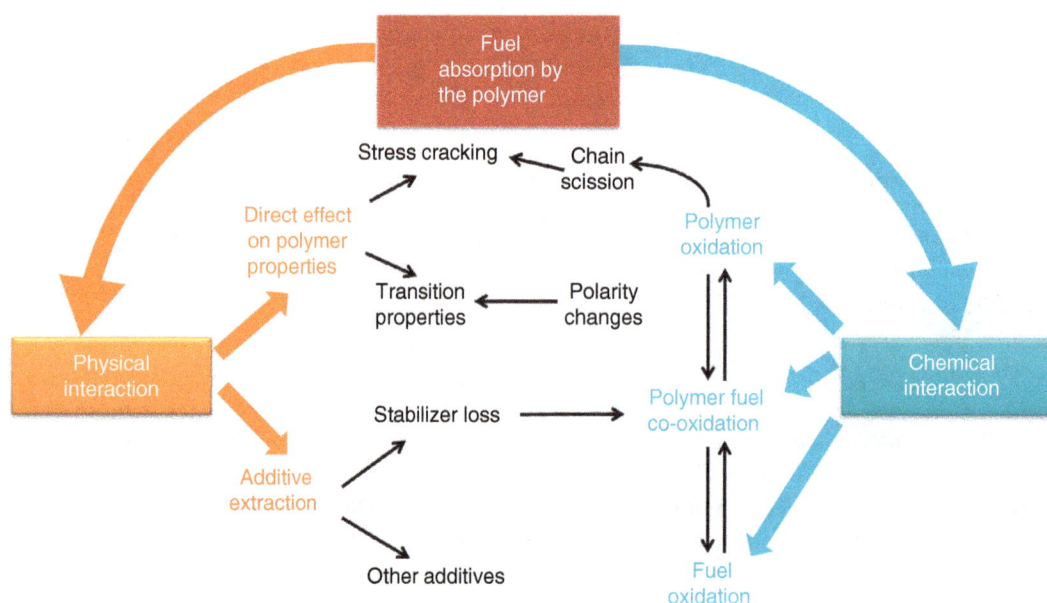

Figure 1

Schematization of possible polymer-fuel interactions having eventual consequences on polymer durability.

Chemical processes are essentially oxidation ones in the automotive conditions where the materials are exposed to temperatures generally lower than 150°C, in relatively dry atmospheres. The main question here is: do biofuels accelerate the polymer oxidation?

One sees many possible interactions between physical and chemical processes: indeed, extraction of stabilizers is expected to accelerate oxidative ageing. Among the consequences of this latter, two are especially important:
- the polymer resistance to stress cracking is sharply linked to its molar mass. Since oxidation induces chain scission, it is expected to decrease the polymer resistance to stress cracking;
- when the polymer is initially almost apolar (case of polyolefins), oxidation increases polarity, that can change the polymer-fuel interaction parameter(s) and have eventual consequences on polymer permeability.

The aim of this article is to illustrate the possibilities of kinetic modeling for the study of oxidation effects by an application to results obtained on some model systems. All the experimental results presented here were published elsewhere [1-4].

1.2 Diffusion Controlled Steps of Chemical Ageing

In a first approach, it will be considered that three molecular species, are able to be exchanged between atmosphere and the material: oil, oxygen (ox) and stabilizer (stab). Here, reactions between oil and polymer are not considered, only the physical role of oil is taken into account. The polymer-oil mixture is considered to react as a single substrate towards oxidation. There are thus two reactive species: oxygen and stabilizer (a single stabilizer is considered for the sake of simplicity) characterized by their equilibrium concentration C_{ox} and C_{stab} and by their diffusivity D_{ox} and D_{stab}. These quantities can, indeed, vary with the degree of oxidation. The chemical analysis must lead to a mechanistic scheme containing all the important elementary steps among which those involving directly oxygen and stabilizer. The kinetic scheme derived from this mechanistic scheme must contain one equation per reactive species, i.e. typically 6 or 7 equations for a case of "homo-oxidation", among which the two ones relative to oxygen and stabilizer of which the specificity is to contain diffusion terms, in order to take into account the atmosphere ↔ material exchanges. Their simplest expression would be:

$$\frac{\partial [O_2]}{\partial t} = D_{ox} \frac{\partial^2 [O_2]}{\partial z^2} - r_{ox}$$

and

$$\frac{\partial [stab]}{\partial t} = -D_{stab} \frac{\partial^2 [stab]}{\partial z^2} - r_{stab}$$

where z is the layer depth in the sample thickness, r_{ox} is the rate of oxygen consumption and r_{stab} the rate of stabilizer consumption, both expressed in function of reactive species concentrations, D_{ox} and D_{stab} the respective diffusivity for oxygen and stabilizer. The boundary conditions are the oxygen equilibrium concentration and the initial stabilizer concentration.

The first step of the research, here, would consist to write the full expressions of r_{ox} and r_{stab}. Then, we will face two problems linked to the fact that neither diffusivities D_{ox} and D_{stab} nor initial conditions are constant. The first reason is linked to the oil penetration in the polymer. Oil plasticizes the polymer, increases its segmental mobility, which presumably increases the diffusivity of small molecules. From this point of view, two situations can be found:
- if the characteristic time of oil diffusion is considerably shorter than oxidation induction time and characteristic time of stabilizer diffusion, then one can consider that the oxidation process occurs when the polymer is saturated by oil, D_{ox} and D_{stab} are the diffusivity values for the plasticized polymer, they are independent of oil transport parameters;
- if, in contrast, oil diffusion, polymer oxidation and thus stabilizer consumption occur in the same timescale, the above equations have to be coupled with oil diffusion one, which is a more complex situation.

From the experimental point of view, one can remark that both above equations are to be solved in time (t) and in space (z). The knowledge of depth distributions of stabilizer and oxidation products concentrations in sample thickness is especially interesting. The kinetic models are complex and cannot be validated only by results relative to global (average) concentrations. Experimental determinations of thickness reaction profiles will bring very useful complementary information.

1.3 From Oxidation to Physical Properties

If oxidation induces changes of mechanical properties, this is through molecular weight changes. In the context of oxidation of saturated hydrocarbon polymers, such changes result essentially from random chain scissions. The first step in a study of chain scission kinetics consists to determine the elementary step(s) of the mechanistic scheme in which chains are broken. The most common precursor of chain scission in radical oxidation is a secondary or tertiary alkoxy radical (PO°) able to rearrange by beta scission (Fig. 2).

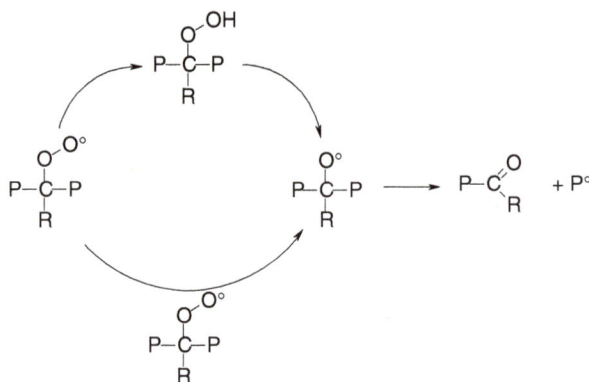

Figure 2

Main routes for oxidative chain scission in conditions of low temperature oxidation in dark.

The precursors of alkoxyls are peroxyls which are the chain carriers of radical oxidation. Peroxyls can give hydroperoxides (POOH) in the propagation of radical chains. Hydroperoxides can decompose uni- or bimolecularly but in both cases they give alkoxyls. Peroxyls can also react with themselves. In most cases, the result of these bimolecular combinations is a termination but in some cases, alkoxy radicals can escape from the cage and initiate new radical chains abstracting hydrogens or rearranging by beta scission.

Let us consider both processes of alkoxyl formation:

$$\delta POOH \rightarrow PO^\circ + \alpha HO^\circ + \beta POO^\circ \qquad k_1$$

and

$$2POO^\circ \rightarrow \gamma PO^\circ + products \qquad k_6$$

where $\alpha = 1$ and $\beta = 0$ if $\delta = 1$ (unimolecular POOH decomposition); $\alpha = 0$, $\beta = 1$ if $\delta = 2$ (bimolecular POOH decomposition). γ must be derived from the kinetic analysis (see for instance Khelidj *et al.* [5]).

The beta scission of alkoxyls is in competition with other processes, *e.g.* H abstraction. A general expression of chain scission rate (*s* being the number of chain scissions per mass unit) would be thus:

$$\frac{ds}{dt} = \gamma_1 k_1 [POOH]^\delta + \gamma_6 k_6 [POO^\circ]^2$$

where γ_1 and γ_6 are yields to be determined experimentally. In many cases, the rates of POOH decomposition and POO° bimolecular combination are almost equal (stationary state), so that it is licit to write simply:

$$\frac{ds}{dt} = \gamma' k_1 [POOH]^\delta$$

Since each chain scission event creates a new chain, *s* is linked to the number average molar mass by:

$$s = \frac{1}{M_n} - \frac{1}{M_{n0}}$$

where M_n and M_{n0} are the respective values of number average molar mass after and before ageing.

The link between oxidation kinetics and molar mass changes is now established. It remains to establish the link between molar mass and mechanical properties. What is clear is that chain scission favors fracture through chain disentanglement in the amorphous phase, but the embrittlement mechanism is not the same in amorphous or low crystallinity polar polymers and in non-polar highly crystalline polymers [6]. In the first category of polymers, for instance polycarbonate, poly(ethylene terephthalate) or polyamides, embrittlement results from the destruction of the entanglement network and occurs when the molar mass approaches the entanglement molar mass. In the second category, chain scission induces secondary crystallization (chemicrystallization); embrittlement occurs when the interlamellar spacing l_a becomes lower than a critical value of the order of 6 nm in polyethylene [7] or polyoxymethylene [8]. However, since l_a depends mainly of molar mass, it can be considered that embrittlement of polymers of the second category occurs, as for the first category, when the molar mass reaches a critical value M'_c. Both categories differ by the ratio critical molar mass / entanglement molar mass: $q = M'_c / M_e$.

$q \sim 2$ to 10 for the first category, typically $M'_c \sim 15$ kg/mol for poly(ethylene terephthalate), polycarbonate or polyamide 11 and $q \sim 20$ to 50 for the second category, typically $M'_c \sim 70$ kg/mol for polyethylene and polyoxymethylene and ~ 200 kg/mol for polypropylene [9].

Ageing experiments are generally performed on unloaded samples and mechanical measurements are made at the end of exposure. This approach is justified when polymer parts don't sustain continuous loads in their use conditions. In this case, M'_c constitutes a very good end-life criterion because when the molar mass approaches this value, the polymer toughness (*Fig. 3a*) and ultimate elongation (*Fig. 3b*) decay abruptly. This means that, when M_n becomes lower than M'_c, the probability of part failure under accidental loadings increases suddenly and approaches unity.

The problem is more complicated when the polymer sustains continuous static (creep) or dynamic (fatigue) loading. In such cases, chain disentanglement is favored by stress ("chain pulling"). Oxidative chain scission accelerates stress-cracking, but there is, to our knowledge, no consensus on the relationships between the time to fracture and the exposure and loading conditions.

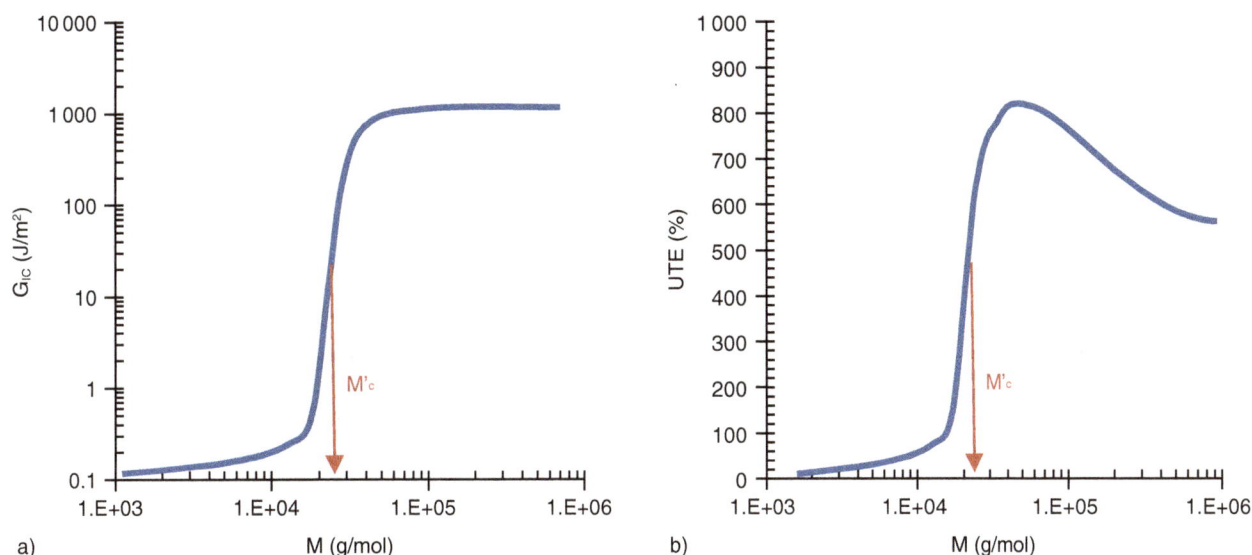

Figure 3

Shape of the molar mass dependance of toughness (here critical rate of elastic energy release for crack propagation in mode 1 a) and shape of the variation of ultimate tensile elongation against molar mass for a ductile polymer b)). The scales are just shown to give the order of magnitude.

Transport properties (oxygen, stabilizers and oils) play a role in ageing kinetics. Do ageing modify transport properties? In principle, the solubility (S) and diffusivity (D) of small molecules in a polymer are not very sensitive to molar mass changes, at least above M'_c. But, in the case of hydrocarbon polymers, oxidation grafts polar groups (alcohols, ketones, acids, etc.) to the polymer. The polarity of this latter and thus its solubility parameter increases, which modifies the strength of polymer-small molecule interaction, with eventual consequences on solubility. Considering first fuel solubility in polyethylene (PE), we can predict opposite consequences of oxidation on hydrocarbon fuels and on ethanol. Hydrocarbon fuels have a solubility parameter close to polyethylene one, an increase of the PE solubility parameter due to oxidation is expected to increase the gap between PE and oil solubility parameters, the oil solubility is expected to decrease. Ethanol has a solubility parameter considerably higher than PE, and then PE oxidation is expected to increase ethanol solubility. Polarity changes are also expected to have consequences on the diffusivity of small molecules, as shown for water in PE by Mc Call et al. [10]. Theoretical tools are lacking to predict the effects of polarity changes on transport properties of all the mobile species under consideration. Here, in a first approach, it will be necessary to make experimental determinations on virgin and oxidized polymers and to derive empirical relationships from the results.

1.4 Polymer – Biofuel Chemical Interaction

When a biofuel is absorbed by a polymer, each BioFuel (BF) molecule is isolated from the others and surrounded by the Polymer Matrix (PM). Two cases can be distinguished:
- BF is considerably less reactive than PM to oxidation, in this case, BF doesn't influence significantly the polymer oxidation rate;
- BF is more reactive than PM. In this case, the question is: do the reactive species (radicals) resulting from BF oxidation attack PM? How to take into account such interactions in a kinetic model? Diesel biofuels are based on unsaturated esters as for instance mixtures of methyl esters of oleic, linoleic and linolenic acids. It is well known that methylenes are considerably more reactive in allylic placement than in a saturated chain. The reactivity is again increased when the methylene is placed between two double bonds, as in linoleic or linolenic esters. One can thus suspect an accelerating effect of these molecules on the oxidation of a saturated polymer matrix.

The problem of "co-oxidation" of mixtures of substrates having reactivities of the same order has been studied one half century ago [11] using solutions of the kinetic problem inspired by radical copolymerization theories. These solutions were based on a set of simplifying hypotheses among which the hypothesis of common initiation and termination and the hypothesis of stationary state. In such cases, the oxidation rate depends only on propagation rate constants (as reactivity ratios in copolymerization) and substrates concentrations. Here, we face more complex situations, for instance the existence of an induction period during which stationary state hypothesis cannot apply, and the complications linked to the presence of stabilizers. In the 1950-60s, the authors were forced to use simplifying hypotheses to obtain kinetic schemes having analytical solutions. Now, we can use numerical tools allowing the resolution of considerably more complex schemes without the recourse to these hypotheses [12]. In the frame of co-oxidation studies, it becomes possible to take into account polymer-biofuel interactions not only in propagation as in classical studies, but also in initiation and termination. An example of this new approach was recently given in the case of radiation initiated oxidation of ethylene-propylene copolymers [13]. In the case under study where a biofuel RH is mixed to a polymer PH, the co-oxidation mechanistic scheme could be, assuming only bimolecular hydroperoxide decomposition and no presence of stabilizers:

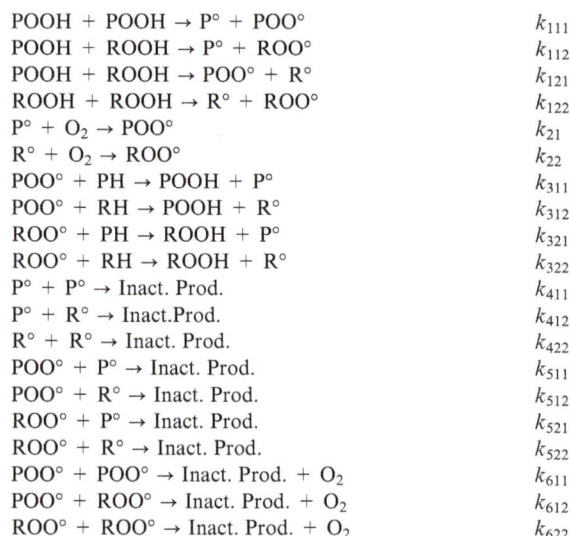

POOH + POOH → P° + POO°	k_{111}
POOH + ROOH → P° + ROO°	k_{112}
POOH + ROOH → POO° + R°	k_{121}
ROOH + ROOH → R° + ROO°	k_{122}
P° + O₂ → POO°	k_{21}
R° + O₂ → ROO°	k_{22}
POO° + PH → POOH + P°	k_{311}
POO° + RH → POOH + R°	k_{312}
ROO° + PH → ROOH + P°	k_{321}
ROO° + RH → ROOH + R°	k_{322}
P° + P° → Inact. Prod.	k_{411}
P° + R° → Inact.Prod.	k_{412}
R° + R° → Inact. Prod.	k_{422}
POO° + P° → Inact. Prod.	k_{511}
POO° + R° → Inact. Prod.	k_{512}
ROO° + P° → Inact. Prod.	k_{521}
ROO° + R° → Inact. Prod.	k_{522}
POO° + POO° → Inact. Prod. + O₂	k_{611}
POO° + ROO° → Inact. Prod. + O₂	k_{612}
ROO° + ROO° → Inact. Prod. + O₂	k_{622}

This scheme involves 21 elementary rate constants plus some stoichiometric yield ratios, it is clearly out of reach of analytical resolutions but it can be solved numerically using commercial solvers. All the rate constants k_{i11} relative to the polymer oxidation are already known [1, 5]. Many rate constants relative to unsaturated fatty esters

can be found in literature; the rest must be experimentally determined. Concerning the rate constants of cross reactions (k_{i12} and k_{i21}) they can be estimated, starting from the idea that they are intermediary between both corresponding "homo-oxidation" rate constants, using for instance the already used geometric average: $k_{i12} = (k_{i11}.k_{i22})^{1/2}$.

In the final step of the model elaboration, stabilization reactions will be added to the above scheme and oxygen and stabilizer(s) diffusion terms will be added to kinetic equations. Indeed, diffusion terms must take into account the eventual role of oil penetration in the polymer. The resulting scheme is among the most complex ones in the field of oxidation of substrates in solid state.

1.5 Principles of Lifetime Prediction

In the case of polymer automotive parts in contact with fuels, not submitted to solar (UV) irradiation, the first factor to consider in a durability analysis is the existence or not of continuous mechanical (static or dynamic) loading.

If the parts are not loaded or sustain low stress levels, for instance ≤ 10% of instantaneous yield stress, failure can only result from a deep embrittlement due to chemical degradation, i.e. oxidation. Then, the parts can undergo fracture under low level stresses linked to small impacts or simply from differential dilatations induced by temperature changes. The most pertinent end-life criterion is then the embrittlement critical molar mass M'_c. From fracture mechanics concepts, one can determine the critical thickness l_c of the brittle layer able to induce crack propagation in the whole sample thickness [14, 15]. The oxidation kinetic model with reaction-diffusion coupling is aimed to predict at every moment the thickness distribution of average molar mass M. The end of life corresponds to the moment where $M = M'_c$ at the depth l_c. This criterion doesn't predict the part fracture but rather the moment at which the probability of part fracture increases suddenly to approach unity.

If the parts sustain continuous loads, two subcases can be envisaged depending on the occurrence or not of chemical degradation (i.e. oxidation). Let us call σ the stress (static creep) or the stress amplitude (fatigue), the curves of lifetime t_f against σ are expected to have the shape of Figure 4.

In the absence of chemical degradation, there is a critical stress σ_c below which there is no fracture. The value of σ_c depends of temperature and fluids in contact. A decrease of σ_c is thus expected in the presence of oil, its amplitude depends mainly of the difference of solubility

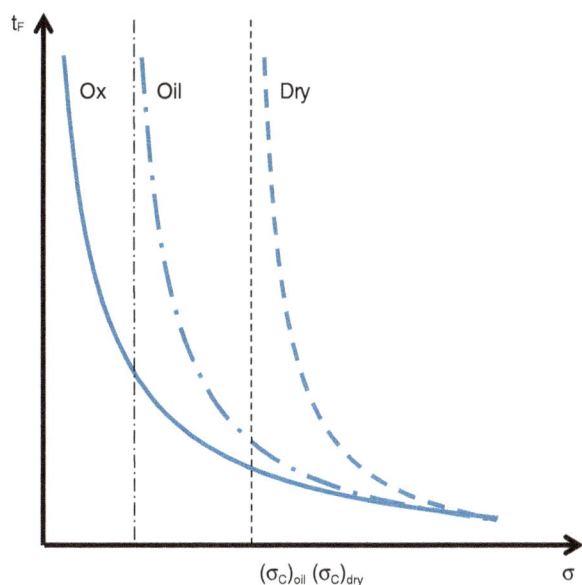

Figure 4

Shape of lifetime-stress curves in the presence ("ox") or the absence of oxygen ("oil" in the presence of oil; "dry" in the absence of oil) of chemical degradation.

parameters between the polymer and oil. The viscosity and surface tension of this latter, which determines the rate at which oil reaches the crack tip, can also play a role. These phenomena are well understood from a qualitative point of view, but there is no, to our knowledge, consensus on the mathematical expression of the curves $t_f = f(\sigma)$.

In the presence of chemical degradation, at high stress levels, failure occurs before significant chemical changes, the curve $t_f = f(\sigma)$ is superposed to the curves corresponding to the absence of degradation (obtained for instance in neutral atmosphere). At low stress levels, i.e. at longer exposure times, degradation affects the polymer strength, the curve diverges from preceding ones and does not display an asymptotic stress. Here, again, there is no wide consensus on the kinetic modeling approach. The following approach has been recently proposed by Colin et al. [16].

Let us consider a polymer sample of thickness L submitted to a constant tensile stress σ in the presence of oxygen at constant temperature T. A model of lifetime prediction can be built associating three "moduli": the first one relative to polymer degradation, aimed to determine the thickness distribution of average molar mass at every time; the second one relative to polymer creep kinetics and the third one relative to the molar mass dependence of ultimate strain ε_R. The principles involved can be described as follows:

The sample undergoes a creep characterized by an anelastic strain rate $d\varepsilon/dt$ depending of the applied stress, temperature and time. A very simple equation was proposed by the authors:

$$d\varepsilon/dt = Ae_T.\sigma.t^{-m}$$

where A is a constant, e_T is a temperature factor obeying for instance Arrhenius law: $e_T = \exp\text{-}(H/RT)$ and m is an exponent expressing the auto-retardated character of creep. The integration of this equation gives the anelastic strain $\varepsilon = f(t)$.

The fracture behavior of the polymer is characterized by the existence of a relatively sharp ductile-brittle transition. It is generally considered that ductility is due to the existence of an entanglement network in the polymer amorphous phase. Under the combined effect of stress and temperature, the chains are mobile enough to disentangle by reptation but the time to disentanglement varies rapidly with molar mass, roughly:

$$t(\text{disent.}) \sim M^3$$

NB: This time is shortened by the plasticizing effect of absorbed fuels.

Thus the sample deforms continuously under the effect of applied stress but becomes abruptly brittle when the time approaches the disentanglement. This disentanglement time decreases rapidly when molar mass decreases as a result of oxidation. As it has been shown above, in a thick sample, fracture will become highly probable when the disentanglement time will be reached at the critical depth l_c depending on polymer fracture properties; in polyethylene for instance, $l_c \sim 100$ μm.

Other approaches, based for instance on considerations of crack propagation, are possible in the case of glassy polymers [17].

2 POLYETHYLENE AGEING IN CONTACT WITH BIOETHANOL

2.1 Polyethylene Stabilization

Stabilizers are incorporated into polymers to reduce oxidative degradation, during processing and in the subsequent service life of the polymer. Processing stabilization of polyethylene is usually done by combination of phenolic and phosphorous antioxidants [18-21]. Fearon et al. [22] attributed the positive effect of phosphite antioxidants to their interaction with peroxides; the trivalent phosphorous additives often help to improve the colour of polymers. Indeed, organic phosphites have been applied as efficient processing stabilizers

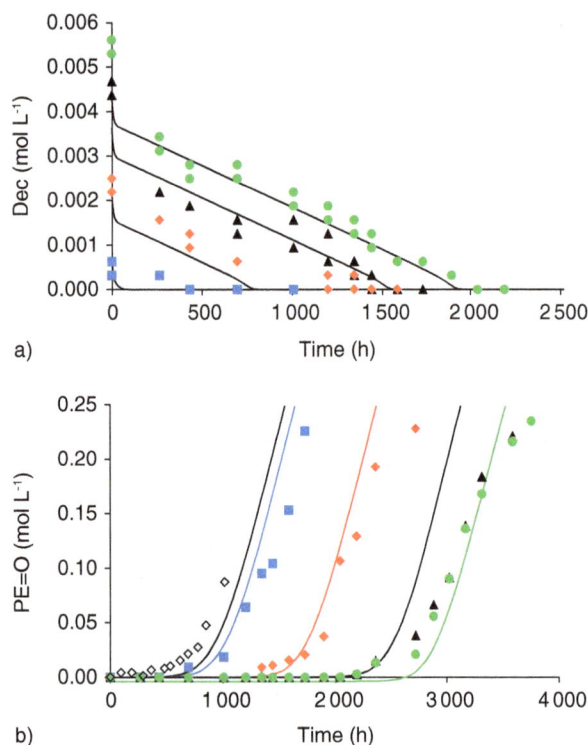

a)

b)

Figure 5

Kinetics of phosphite depletion a) and carbonyl build up b) for pure PE (◇), PE + 0.1% Irgafos 168 (■), PE + 0.2% Irgafos 168 (◆), PE + 0.3% Irgafos 168 (▲), PE + 0.4% Irgafos 168 (●) at 80°C (full lines correspond to kinetic modeling – see [22]).

Figure 6

OIT$_{200}$ variation for the commercial sample exposed in ethanol (◆), E10 (□) and E50 (▲) at 80°C.

in numerous polymers, especially polyolefins. Various mechanisms of phosphite stabilization have been proposed in the literature [23, 24]. Here, in order to put in evidence the major pathway of phosphite consumption, we present some results obtained at moderate temperatures ($\leq 180°C$) for ageing of a pure polyethylene mixed with 0.1, 0.2, 0.3 or 0.4% of an organophosphite stabilizer Irgafos 168 [1]. An FTIR study confirmed that, in the conditions under study, this phosphite is consumed by reducing hydroperoxides and yields a phosphate. When phosphites (or more generally all sacrificial stabilizers) are totally consumed (*Fig. 5b*), the sudden auto-acceleration of oxidation takes place (*Fig. 5a*) and embrittlement occurs shortly after. It was also shown that adding 0.4% (in weight) of phosphite permitted to increase the lifetime of PE at 80°C from less than 1 000 h to more than 2 000 h. The non-linear change of induction period with the concentration in phosphites can be interpreted as the consequence of a partial loss of stabilizer by evaporation, which is completely modeled elsewhere [1].

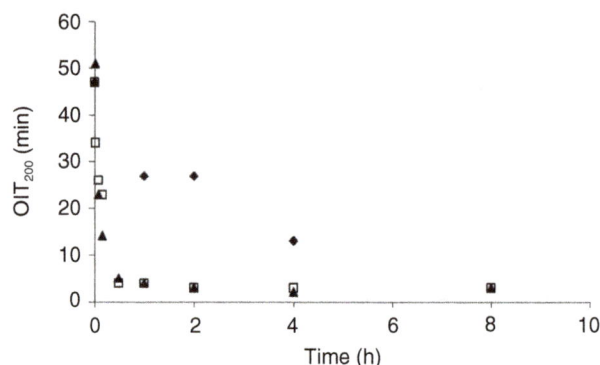

Hindered phenols are intrinsically more efficient than phosphites, for instance, lifetimes of the order of more than 10 000 h can be expected for PE + some ppm of Irganox 1010 [25]. It means that phenol + phosphite stabilized PE is expected to keep its engineering properties over years provided stabilizers disappear only by chemical consumption (radicals or hydroperoxides trapping). Let us now turn to the possible influence of fuel-medium environment likely to promote physical loss.

2.2 Experimental Evidence for Stabilizer Loss in the Presence of Ethanol

It has been tried to appreciate an eventual effect of ethanol based fuels on the oxidative stability of stabilized PE, focusing on stabilizer extraction by the fuel. A commercial HDPE sample stabilized by a phenol-phosphite synergistic blend was immersed in ethanol-cyclohexane mixtures used as model fuels (denoted by E0, E10 or E50, the number expressing the volume ratio of ethanol) at 80°C (*Fig. 6*) [2].

Polyethylene and common hydrocarbon fuels have relatively close solubility parameters, typically of the order of (15.8 to 17.1 MPa$^{1/2}$). Phenolic and phosphites stabilizers have noticeably higher solubility parameters. Ethanol is more polar than the stabilizers under study (26.3 MPa$^{1/2}$). Stabilizers are expected to be more soluble in ethanol than in PE so that one could expect a negative effect of ethanol on PE durability owing to its eventual extractive power on stabilizers.

To check this hypothesis, samples were immersed in various ethanol-cyclohexane mixtures and the stabilizer disappearance was monitored by Oxidation Induction Time (OIT) measurements (length of oxidation period

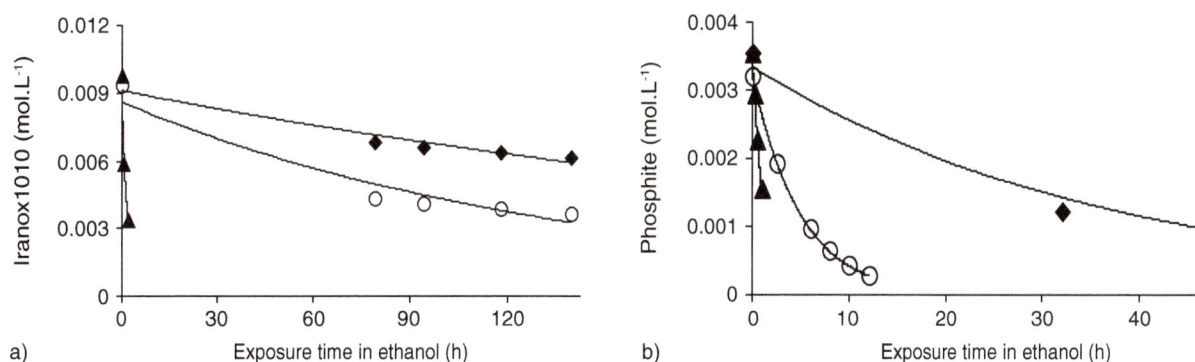

Figure 7

Exposure of laboratory made samples stabilized by Irganox 1010 a) and Irgafos 168 b) stabilizer loss at 40°C (♦), 60°C () and 80°C (▲) in pure ethanol.

TABLE 1

Molar mass and apparent first order rate constants for Irgafos 168 and Irganox 1010 at various temperatures and apparent activation energy (E_a)

		Rate constant β (h$^{-1} \times 10^{-2}$)			
	M (g.mol^{-1})	40°C	60°C	80°C	E_a (kJ.mol^{-1})
Irgafos 168	647	2.7	20	84	90 ± 10
Irganox 1010	1 178	0.33	1.3	25	110 ± 10

recorded *in situ* for an isothermal ageing test under 100% oxygen at atmospheric pressure, at 190 or 200°C in the DSC cell), assumed to be proportional to the residual quantity of stabilizers after ageing. Some results obtained at 80°C are shown in Figure 6. They call for the following comments:

Immersion in fuel drastically increases the stabilizer loss rate. Phosphite and phenol concentrations tend towards 0 after respectively c.a. 50 and 100 h. Thus alcohol-hydrocarbon fuels are expected to decrease the PE oxidative stability in proportion depending, indeed, of temperature and sample thickness.

Stabilizer extraction appears considerably slower in pure ethanol than in ethanol-cyclohexane mixtures.

Concerning the latter, no significant difference was found between E10 and E50. Extraction rate was in any case lowered when fuel is mixed with ethanol. One can thus conclude that bioethanol doesn't negatively influence the ageing behavior of polyethylene parts.

2.3 Modeling Aspects for Stabilizer Depletion

It seemed to us interesting to appreciate the extractive power of pure ethanol for samples containing a single stabilizer (here made of an additive free PE grade mixed with 0.3% Irganox 1010 or 0.3% Irgafos 168 prepared as described in [1, 2]). For those stabilizers, the concentration

can be easily monitored by FTIR [1, 2]. The results of immersion tests at 40, 60 and 80°C showed that stabilizer loss obeys first order kinetics:

$$\frac{dOIT}{dt} = -\beta(OIT - OIT_\infty)$$

where OIT_∞ is the induction time of the non-stabilized polymer and β is a first order rate constant.

The loss rate is significantly higher for Irgafos 168 than for Irganox 1010 that can be explained by the well-known effect of molar mass on migration rate [26].

Then the work is focused on the study of the depletion of each stabilizer separately. The results argue for a first order kinetics (*Fig. 7*).

Apparent first-order rate constant values β for stabilizer loss at 40, 60 and 80°C were determined. They are compiled in Table 1.

It can be verified that β obeys Arrhenius law, which allows extrapolating at lower temperature to perform some prediction of extraction kinetics in the device temperature range. Those values can now be used in a model coupling extraction of stabilizer in the sample superficial layer and diffusion from the bulk.

By comparing a characteristic time for extraction $\tau_E = 1/\beta$ with the diffusion characteristic time $\tau_D = L^2/D$, L being the sample thickness, it seems however, that stabilizer migration in the external fuel media is a diffusion controlled phenomena [2].

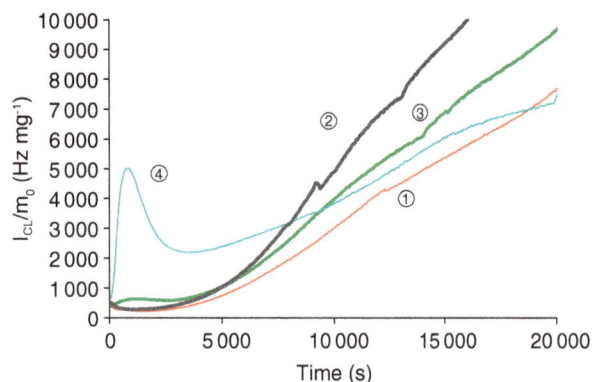

Figure 8

Kinetic curves of chemiluminescence emission for thermal oxidation at 150°C of pure PE ①, PE + methyl oleate ②, PE + methyl linoleate ③ and PE + methyl linolenate ④.

the early times of exposure can be suspected for PE + methyl linoleate;

– the trends observed for the PE + methyl linoleate are strongly exaggerated for the PE + methyl linolenate system where an intense peak develops in the early hours of exposure and where the light emission in the first 5 000 seconds is considerably stronger than for the other samples.

It can be shown that pure methyl esters of unsatturated fatty acid oxidize faster than PE. Figure 8 could be considered as the overlap of PE oxidation curve and UFE one. The shift of the part of curve ascribed to PE observed when this latter has been impregnated by UFE militates in favor of a cooxidation process, *i.e.* that UFE oxidation generates some radical species attacking PE chains. Some complementary experiments made on PE in solid state ($T < 130°C$) were aimed at deconvoluting the overall carbonyl concentration into the part generated from UFE and the part attibuted to pure PE oxidation [4]. The results clearly showed that the presence of unsaturated fatty esters accelerates the oxidation kinetics of PE matrix because the UFE are more oxidizable than PE due to their allylic hydrogens. We will now try to derive a kinetic modeling permitting to simulate this striking fact.

3 POLYETHYLENE AGEING IN CONTACT WITH BIODIESELS

3.1 Experimental Evidence for Co-Oxidation

Let us first recall that biodiesel from vegetable source are mixtures of 16 or 18 carbon fatty acids with 0, 1, 2 or 3 doubles bonds [27]. It seemed to us interesting to study polymer-biodiesel interaction through the case of PE ageing in presence of methyl oleate, linoleate or linolenate chosen as model systems. Stabilizer free PE films were impregnated with those methyl esters at room temperature. Let us precise that this study was just aimed at highlighting a possible co-oxidation, that leads to the choice of a non-stabilized PE rather than a commercial one. These films were then submitted to thermo-oxidative ageing at 150°C and the oxidation rate was monitored *in situ* by ChemiLuminescence (CL) measurements. Typical CL results are presented in Figure 8. They call for the following comments:

– for pure PE: the curve has the classical sigmoidal shape. A small shoulder at time ~ 10 000 s is noticed. According to Broska and Rychly [28], it could be due to the existence of structural irregularities in PE. This interpretation is, as it will be seen after, in good agreement with the proposal of kinetic model for co-oxidation involving the role of double bonds;

– for PE + methyl oleate and PE + methyl linoleate, the CL kinetic curve is progressively shifted towards shorter times, suggesting that PE matrix is oxidized faster in the presence than in the absence of unsaturated ester (UFE). A small CL peak appearing in

3.2 Co-Oxidation Modeling Aspects

It is now well established that the oxidative degradation of additive free PE can be described by the following kinetic model [5]:

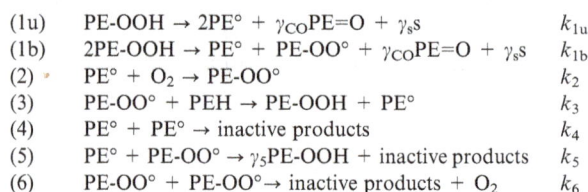

(1u)	PE-OOH → 2PE° + γ_{CO}PE=O + γ_ss	k_{1u}
(1b)	2PE-OOH → PE° + PE-OO° + γ_{CO}PE=O + γ_ss	k_{1b}
(2)	PE° + O₂ → PE-OO°	k_2
(3)	PE-OO° + PEH → PE-OOH + PE°	k_3
(4)	PE° + PE° → inactive products	k_4
(5)	PE° + PE-OO° → γ_5PE-OOH + inactive products	k_5
(6)	PE-OO° + PE-OO° → inactive products + O₂	k_6

where: k_{1u} is the rate constant for unimolecular hydroperoxide decomposition (s^{-1}); k_{1b}, ..., k_6 are second order rate constants for bimolecular processes ($L.mol^{-1}.s^{-1}$); γ_{CO}, γ_s and γ_5 are respectively yields in carbonyl, chain scission and hydroperoxides for (1u), (1b), (5) and (6) equations. Here, reaction 6 is a virtual balance equation kinetically equivalent to several coexisting termination reactions (coupling, disproportionation, etc.) [5].

The ratio k_3^2/k_6 expresses the intrinsic substrate oxidizability, independent of the initiation mode (peroxide decomposition, polymer radiolysis or photolysis, etc.) [1]. In the case of ethylene propylene co-oxidation, Decker *et al.* [29] showed that the oxidizability ratio varies with ethylene molar fraction e according to a pseudo-hyperbolic curve which could be approximated by the following function:

$$\frac{k_3}{\sqrt{k_6}} = 10^{-4} \times \frac{2 - 1.63 \times e}{1 + 3.13 \times e}$$

where k_3 and k_6 can be defined as the rate constants characteristic of a virtual homopolymer which would have the same kinetic behavior as the copolymer under study. An approach in which the kinetic behavior of the copolymer would be predicted from the characteristics of the corresponding homopolymers would be, indeed, more satisfactory. Such an approach needs to solve the co-oxidation kinetic scheme in which there are two distinct reactive sites: here tertiary carbons present only in propylene units and secondary carbons present in both comonomers are simultaneously oxidized (oxidation of primary carbons of propylene units can be neglected).

The presence of two reactive sites needs to take into account supplementary reaction for cross initiation, propagation, termination to be added to the self-initiation, propagation and termination reactions for pure substrates. If PE represents the aliphatic (-CH$_2$-) substrate (for polymer matrix) and UFE the allylic >C=CH-CH$_2$- one (for methyl ester), the kinetic scheme could be written:

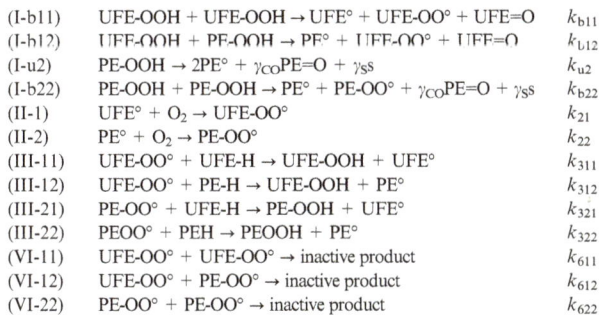

(I-b11)	UFE-OOH + UFE-OOH → UFE° + UFE-OO° + UFE=O	k_{b11}
(I-b12)	UFE-OOH + PE-OOH → PE° + UFE-OO° + UFE=O	k_{b12}
(I-u2)	PE-OOH → 2PE° + γ_{co}PE=O + γ_ss	k_{u2}
(I-b22)	PE-OOH + PE-OOH → PE° + PE-OO° + γ_{co}PE=O + γ_ss	k_{b22}
(II-1)	UFE° + O$_2$ → UFE-OO°	k_{21}
(II-2)	PE° + O$_2$ → PE-OO°	k_{22}
(III-11)	UFE-OO° + UFE-H → UFE-OOH + UFE°	k_{311}
(III-12)	UFE-OO° + PE-H → UFE-OOH + PE°	k_{312}
(III-21)	PE-OO° + UFE-H → PE-OOH + UFE°	k_{321}
(III-22)	PEOO° + PEH → PEOOH + PE°	k_{322}
(VI-11)	UFE-OO° + UFE-OO° → inactive product	k_{611}
(VI-12)	UFE-OO° + PE-OO° → inactive product	k_{612}
(VI-22)	PE-OO° + PE-OO° → inactive product	k_{622}

Similar models were presented in literature for extrinsically initiated oxidation. They were analytically solved using *ad hoc* hypothesis (for example equality of cross propagation rates $r_{312} = r_{321}$) and under the assumption of constant initiation rate (which is not suitable for thermal oxidation). Here, we will check this model on PE + methyl esters of unsaturated esters of oleic, linoleic or linolenic acids, these latter being expected to have distinct oxidizabilities and also to oxidize faster than PE. The use of a numerical tool will permits to solve the system of differential equations without using questionable hypotheses and also to generate a wide variety of simulations to be compared with our experimental results, in particular here the non-monotonous shape of CL curves.

The simulations runs were done using the following hypotheses:

– k_{21} was chosen equal to 10^7 L.mol^{-1}.s^{-1}, having in mind that variations of this value have a negligible influence on oxidation kinetics. It seems reasonable to assume $k_{21} < k_{22}$ because of the difference of reactivity between PE° and UFE° alkyl radicals;

– k_{b11} and k_{611} have been here fixed respectively equal to 10^{-2} and 10^8 L.mol^{-1}.s^{-1}. Their precise adjustment from the CL curves of pure methyl esters oxidation is under study in our lab;

– Cross initiation and cross termination rate constants were calculated under the assumption of geometrical means:

$$k_{612}{}^2 = k_{611} \times k_{622}$$

$$k_{b12}{}^2 = k_{b11} \times k_{b22}$$

– $k_{31}, k_{312}, k_{32}, k_{321}$ for the reaction:
ROO° + PH → ROOH + P°
propagation reactions can be calculated by the relationships established by Korcek *et al.* [30]:

$$\log_{10} k_p{}^{\text{sec-ROO°}}(30°C) = 16.4 - 0.2 \times \text{BDE(C-H)}$$

$$E_P = 0.55 \times (\text{BDE(C-H)} - 62.5)$$

where $^{\text{sec-ROO°}}$ denotes a secondary peroxy radical, BDE(C-H) is the bond dissociation energy (in kcal. mol^{-1}) of an abstractable hydrogen hold by a P-H substrate.

Using these relationships together with BDE values reported by Denisov [31] for several unsaturated hydrocarbons, propagation rate constants values at 150°C can be proposed (*Tab. 2*):

Since the propagation rate constants depend only on the bond dissociation energy of broken C-H bond, we will first assume:

$$k_{312} = k_{322}$$

$$k_{321} = k_{311}$$

(instead of equality of the rates:
k_{312}[UFE-OO°][PE-H] = k_{321}[PE-OO°][UFE-H]
as done in the original paper by Russell [11]).

The following initial conditions were chosen:

$$[PE°]_0 = [PEOO°]_0 = [UFE°]_0 = [UFE-OO°]_0 = 0$$
$$[PEOOH]_0 = 10^{-4} \text{ mol.L}^{-1}, [UFEOOH]_0 = 10^{-2} \text{ mol.L}^{-1}$$
$$[PH]_0 = 60 \text{ mol.L}^{-1}, [UFE-H]_0 = 0.3 \text{ mol.L}^{-1}$$

calculated as the number of moles of the more reactive hydrogens present in the c.a. 5% fatty ester absorbed

TABLE 2

Kinetic parameters of propagation reactions in saturated and unsaturated substrates

	BDE(C-H) (kJ.mol^{-1})	E_3 (kJ.mol^{-1})	k_3 (30°C) (L.mol^{-1}.s^{-1})	k_3 (150°C) (L.mol^1.s^{-1})
CH-H (cyclohexane)	395.5	73.7	0.0	12.6
CH$_2$=CHCH$_2$-H	368	58.5	0.1	47.2
CH$_2$=CH(CH-H)Me	349.8	48.5	0.5	113.4
Z-MeCH=CH(CH-H)Me	344	45.3	0.9	149.9
Me$_2$C=CH(CH-H)Me	332	38.7	3.4	267.2
Me$_2$C=CH(C-H)Me$_2$	322.8	33.7	9.4	416.2
CH-H (cyclohexene)	341.5	44.0	1.2	169.1
CH-H (cyclohexadiene)	330.9	38.1	3.9	281.8
CH-H (cyclohexadiene)	312.6	28.1	28.9	680.2
CH-H (cycloheptatriene)	301	21.7	103.6	1 189.1

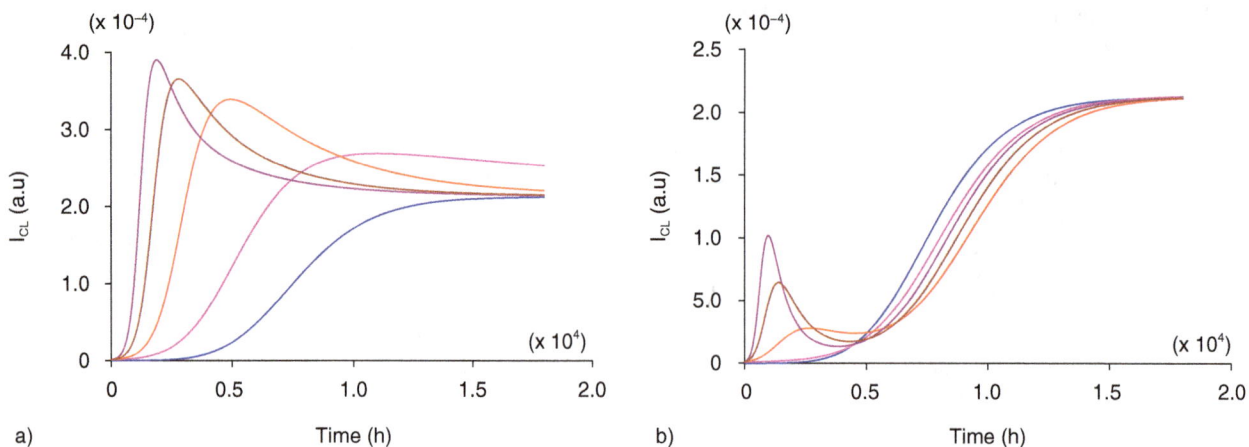

a) b)

Figure 9

Simulation of CL runs with $k_{b11} = 10^{-2}$ L.mol^{-1}.s^{-1}, $k_{611} = 10^8$ L.mol^{-1}.s^{-1}, [POOH]$_0$ = 10^{-2} mol.L^{-1}, $k_{312} = k_{322}$ a) or $k_{312} = 0$ b) for pure PE, PE + ME with k_{311} = 100, 500, 1 000 and 1 500 L.mol^{-1}.s^{-1}.

in PE at room temperature (see 'Oils Permeability in Polyethylene' section).

The following expression of chemiluminescence intensity can be derived:

$$I_{CL} = \Phi \times (k_{b11}.[\text{UFE-OOH}]^2 + k_{b12}.[\text{UFE-OOH}].[\text{PE-OOH}] + k_{b22}.[\text{PE-OOH}]^2)$$

Simulations for CL curves are given in Figure 9 for several sets of rate constants differing only by k_{311} and k_{321} value. It is noteworthy that the best simulations were obtained using $k_{312} = 0$ (instead of $k_{312} = k_{322}$). The shape of simulated CL curves is in reasonable agreement with experimental observations.

It seems clear that whatever the k_{311} value extrapolated at 150°C, the CL curves can be simulated provided that the following inequality is verified:

$$k_{311}{}^2/k_{611} \gg k_{322}{}^2/k_{622}$$

i.e. that both compounds have a significantly different reactivity. The next step of the approach is to investigate if, in real conditions, the co-oxidation phenomenon is limited to the superficial layers of a thick PE tank (here represented by a thin film) or on the contrary if UFE methyl esters migrate into the PE bulk and promote its oxidation.

3.3 Oils Permeability in Polyethylene

Vegetable oils are more or less complex mixtures of fatty compounds. We focus here on their permeation properties (diffusion and solubility) in a general purpose grade of PE. It seemed to us interesting to compare rapeseed and soy methyl esters with their major components *i.e.* methyl oleate (C18:1) and methyl linoleate (C18:2). The gravimetric sorption curves display an equilibrium plateau at a weight gain c.a. 4-5% at room temperature (*Fig. 10*), which is almost independent of the ester nature and having the same value for the vegetable oils and for their major component confirming thus the pertinence of its choice as a model compound.

Since oils are supposed to penetrate only into the amorphous phase of the polymer, it seemed to us interesting to determine the oil equilibrium concentration c in the amorphous phase from the equilibrium mass gain w_{eq}

using the following relationship (assuming that densities of PE amorphous phase and methyl ester are equal):

$$c = \frac{w_{eq} \cdot \rho_a^{PE}}{M_{ester} \cdot (1 - x_C) \cdot (1 + w_{eq})}$$

where: ρ_a^{PE} is the density of PE amorphous phase ($\rho_a^{PE} = 850$ g.L^{-1}), M_{ester} is the molar mass of the methyl ester (taken equal to the average molar mass for the soy and rapeseed methyl esters), x_C is the crystallinity ratio (here $x_C \sim 0.5$).

Equilibrium concentrations determined at 23, 45, 60 and 75°C are listed in Table 3. They call for the following comments:

- methyl linoleate is slightly but significantly less soluble than methyl oleate;
- vegetable methyl esters do not differ strongly from their major component;
- the solubility is an increasing function of temperature.

The relative mass uptake was plotted against square root of time in Figure 11 for rapeseed oil at the four temperatures under investigation. For the temperatures of 23°C, 45°C and 60°C, the plots are linear ($R^2 \geq 0.977$) in the domain of low mass uptake indicating that diffusion obeys Fick's law. At the highest temperature (75°C), the plots are clearly sigmoidal for all the samples under study. The diffusion coefficient D was calculated from the slope of the straightline using:

$$D = \frac{\pi L^2}{16} \cdot \left(\frac{dw/w_{eq}}{d\sqrt{t}} \right)^?$$

Its values are listed in Table 4. For the temperature of 75°C, we have taken the average slope but the corresponding D values must be considered cautiously.

Data are well fitted by Arrhenius law.

In conclusion, the presented experimental results show that methyl esters derived from soy or rapeseed oil are relatively soluble in PE (roughly 5 to 10% in weight depending on temperature) together with a high diffusion rate (10^{-13} m^2.s^{-1} corresponds to a time to reach equilibrium of c.a. 3 years for a 3 mm thick sample). In other words, fatty esters of the soy or rapeseed type can easily migrate towards PE bulk and promote its oxidative degradation (see above).

Let's us now turn to possible modeling for diffusivity prediction. The diffusivity can be considered as resulting from the balance between penetrant size (V^*) and free volume (V_f) permitting molecular jumps, as expressed by Cohen and Turnbull [32]:

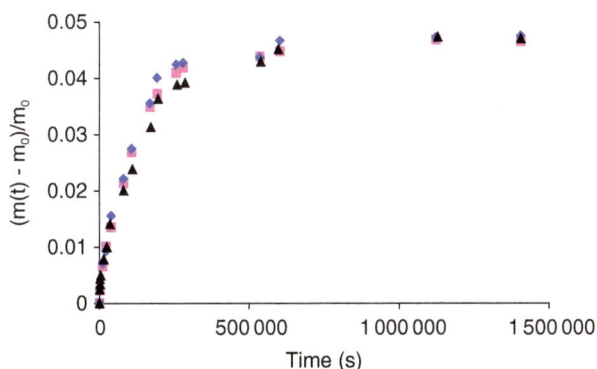

Figure 10

Mass uptake in polyethylene immersed in rapeseed methyl ester (measurement is done in triplicate).

$$D = A. \exp \left(-\gamma \frac{V^*}{V_f} \right)$$

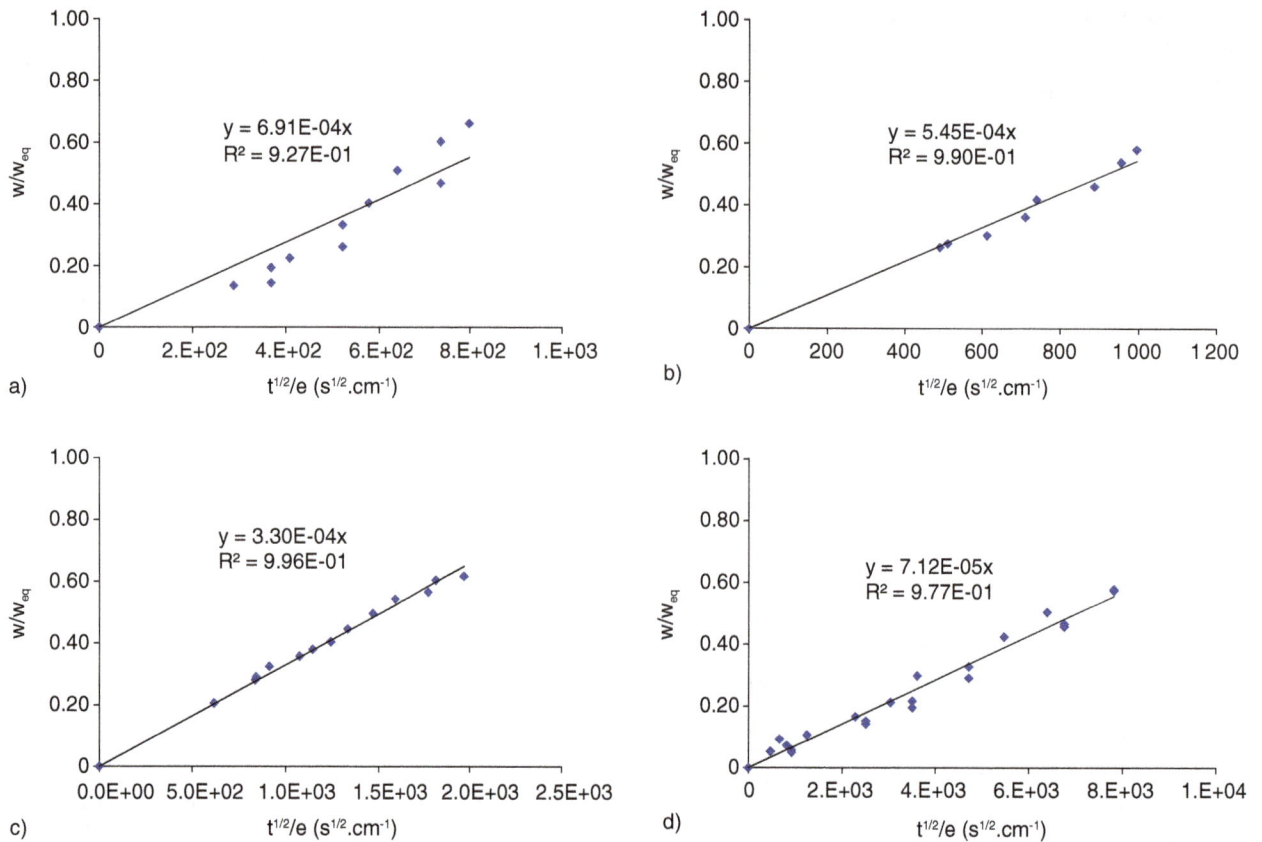

a)
b)
c)
d)

Figure 11

Relative mass uptake against square root of time for polyethylene films immersed in rapeseed methyl ester at 75°C a), 60°C b), 45°C c) and room temperature d).

TABLE 3

Equilibrium mass gain and corresponding concentration at the four temperatures under study

T (°C)	Methyl oleate		Methyl linoleate		Rapeseed methyl ester		Soy methyl ester	
	w_{ep}	c (mol.L^{-1})	w_{ep}	c (mol.L^{-1})	w_{ep}	c (mol.L^{-1})	w_{ep}	c (mol.L^{-1})
75	0.116	0.614	0.092	0.503	0.095	0.512	0.092	0.497
60	0.070	0.389	0.064	0.358	0.071	0.393	0.069	0.382
45	0.056	0.316	0.051	0.289	0.055	0.309	0.054	0.303
23	0.047	0.265	0.042	0.238	0.047	0.267	0.047	0.268

In which γ is a parameter ranging between 0.5 and 1. This theory has led to two sorts of models:
— non empirical models. The most sophisticated one was established by Vrentas and Duda [33]:

$$D_1 = D_{01} \cdot \exp\left(-\frac{E}{RT}\right)$$

$$\times \exp\left(-\frac{\omega_1 \hat{V}_1^* + \omega_2 \xi \hat{V}_2^*}{\frac{K_{11}\omega_1\left(K_{21}-T_{g1}+T\right)}{\gamma_1} + \frac{K_{12}\omega_2\left(K_{22}-T_{g2}+T\right)}{\gamma_2}}\right)$$

where D_1 is the penetrant self-diffusion coefficient, E is the activation energy for a penetrant jump corresponding to the energy which is necessary for a molecule to overcome the attraction of neighboring molecules, ω_1 and ω_2 are the weight fractions of polymer and penetrant, V_i^* is the specific volume necessary for a penetrant molecule or polymer segment jump, and ξ is the ratio of penetrant and polymer jumping unit critical volumes.

This relationship is especially designed for cases where penetrant brings its own free volume so that diffusion is

TABLE 4

Diffusion coefficients (in $m^2.s^{-1}$) pre-exponential factor and activation energy for all the samples under study

	75°C	60°C	45°C	23°C	ln D_0 (D_0 in $m^2.s^{-1}$)	ΔH_D ($kJ.mol^{-1}$)
Methyl oleate	2.40×10^{-11}	8.21×10^{-12}	2.36×10^{-12}	1.39×10^{-13}	5.18	85.2
Methyl linoleate	1.96×10^{-11}	5.02×10^{-12}	2.30×10^{-12}	1.13×10^{-13}	4.15	83.1
Rapeseed methyl ester	1.82×10^{-11}	6.22×10^{-12}	2.12×10^{-12}	1.26×10^{-13}	3.69	81.8
Soy methyl ester	2.56×10^{-11}	6.48×10^{-12}	2.10×10^{-12}	1.25×10^{-13}	5.63	86.7

TABLE 5

Model parameters from Begley et al. [35]

Polymer	A_P	τ	Temperature
LDPE	11.5	0	<90°C
HDPE	14.5	1577	<100°C
PP homopolymer	13.1	1577	<120°C
PP rubber	11.5	0	<100°C
PS	0.0	0	<70°C
HIPS	1.0	0	<70°C
PET	6.0	1577	<175°C
PEN	5.0	1577	<175°C
PA66	2.0	0	<100°C

auto-facilitated. However, its implementation required the determination of several coefficients *a priori* unknown.

Molecular (empirical) models also express diffusion coefficient in function of penetrant size (the most often its molar mass), temperature, and polymer structure dependant parameters. They were developed for predicting migration of chemicals from food packaging. They work especially well for linear molecules [34]. An example of these models is the following one:

$$D = 10^4 . \exp\left(A_p - 0.1351.M_r^{2/3} + 0.003.M_r - \frac{10\,454 + \tau}{T}\right)$$

in which: D is expressed in $cm^2.s^{-1}$, M_r being the molar mass of the penetrant expressed in $g.mol^{-1}$, A_P and τ depend on polymer matrix nature.

The parameter values determined by Begley et al. [35] for some polymers are listed in Table 5.

It is easy to verify that despite its simplicity:
- this model permits a fair prediction of diffusivity values,
- the prediction of temperature effect is excellent.

This model can be thus applied in the case under study, to predict the diffusivity values of biofuels in PE.

CONCLUSIONS

This paper provides an overview perspective how lifetime of polyethylene in contact of ethanol based or biofuel based fuels can be predicted. Indeed, embrittlement time of polyethylene parts is governed by several processes such as physical interactions (stress cracking, additive extraction) or chemical interactions (polymer oxidation promoted by fuel oxidation). Here, Arrhenius law cannot be applied to the whole ageing process owing to the complexity of mechanisms involving several processes of diffusion (oil, oxygen and stabilizers with distinct timescales) and chemical processes (oxidation). The classical method consisting of using Arrhenius law to predict lifetime from experimental results obtained at high temperature (typically $T \geq 100$°C) is thus questionable here.

Thanks to numerical methods, a non-empirical kinetic model can be derived from a realistic mechanistic scheme to simulate property changes, for instance carbonyl growth. The main advantage of this analytical approach is that stabilizers which are frequently present in commercial polyethylene can be included into the simulation since kinetic parameters are already known for the pure polymer. Some simulation results have been reported for stabilized polyethylene, it has been shown how they may guide extrapolations towards lower temperatures. Furthermore, this kinetic approach can include physical phenomena occurring during the degradation process as stabilizers physical loss. It has been found that the stabilizer depletion rate is reduced when ethanol content increases and obeys Arrhenius law.

A specific attention has been paid to chemical interaction between oil and PE. It has been shown that unsaturated fatty esters can penetrate into PE and thus promotes PE oxidation. A kinetic model involving all these processes has been proposed. The predictive value of this approach has been partially checked from chemiluminescence experiments.

REFERENCES

1 Djouani F., Richaud E., Fayolle F., Verdu J. (2011) Modelling of thermal oxidation of phosphite stabilized polyethylene, *Polym. Degrad. Stab.* **96**, 7, 1349-1360.

2 Djouani F., Patel V., Richaud E., Fayolle B., Verdu J. (2012) Antioxidants loss kinetics in polyethylene exposed to model ethanol based biofuels, *Fuel* **93**, 502-509.

3 Richaud E., Flaconnèche B., Verdu J. (2012) Biodiesel permeability in polyethylene, *Polymer Testing* **31**, 8, 1070-1076.

4 Richaud E., Fayolle B., Verdu J., Rychlý J. (2013) Co-oxidation kinetic model for the thermal oxidation of polyethylene-unsaturated substrate systems, *Polym. Degrad. Stab.* **98**, 5, 1081-1088.

5 Khelidj N., Colin X., Audouin L., Verdu J., Monchy-Leroy C., Prunier V. (2006) Oxidation of polyethylene under irradiation at low temperature and low dose rate. Part II. Low temperature thermal oxidation, *Polym. Degrad. Stab.* **91**, 7, 1598-1605.

6 Fayolle B., Richaud E., Colin X., Verdu J. (2008) Review: Degradation induced embrittlement in semi-crystalline polymers having their amorphous phase in rubbery state, *J. Mater. Sci.* **43**, 6999-7012.

7 Kennedy M.A., Peacock A.J., Mandelkern L. (1994) Tensile properties of crystalline polymers: Linear polyethylene, *Macromolecules* **27**, 19, 5297-5310.

8 Fayolle B., Verdu J., Piccoz D., Dahoun A., Hiver J.M., G'Sell C. (2009) Thermooxidative ageing of polyoxymethylene Part 2: Embrittlement mechanisms, *J. Appl. Polym. Sci.* **111**,1, 469-475.

9 Fayolle B., Audouin L., Verdu J. (2004) A critical molar mass separating the ductile and brittle regimes as revealed by thermal oxidation in polypropylene, *Polymer* **45**,12, 4323-4330.

10 McCall D.W., Douglass D.C., Blyler Jr L.L., Johnson G. E., Jelinski L.W., Bair H.E. (1984) Solubility and diffusion of water in low-density polyethylene, *Macromolecules* **17**, 9, 1644-1649.

11 Russell G.A., Williamson Jr R.C. (1964) Directive effects in aliphatic substitutions. XXV. Reactivity of aralkanes, aralkenes, and benzylic ethers toward peroxy radicals, *J. Am. Chem. Soc.* **86**, 12, 2364-2367.

12 Rincon-Rubio L.M., Fayolle B., Audouin L., Verdu J. (2001) A general solution of the Closed-loop kinetic scheme for the thermal oxidation of polypropylene, *Polym. Degrad. Stab.* **74**, 1, 177-188.

13 Colin X., Richaud E., Verdu J., Monchy-Leroy C. (2010) Kinetic modelling of radiochemical ageing of ethylene–propylene copolymers, *Radiat. Phys. Chem* **79**, 3, 365-370.

14 So P.K., Broutman L.J. (1986) The fracture behavior of surface embrittled polymers, *Polym. Eng. Sci.* **26**, 17, 1173-1179.

15 Schoolenberg G.E. (1988) A fracture mechanics approach to the effects of UV-degradation on polypropylene, *J. Mater. Sci.* **23**, 5, 1580-1590.

16 Colin X., Audouin L., Verdu J., Rozental-Evesque M., Rabaud B., Martin F., Bourgine F. (2009) Aging of polyethylene pipes transporting drinking water disinfected by chlorine dioxide. Part II — Lifetime prediction, *Polym. Eng. Sci.* **49**, 8, 1642-1652.

17 Verdu J. (2012) Oxidative Ageing of Polymers, John Wiley & Sons, Inc., Hoboken, NJ, USA.

18 Chirinos-Padrón A.J., Hernández P.H., Allen N.S., Vasilion C., Marshall G.P., de Poortere M. (1987) Synergism of antioxidants in high density polyethylene, *Polym. Degrad. Stab.* **19**, 2, 177-189.

19 Al-Malaika S., Goodwin C., Issenhuth S., Burdick D. (1999) The antioxidant role of α-tocopherol in polymers II. Melt stabilising effect in polypropylene, *Polym. Degrad. Stab.* **64**, 1, 145-156.

20 Verdu J., Rychly J., Audouin L. (2003) Synergism between polymer antioxidants-kinetic modelling, *Polym. Degrad. Stab.* **79**, 3, 503-509.

21 Zweifel H. (1996) Effect of stabilization of polypropylene during processing and its influence on long term behavior under thermal stress, *Polymer Durability. Advances in Chemistry Series 249*, Clough R.L., Billingham N., Gillen K.T. (eds), American Chemical Society, Washington

22 Fearon P.K., Phease T.L., Billingham N.C., Bigger S.W. (2002) A new approach to quantitatively assessing the effects of polymer additives, *polymer* **43**, 17, 4611-4618.

23 Schwetlick K., Pionteck J., Winkler A., Hähner U., Kroschwitz H., Habicher W.D. (1991) Organophosphorus antioxidants: Part X – Mechanism of antioxidant action of aryl phosphites and phosphonites at higher temperatures, *Polym. Degrad. Stab.* **31**, 2, 219-228.

24 Peinado C., Corrales T., García-Casas M.J., Catalina F., Ruiz Santa Quiteria V., Parellada M.D. (2006) Chemiluminescence from poly(styrene-b-ethylene-co-butylene-b-styrene) (SEBS) block copolymers, *Polym. Degrad. Stab.* **91**, 4, 862-874.

25 Mallégol J., Carlsson D.J., Deschênes L. (2001) A comparison of phenolic antioxidant performance in HDPE at 32–80°C, *Polym. Degrad. Stab.* **73**, 2, 259-267.

26 Vitrac O., Mougharbel A., Feigenbaum A. (2007) Interfacial mass transport properties which control the migration of packaging constituents into foodstuffs, *J. Food Eng.* **79**, 3, 1048-1064.

27 Demirbas A. (2005) Biodiesel production from vegetable oils *via* catalytic and non-catalytic supercritical methanol transesterification methods, *Progr. Energ. Combust. Sci.* **31**, 5-6, 466-487.

28 Broska R., Rychlý J. (2001) Double stage oxidation of polyethylene as measured by chemiluminescence, *Polym. Degrad. Stab.* **72**, 2, 271-278.

29 Decker C., Mayo F.R., Richardson H. (1973) Aging and degradation of polyolefins - 3. Polyethylene and ethylene-propylene copolymers, *J. Polym. Sci. Part A-1: Polym. Chem.* **11**, 11, 2879-2898.

30 Korcek S., Chenier J.H.B., Howard J.A., Ingold K.U. (1972) Absolute Rate Constants for Hydrocarbon Autoxidation. XXI. Activation Energies for Propagation and the Correlation of Propagation Rate Constants with Carbon-Hydrogen Bond Strengths, *Can. J. Chem.* **50**, 2285-2297.

31 Denisov E.T., Afanas'ev I.B. (2005) Oxidation and Antioxidants in Organic Chemistry and Biology, CBC Taylor & Francis Group, Boca Raton, London, New York, Singapore.

32 Cohen M.H., Turnbull D. (1959) Molecular transport in liquids and glasses, *J. Chem Phys.* **31**, 1164-1169.

33 Vrentas J.S., Duda J.L. (1977) Diffusion in polymer – Solvent systems – 2. A predictive theory for the dependence of diffusion coefficients on temperature, concentration and molecular weight, *J. Polym. Sci. Polym. Phys. Ed* **15**, 3, 417-439.

34 Brandsch B., Mercea P., Piringer O. (2000) Modeling of Additive Diffusion Coefficients in Polyolefins, in *Food Packaging Testing Methods and Applications. ACS Symposium Series 753*, Risch S.J. (ed.). Washington DC.

35 Begley T., Castle L., Feigenbaum A., Franz R., Hinrichs K., Lickly T., Mercea P., Milana M., O'Brien A., Rebre S., Rijk R., Piringer O. (2005) Evaluation of migration models that might be used in support of regulations for food-contact plastics, *Food Addit. Contam.* **22**, 1, 73-90.

9

Combined Effect of Pressure and Temperature on the Viscous Behaviour of All-Oil Drilling Fluids

J. Hermoso, F. Martínez-Boza* and C. Gallegos

Departamento de Ingeniería Química, Universidad de Huelva, Facultad de Ciencias Experimentales, Campus del Carmen, 21071 Huelva - Spain
e-mail: martinez@uhu.es

* Corresponding author

Abstract — *The overall objective of this research was to study the combined influence of pressure and temperature on the complex viscous behaviour of two oil-based drilling fluids. The oil-based fluids were formulated by dispersing selected organobentonites in mineral oil, using a high-shear mixer, at room temperature. Drilling fluid viscous flow characterization was performed with a controlled-stress rheometer, using both conventional coaxial cylinder and non-conventional geometries for High Pressure/High Temperature (HPHT) measurements. The rheological data obtained confirm that a helical ribbon geometry is a very useful tool to characterise the complex viscous flow behaviour of these fluids under extreme conditions.*

The different viscous flow behaviours encountered for both all-oil drilling fluids, as a function of temperature, are related to changes in polymer-oil pair solvency and oil viscosity. Hence, the resulting structures have been principally attributed to changes in the effective volume fraction of disperse phase due to thermally induced processes. Bingham's and Herschel-Bulkley's models describe the rheological properties of these drilling fluids, at different pressures and temperatures, fairly well. It was found that Herschel-Bulkley's model fits much better B34-based oil drilling fluid viscous flow behaviour under HPHT conditions.

Yield stress values increase linearly with pressure in the range of temperature studied. The pressure influence on yielding behaviour has been associated with the compression effect of different resulting organoclay microstructures. A factorial WLF-Barus model fitted the combined effect of temperature and pressure on the plastic viscosity of both drilling fluids fairly well, being this effect mainly influenced by the piezo-viscous properties of the continuous phase.

Résumé — **Effet combiné de la pression et de la température sur le comportement visqueux des boues de forage non-aqueuses** — L'objectif principal de cette recherche était d'étudier l'influence combinée de la pression et de la température sur le comportement visqueux de deux boues de forage à base d'huile.

Les boues de forage à base d'huile ont été formulées par dispersion d'organo-bentonites dans de l'huile minérale à température ambiante, en utilisant un mélangeur à fort cisaillement. La caractérisation de l'écoulement visqueux des boues de forage a été réalisée avec un rhéomètre à contrainte imposée, utilisant un cylindre coaxial conventionnel et les géométries non conventionnelles pour des mesures à Hautes Pressions/Hautes Températures (HPHT). Les données rhéologiques obtenues confirment que la géométrie en « ruban hélicoïdal » est un outil très utile pour la caractérisation du comportement de l'écoulement des fluides complexes dans des conditions extrêmes.

Les différents comportements visqueux rencontrés pour les deux types de boues à base d'huile, en fonction de la température, sont liés aux changements de solvabilité du couple polymère-huile et de la viscosité de l'huile. Par conséquent, les structures qui en résultent sont principalement dues aux changements de la fraction volumique effective de la phase dispersée due à des procédés induits thermiquement.

Les modèles de Bingham et Herschel-Bulkley décrivent avec précision les propriétés rhéologiques de ces fluides de forage, à différentes pressions et températures. Le modèle de Herschel-Bulkley s'adapte mieux au comportement d'écoulement de la boue de forage B34 dans les conditions HPHT.

Les valeurs de contraintes seuil augmentent linéairement avec la pression dans le domaine de température étudiées. L'influence de la pression sur le rendement a été associée à l'effet de compression des différentes microstructures d'organo-argiles qui en résultent. Le modèle factoriel WLF-Barus s'ajuste bien à l'effet combiné de la température et de la pression sur la viscosité plastique des deux fluides de forage. Cet effet est principalement influencé par les propriétés piezo-visqueuses de la phase continue.

SYMBOLS

c_1	Experimental constant
c_2	Experimental constant, T, (°C)
HPHT	High Pressure/High Temperature
K	Consistency index
n	Flow index
p	Pressure
p_0	Pressure of reference
T	Temperature
T_0	Temperature of reference
β	Piezo-viscous coefficient
β_0	Empirical parameter
β_1	Empirical parameter
β_τ	Piezo-yield stress coefficient
Δp	Differential pressure
$\dot{\gamma}$	Shear rate
η	Viscosity
η_0	Viscosity or plastic viscosity at 1 bar and 40°C
η_p	Plastic viscosity
τ	Shear stress
τ_c	Yield stress
τ_B	Bingham's yield stress
τ_H	Herschel-Bulkley's yield stress
τ_{c0}	Bingham's or Herschel-Bulkley's yield stress, at 1 bar and 40°C
χ	Flory-Huggins interaction parameter for the polymer-solvent pair

INTRODUCTION

The oil industry is increasingly drilling more technically challenging and difficult wells. Drilling of deeper wells worldwide requires a constant searching for adequate drilling muds to overcome extreme conditions (Williams et al., 2011). The successful completion of an oil well and its cost depend, on a considerable extent, on the properties of drilling fluids. Hence, the proper selection of the drilling fluid during rotary drilling process is essential for dealing with the wide range of challenges encountered, particularly in high angle drilling, completion and workover operations.

Most of the worldwide drilling operations use water-based muds because of their lower environmental impact and cheaper cost (Caenn and Chillingar, 1996). However, water muds also can give wellbore instability problems, such as bit balling, wash out, high torque and drag, due to clay swelling at extreme temperature conditions. On the contrary, non aqueous drilling fluids (oil based and synthetic based), owing to their excellent lubricity, high rate of penetration and outstanding thermal stability, are often used to drill difficult wells, such as long sections of high angle and/or HPHT wells (Ghalambor et al., 2008).

Drilling fluid viscous behaviour is a critical issue in the success of drilling operations, particularly for drill cuttings removal. The properties that drilling fluids should possess are appropriate viscosity, high-shear thinning behaviour and a finite yield stress for suspending and transferring drill cuttings to the surface (Kelessidis et al., 2007). Nevertheless, the rheological characterization of these systems is not a trivial task because of the inherent heterogeneous nature of the system. The use of non-conventional geometries, such as helical ribbons and blade turbines, has become valuable tools for characterizing the viscous flow behaviour of disperse systems, mainly due to the elimination of serious wall slip effects of apparent yield stress materials (Barnes and Nguyen, 2001; O'Shea and Tallon, 2011). Taking into account the particular advantages of this mixing-rheology technique, a better rheological characterization of drilling fluids can be achieved, mainly at low shear rates.

In a previous work (Hermoso et al., 2014), the influence of both pressure and organoclay concentration, at a given constant temperature, on the viscous flow behaviour of all-oil drilling fluids, formulated with two organobentonites, using an extended Sisko model, was highlighted. This model fitted the flow behaviour of these suspensions, in the range of intermediate to high shear rates, fairly well. However, Sisko's model is not useful to estimate an apparent yield stress. This is why, in this work, two well-known yield-stress models used in drilling mechanics, Bingham and Herschel-Bulkley models, were tested for describing the viscous flow behaviour of all-oil drilling fluids.

On the other hand, data concerning HPHT influence on the yielding behaviour of oil drilling suspensions are very limited (Lu et al., 2008; Zhao et al., 2009). However, knowledge of the flow behaviour of clay-based drilling fluid, as function of temperature and pressure, is decisive in order to deal with different important issues, such as excessive torque, gelation, hole cleaning or barite sag (Shahbazi et al., 2007). In this sense, one of the objectives of this study was to define the temperature-pressure-shear stress relationship for two organoclays suspensions, aiming to support a physical interpretation of the rheological behaviour obtained. In addition, this work also aims to check the accuracy of two rheological models that describe the combined effect of both pressure and temperature on the viscous flow behaviour of theses suspensions. Particular efforts has been dedicated to the yielding transition, measured by using non-conventional geometries (double helical ribbon), aiming to get further insight into the complex rheology of these fluids in extreme HPHT conditions.

1 MATERIALS AND PROCEDURES

1.1 Materials

The organoclays used in the present study (B34 and B128) were purchased from *Elementis* (Belgium) and used without further purification. These organoclays are industrial powdered products, typically used in oil well drilling. The original organoclays were modified by a cationic exchange reaction between bentonite clay and different quaternary ammonium chloride ions, dimethyl-dihidrogenated tallow ion for B34 and dimethyl-benzyl-hydrogenated tallow for B128. The particle size distributions of dry powders were measured by light scattering (Mastersizer Scirocco 2000, *Malvern Instruments Ltd*, UK). Sauter's mean diameters were 9.81 μm (B34) and 6.17 μm (B128), respectively.

The SR-10 oil was a naphthenic lubricating oil, supplied by *Verkol* (Spain), with a viscosity of 115 cSt and a specific gravity of 0.916 g/cm^3, at 40°C.

1.2 Drilling Fluid Manufacture

For the preparation of the drilling fluids, 5 wt% organoclay was mixed with SR-10 base oil, using a high shear mixer (*Ultraturrax*, Ika, Germany), at 9 000 rpm, for 5 minutes and room temperature. Before mixing, both organoclays were wetted with oil, for an hour, to guarantee good organoclay dispersion into the oil base. When the dispersion process was finished, the resulting suspensions became viscous and stable. The all-oil drilling fluids formulated were stored at room temperature.

1.3 Rheological Characterization

Traditional procedures to characterize drilling mud rheology follow the API recommendations using the FANN rheometer for a field characterization. In this work, in order to get a better accuracy in the low shear-rate region, viscous flow measurements were carried out using a controlled stress rheometer, MARS II, from *Thermo Scientific* (Germany). Four geometries were used: a conventional coaxial cylinder geometry (41 mm inner diameter, 1 mm gap, 60 mm length) and a serrated plate-and-plate geometry (35 mm diameter, 1 mm gap) for atmospheric pressure rheological tests; whilst a coaxial cylinder geometry (38 mm diameter, 80 mm length) and a double helical ribbon geometry (36 mm diameter, 78 mm length), in a pressure cell D400/200 (*Thermo Scientific*, Germany), were used for HPHT measurements. The latter sensor, called DHR, has a non-conventional geometry, previously calibrated with a Newtonian fluid and several shear-thinning fluids, in the pressure range used in the present study (Hermoso et al., 2012). The cell D400/200 is a pressure vessel of 39 mm inner diameter. Inside the cell, the double helical ribbon geometry was put in contact with a sapphire surface, at the bottom of the vessel, by a steel needle. This inner cylinder was equipped, at the top, with a secondary magnetic cylinder (36 mm diameter, 8 mm length), magnetically coupled to a tool outside the cell, which was connected to the motor-transducer of the rheometer. The pressure cell was connected to a hydraulic pressurization system, which consists of two units, a high pressure valve and a hand pump (*Enerpac*, USA), connected by a high pressure line. The pressure cell was pressurized using, as pressurizing liquid, the same fluid to be tested. A pressure transducer GMH 3110 (*Gresingeg Electronic*, Germany), able to measure differential pressures ranging from 0 to 400 bar (0.1 bar resolution), was used.

Steady-state viscosity measurements, at different differential pressures (0, 100, 200, 300 and 390 bar) and temperatures (40, 80, 100, 120 and 140°C), were performed in a shear rate range dependent on sample viscosity. Temperature was fixed with a circulating silicone bath (DC30 *Thermo Scientific*, Germany). In order to guarantee the same conditions of the drilling fluids at the starting point, all samples were pre-sheared at 100 s^{-1}, for 25 s, before the rheological tests. Steady-state viscosities were obtained by performing step increases in shear stress. Each shear stress was applied on the sample during 120 s. After this elapsed time, a constant shear rate was always obtained.

Downward shear rate sweep flow curves were used to compare drilling fluids viscous behaviour. This type of procedure is highly recommend when viscous flow parameters are to be used for the purpose of calculating pressure drop in a condition corresponding to that prevailing at the point of interest in the well (Darley and Gray, 1988). According to this procedure, the yield stresses estimated from both models are defined as dynamic yield stresses.

Before performing the rheological tests, all samples were submitted to storage stability tests. In this sense, closed non-agitated containers were introduced in an oven, at temperatures between 40°C and 140°C. No particle separation was observed after 48 h ageing.

On the other hand, at least, two replicates of each rheological test were performed on fresh samples. The results shown correspond to the mean values obtained. Standard deviations among replicates were always inferior to ±5%.

1.4 Yield Stress Models

The Bingham model is the most traditional model used to fit the viscoplastic behaviour of drilling fluids (Rossi et al., 2002):

$$\tau = \tau_B + \eta_P \dot{\gamma} \tag{1}$$

where τ_B is the apparent Bingham yield stress and η_p is the plastic viscosity. The yield stress values can be obtained by extrapolating the flow curve to zero shear rate. On the other hand, the plastic viscosity is the slope of the stress-shear rate curves at high shear rates. Most drilling fluids are not exactly described by this model, although drilling fluid behaviour can be predicted reasonably well, for practical purposes (Harris and Osisanya, 2005).

However, the three-parameter Herschel-Bulkley model is in much stronger agreement with drilling fluids experimental rheological data, especially at low shear rate:

$$\tau = \tau_H + K\dot{\gamma}^n \tag{2}$$

where τ_H is the apparent Herschel-Bulkley yield stress, K is the consistency index and n is the flow index. This model accounts for both the yielding behaviour as well as for the further non-linear relationship between shear stress and shear rate of *i.e.* bentonite suspensions (Kelessidis et al., 2009), and most drilling fluids (Munawar and Jan, 2012a,b). For oil muds, Herschel-Bulkley's model has been used to describe the influence of both temperature and pressure on the complex flow behaviour at intermediate and high shear rates fairly well (Herzhaft et al., 2003).

All the rheological parameters of these two models were estimated by using a curve-fitting technique with a weighted-average method.

2 RESULTS AND DISCUSSION

2.1 Effect of Temperature at Atmospheric Pressure

Figure 1 shows a comparison between the flow curves obtained from downward shear rate sweep tests using the serrated plate-plate geometry (PR35) and the Double

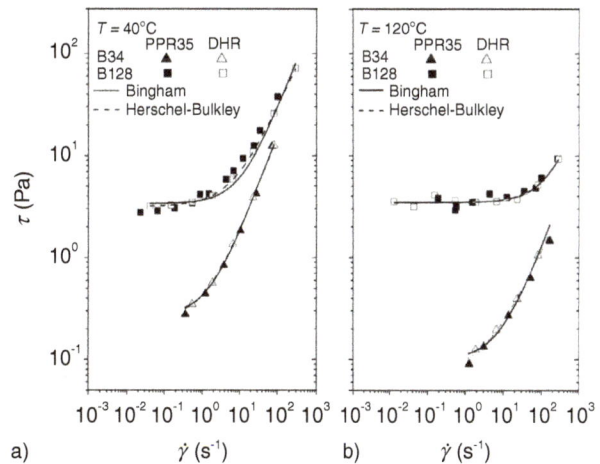

Figure 1

Viscous flow curves, obtained by using serrated plate-plate and double helical ribbon geometries, for both oil-drilling fluids at a) 40°C and b) 120°C.

Helical Mixer (DHR), for both drilling fluids, at 40°C (*Fig. 1a*) and 120°C (*Fig. 1b*). It can be seen that measurements performed using the non-conventional geometry (DHR) are in good agreement with those obtained using the rough plate-plate geometry. Consequently, in the temperature range tested, both geometries give a reliable measurement of the viscous flow behaviour of these fluids in the low shear rate region. These results confirm that the helical ribbon geometry is an appropriate tool to cover accurately a wide shear rate range, especially for high-viscosity fluids (Hermoso *et al.*, 2012). Likewise, the viscous flow curves correspond to a non-Newtonian flow behaviour over a wide range of shear rate, showing an apparent yield stress, mainly for sample B128. From these results, it can be deduced that surface modification of bentonite particles by different organic groups modifies the flow behaviour of these suspensions (Tropea *et al.*, 2007). In addition, the viscous flow behaviour of both drilling fluids drastically changes as temperature increases. Thus, the drilling fluid B34 exhibits significantly lower shear stresses, in the low shear rate region, as temperature increases. Meanwhile, drilling fluid B128 viscous flow curve exhibits a well-defined yield stress in the low shear rate region at high temperature. This temperature-dependent behaviour was reported by Rossi *et al.* (1999) for water-based muds. In general, at 40°C, Herschel-Bulkley's model fits the experimental data better than Bingham's model, particularly for B128-based drilling fluid, whilst, at 120°C, both models predict the flow behaviour of the two all-oil drilling fluids fairly well, being the deviations between predicted and calculated shear stresses within the order of the experimental error ($\pm 10\%$).

Figure 2 depicts upward and downward viscous flow curves for the all-oil drilling fluids studied, at atmospheric pressure and 40°C. Upward and downward shear sweep tests were performed with the double helical ribbon geometry after suspension manufacture. At this temperature, no hysteresis was observed for B34-based drilling fluid (*Fig. 2a*). Nevertheless, as can be observed in Figure 2b, the shear stress values obtained in the upward shear rate ramp are, in general, slightly higher than those in the downward shear rate ramp for B128-based drilling fluid. This behaviour is typical of clay suspensions, such as drilling fluids, as the interparticle network breaks down in separate flocs, which decrease further in size, when shear rate is increased (Abu-Jdayil, 2011). The slightly different hysteresis degrees of this suspension, in the shear rate range studied, could be explained on the basis of a competition between breakup and recombination of structural elements (flocs). Besides, it should be mentioned that these systems do not display significant thixotropic effects,

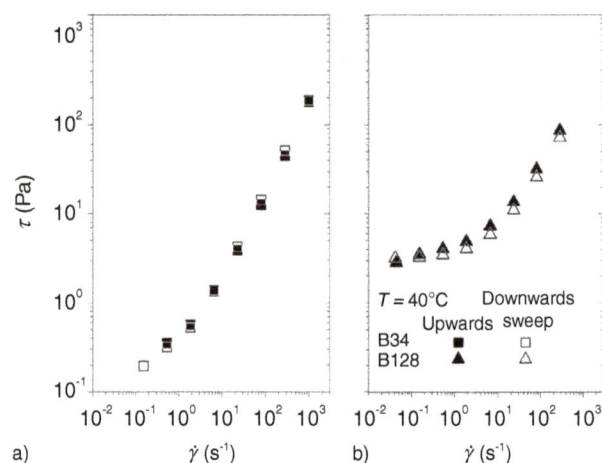

Figure 2

Upward and downward viscous flow curves for B34-based a) and B128-based b) drilling fluids at 40°C.

at temperatures above 40°C, under these experimental conditions. On the other hand, both organoclay suspensions exhibit a shear-thinning behaviour, due to the alignment of basal faces of particles in the direction parallel to the applied shear (Massinga *et al.*, 2010). Additionally, the B128-based suspension presents higher yielding point than B34-based suspensions, indicating that the structure of the former is more resistant to large rearrangements responsible for the yielding behaviour and the destruction of such microstructure (Møller *et al.*, 2006). This behaviour, precisely a high apparent yield stress and weak thixotropy, seems to be related to strongly connected clay aggregates, due to a high degree of swelling, developed by interactions between the benzyl nature of the covered surface B128 clays and the naphthenic oil (Burgentzlé *et al.*, 2004). Conversely, the resulting structures, for B34-based drilling fluid, showing relatively low yield stress values, suggest less connected formations, which are easier to orient in the flow field.

Figure 3a displays the evolution of the apparent yield stress with temperature for the drilling fluids studied, at atmospheric pressure, obtained from Bingham's model fitting. As can be observed, the yield stress values, and their evolution with temperature, are different depending on the sample studied. Thus, for B34-based drilling fluid, the yield stresses in the temperature range studied show similar values for 40°C and 140°C, with a minimum value at 80°C. Otherwise, the yield stress values encountered for B128-based suspension are almost constant up to 120°C, showing a noticeable increase at

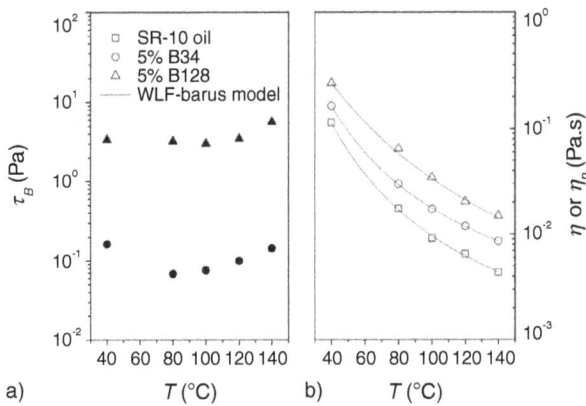

a) b)

Figure 3

Temperature dependence of Bingham's yield stress a) and viscosity (SR-10)/plastic viscosity b) for both drilling fluids.

higher temperature. In addition, significantly higher yield stress values for B128 samples (at least one order of magnitude) are observed.

The precise physical origin of the influence of temperature on drilling fluid yield stress is not clear. A possible explanation for the different behaviour of the oil drilling fluids studied may be related to changes in polymer-oil pair solvency with temperature. Thus, in the case of polymer-covered particles, the type of interaction is governed by the Flory-Huggins interaction parameter, χ if $\chi < 0.5$ (good solvents), the interaction is repulsive, and, if $\chi > 0.5$ (bad solvents), the interaction is attractive (Quemada and Berli, 2002). The low yield stress values for B34-based drilling fluid suggest a short-range attractive force between the particles, which develop a network of aggregates weakly interconnected (Moraru, 2001).

On the other hand, it is worth noting that the unusual temperature dependence for B128-based suspension suggests a thermal thickening above a threshold temperature ($T > 120°C$). The possible thermo-thickening phenomenon of organoclay suspensions in oil is still unclear. Apparently, the rise in temperature probably leads to stronger repulsive interactions between flocs, due to the improvement in the solvency between the aromatic group of the clay and the naphthenic oil. Then, above the threshold temperature, particle associations might undergo an increase in the effective volume. Thus, an increase in volume fraction leads to an increase in yield stress value, as has been observed by other authors for clay suspensions (Minase et al., 2008). Interestingly, this particular behaviour has been recently reported by

Wu et al. (2012), who also suggested an increase in the effective volume fraction with increasing temperature for particulate dispersion in non-aqueous media.

Figure 3b shows the evolution of plastic viscosity with temperature for the drilling fluids studied, at atmospheric pressure. The Newtonian viscosities for the base oil (SR-10) have been also included as reference. A factorial WLF-Barus model fits the above-mentioned dependence fairly well (average error less than 10%).

As expected, drilling fluid plastic viscosity always decreases with temperature (Joshi and Pegg, 2007), being its dependence very similar to that of the base oil. These results suggest that the viscous flow behaviour of these fluids is largely governed by the viscosity of the base oil, as has been reported elsewhere (Herzhaft et al., 2001). Nevertheless, drilling fluids show significantly higher viscosities than the naphthenic oil media, in the whole range of temperature tested, as shown in Figure 3b. In addition, the different plastic viscosities obtained for both all-oil drilling fluids, formulated with the same thickener concentration, suggest the development of different microstructures, being the strength of the covered particle-particle interactions stronger for B128-based fluid.

2.2. Influence of Temperature and Pressure on Flow Behaviour of All-Oil Drilling Fluids

Figure 4 displays the viscous flow curves (shear stress versus shear rate and viscosity versus shear rate) for B128-based oil drilling fluid, as a function of pressure, at 40°C (Fig. 4a, c) and at 140°C (Fig. 4b, d), obtained by using the non-conventional geometry. As can be observed, the yield stress increases with temperature in the pressure range studied. On the other hand, the critical shear rate for the onset of the liquid-like regime also increases with temperature ($\dot{\gamma} \sim 0.1$ s^{-1} at 40°C, and $\dot{\gamma} \sim 20$ s^{-1} at 140°C). This increment can be attributed to thermally-induced changes in particle-particle interactions, as has been pointed out elsewhere (Briscoe et al., 1994). Besides, it is worth mentioning that the influence of pressure on the yield stress values is only significant at the highest temperature studied, resulting in an increase in yield stress with pressure.

Figures 4a-d also show the fitting of both Bingham's and Herschel-Bulkley's models to the experimental flow curves. As can be seen, both models predict the viscous flow behaviour of this oil drilling fluid, at 40°C and 140°C and in the range of pressure selected, fairly well.

Bingham's and Herschel-Bulkley's parameters are listed, in Tables 1 and 2, for B34-based and B128-based drilling fluids, respectively. In these tables, the values of the Average Absolute Relative Deviation percentage

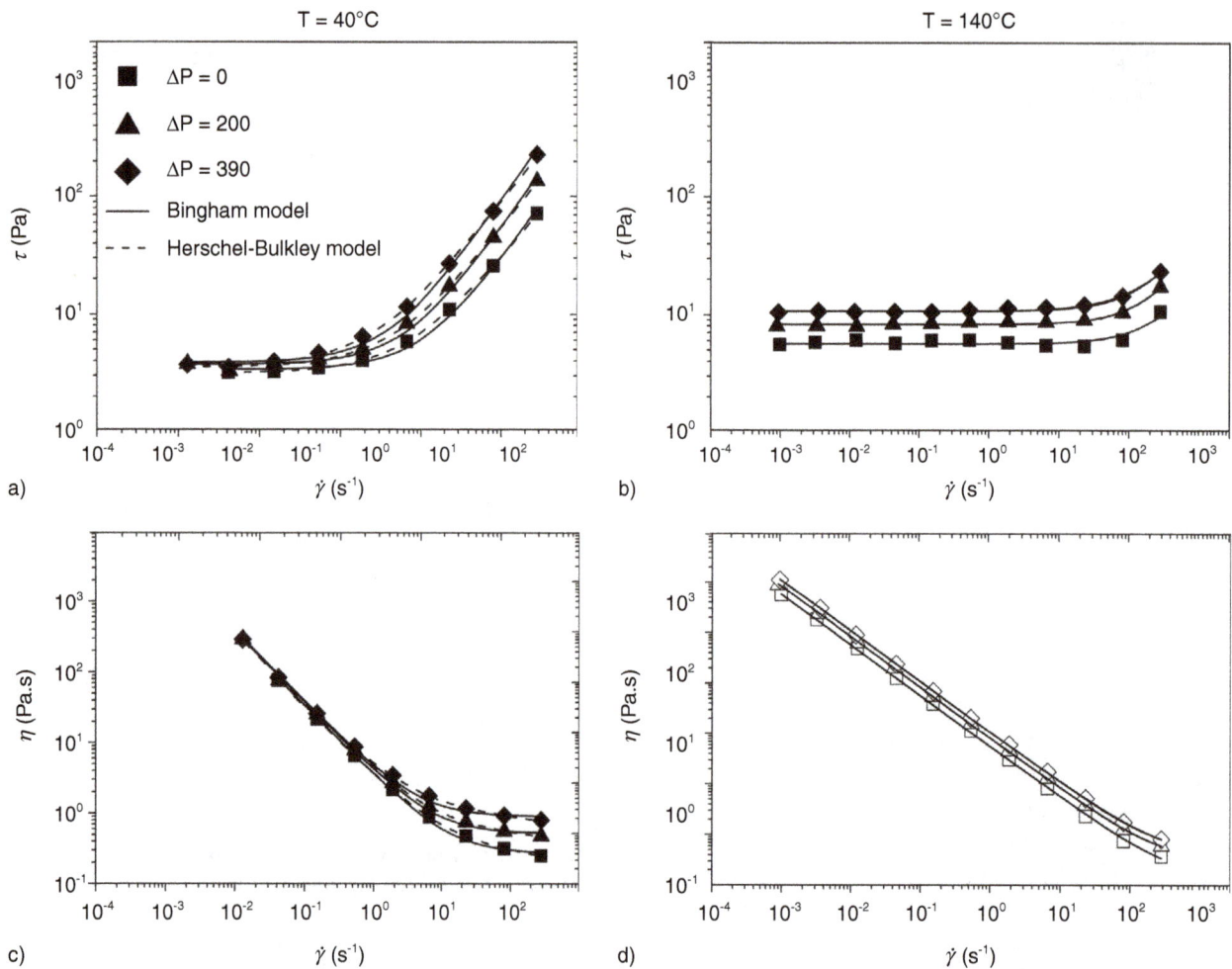

Figure 4

Experimental viscous flow curves a) and b): shear stress *versus* shear rate: c) and d): viscosity *versus* shear rate), at three differential pressures ($\Delta p = p - p_0$) and two temperatures (40 and 140°C), and Herschel-Bulkley's and Bingham's models fittings for B128-based drilling fluid.

(%AARD) have been also included. As shown in Table 1, the values of %AARD are always lower than 10%, indicating that both models are suitable for describing the viscous flow properties of B34-based oil drilling fluid, in the range of temperature and pressure studied, from an engineering point of view. In Tables 3 and 4, the correlation coefficients (R^2) and Residual Sum of Squares (*RSS*) are reported. It is worth noting that Herschel-Bulkley's model always shows significantly lower *RSS* values and R^2 values closer to 1 for B34-based drilling fluid. Hence, this model seems to be more reliable to describe the viscous flow behaviour of this drilling fluid. Otherwise, the slight differences in *RSS* and R^2 values found with both models, for B128-based drilling fluid, would indicate that they are equally valid to model its complex rheological behaviour, under extreme pressure-temperature environments, from an engineering point of view.

2.2.1 Combined Effect of Pressure and Temperature on Yield Stress

Figures 5 and 6 display the evolution of the Herschel-Bulkley yield stress with pressure, in the temperature range comprised between 40°C and 140°C, for B34 and B128-based drilling fluids, respectively. As can be seen, yield stress always linearly increases with pressure in the range of pressure tested. Consequently, the isothermal yield stress behaviour can be modelled by a simple equation, as follows:

$$\tau_c = \tau_{c0} + \beta_\tau(p - p_0) \tag{3}$$

where τ_c is the Herschel-Bulkley yield stress, τ_{c0} is the yield stress at the reference pressure for each temperature, β_τ is a isothermal piezo-yield stress coefficient, p is the applied pressure, and p_0 is the pressure of reference (1 bar). The values of the piezo-yield stress

TABLE 1

Bingham's and Herschel-Bulkley's parameters and %AARD for B34-based drilling fluid, at different temperatures and pressures

Bingham's model

Δp (bar)	0			200			390		
T (°C)	η_p (Pa·s)	τ_B (Pa)	%AARD	η_p (Pa·s)	τ_B (Pa)	%AARD	η_p (Pa·s)	τ_B (Pa)	%AARD
40	0.164	0.16	2.9	0.301	0.41	9.5	0.502	0.74	7.3
80	0.030	0.07	6.3	0.043	0.11	6.3	0.063	0.20	5.2
100	0.017	0.08	6.7	0.023	0.11	7.3	0.028	0.17	5.6
120	0.012	0.01	6.4	0.014	0.12	4.5	0.020	0.15	1.9
140	0.009	0.14	5.6	0.017	0.18	5.9	0.019	0.21	4.9

Herschel-Bulkley's model

Δp (bar)	0				200				390			
T (°C)	K (Pa·s^{-n})	n	τ_H (Pa)	%AARD	K (Pa·s^{-n})	n	τ_H (Pa)	%AARD	K (Pa·s^{-n})	n	τ_H (Pa)	%AARD
40	0.188	0.97	0.12	0.4	0.436	0.92	0.27	3.3	0.668	0.94	0.56	2.6
80	0.044	0.89	0.06	1.6	0.064	0.89	0.09	1.8	0.091	0.90	0.17	0.8
100	0.033	0.84	0.05	1.8	0.038	0.85	0.08	3.1	0.042	0.88	0.14	1.8
120	0.025	0.82	0.08	2.0	0.023	0.87	0.10	1.2	0.022	0.95	0.15	0.8
140	0.038	0.69	0.12	0.1	0.048	0.67	0.15	1.1	0.050	0.71	0.18	0.2

%AARD = $1/N \left(\Sigma |(\tau_{cal} - \tau_{exp})/\tau_{exp}| \right) \times 100$

TABLE 2

Bingham's and Herschel-Bulkley's parameters and %AARD for B128-based drilling fluid, at different temperatures and pressures

Bingham's model

Δp (bar)	0			200			390		
T (°C)	η_p (Pa·s)	τ_B (Pa)	%AARD	η_p (Pa·s)	τ_B (Pa)	%AARD	η_p (Pa·s)	τ_B (Pa)	%AARD
40	0.270	3.38	3.8	0.517	3.74	6.8	0.899	3.87	8.5
80	0.065	2.72	4.2	0.092	3.01	3.0	0.132	3.60	3.5
100	0.034	3.02	4.3	0.049	3.39	3.7	0.066	4.04	5.2
120	0.020	3.48	4.7	0.029	4.81	5.0	0.042	5.86	3.4
140	0.015	5.64	5.6	0.031	8.26	2.5	0.045	10.81	2.1

Herschel-Bulkley's model

Δp (bar)	0				200				390			
T (°C)	K (Pa·s^{-n})	n	τ_H (Pa)	%AARD	K (Pa·s^{-n})	n	τ_H (Pa)	%AARD	K (Pa·s^{-n})	n	τ_H (Pa)	%AARD
40	0.519	0.86	3.16	0.3	0.854	0.89	3.55	3.3	1.508	0.88	3.62	2.3
80	0.068	0.99	2.71	4.7	0.096	0.99	3.01	3.1	0.137	0.99	3.59	3.5
100	0.036	0.99	3.05	4.5	0.053	0.98	3.39	4.3	0.069	0.99	4.04	5.6
120	0.022	0.98	3.47	4.7	0.032	0.98	4.80	5.4	0.043	0.99	5.86	3.3
140	0.016	0.99	5.64	5.6	0.034	0.98	8.25	2.5	0.081	0.89	10.74	1.9

TABLE 3

Correlation coefficient (R^2) and Residual Sum of Squares (RSS) of models used for B34-based drilling fluid, at different temperatures and pressures

Bingham's Model

Δp (bar)	0		200		390	
T (°C)	R^2	RSS (Pa^2)	R^2	RSS (Pa^2)	R^2	RSS (Pa^2)
40	0.9992	0.0066	0.9919	0.0857	0.9951	0.0511
80	0.9945	0.0243	0.9944	0.0244	0.9958	0.0167
100	0.9933	0.0209	0.9904	0.0276	0.9933	0.0178
120	0.9927	0.0180	0.9963	0.0086	0.9993	0.0016
140	0.9864	0.0138	0.9851	0.0146	0.9889	0.0109

Herschel-Bulkley's Model

Δp (bar)	0		200		390	
T (°C)	R^2	RSS (Pa^2)	R^2	RSS (Pa^2)	R^2	RSS (Pa^2)
40	0.9999	1.5E-4	0.9985	0.0112	0.9994	0.0065
80	0.9996	0.0017	0.9995	0.0020	0.9999	4.5E-4
100	0.9995	0.0016	0.9984	0.0047	0.9994	0.0016
120	0.9992	0.0020	0.9997	0.0007	0.9999	0.0003
140	0.9999	4E-7	0.9994	0.0006	0.9999	1.4E-5

$RSS = \Sigma |(\tau_{cal} - \tau_{exp})^2|$

coefficient in a temperature range comprised between 40°C and 140°C, for both oil drilling fluids, are shown in Table 5. As can be observed in these figures, the influence of pressure on the yield stress is more significant for B128-based drilling fluid, which also shows the highest values of yield stress. Similar behaviour, in the low shear-rate region, has been reported by Gandelman et al. (2007).

Changes in B34-based drilling fluid piezo-yield stress could be associated with the thermal weakening of interparticle interactions in the agglomerates and reduction in the effective agglomerate concentration, being the yielding behaviour less sensitive to pressure changes as temperature increases, even enhancing its compressibility with temperature (Paredes et al., 2012). In addition, this decrease in yield stress with an increase in temperature may be related to a decrease in oil viscosity (Houwen and Geehan, 1986).

On the contrary, for B128-based drilling fluid, the pressure/temperature dependence of the yield stress is quite different. As can be seen in Figure 6, yield stress dependence on pressure is roughly independent of temperature below 120°C. In addition, the larger influence of pressure on this parameter could be related to the compression of a thermally-modified denser structure, as have been pointed out by Politte (1985) for inverted oil muds.

From the engineering point of view, the temperature and pressure dependence of the yield stress is of major interest for the design of transport operations in the oil drilling industry (Darley and Gray, 1988). In this sense, drilling fluids are mainly designed to be able of suspending drill cuttings under static conditions. According to the results obtained in this study, the B34-based oil drilling fluid presents fairly low cutting carrying capacity, but also not excessive horse-power pump requirements in the range of pressure and temperature tested (see yield stress values in Tab. 1). On the contrary, B128 organoclay can be considered a potential candidate to prepare suspensions able to be used for drilling fluid applications, with yield stress values below 15 Pa (Munawar and Jan, 2012a) in the whole range of pressure and temperatures studied.

2.2.2 Temperature-Pressure-Plastic Viscosity Relationship

The combined effect of temperature and pressure on the viscous flow properties of many materials has been

TABLE 4

Correlation coefficient (R^2) and Residual Sum of Squares (RSS) of models used for B128-based drilling fluid, at different temperatures and pressures

Bingham's Model

Δp (bar)	0		200		390	
T (°C)	R^2	RSS (Pa^2)	R^2	RSS (Pa^2)	R^2	RSS (Pa^2)
40	0.9769	0.0520	0.9756	0.0662	0.9726	0.0860
80	0.9340	0.0299	0.9784	0.0140	0.9960	0.0029
100	0.8878	0.0305	0.9364	0.0206	0.9152	0.0310
120	0.9199	0.0391	0.9380	0.0289	0.9674	0.0184
140	0.7898	0.0469	0.9700	0.0092	0.9811	0.0065

Herschel-Bulkley's Model

Δp (bar)	0		200		390	
T (°C)	R^2	RSS (Pa^2)	R^2	RSS (Pa^2)	R^2	RSS (Pa^2)
40	0.9998	0.0005	0.9939	0.0165	0.9978	0.0069
80	0.9418	0.0308	0.9783	0.0141	0.9963	0.0027
100	0.8833	0.0317	0.9329	0.0218	0.9133	0.0317
120	0.9187	0.0397	0.9335	0.0310	0.9668	0.0187
140	0.7855	0.0479	0.9697	0.0094	0.9846	0.0053

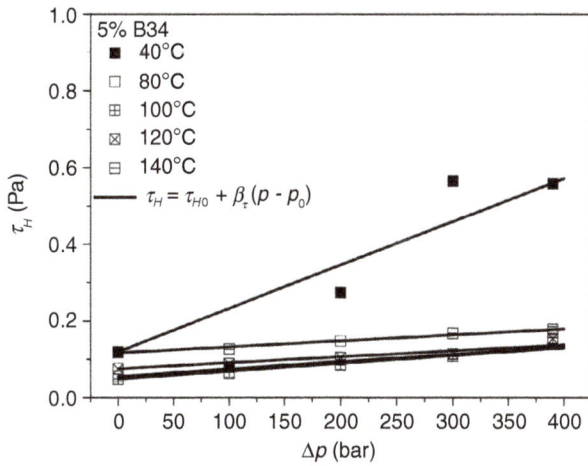

Figure 5

Effect of pressure on Herschel-Bulkley's yield stress for B34-based drilling fluid, as a function of temperature.

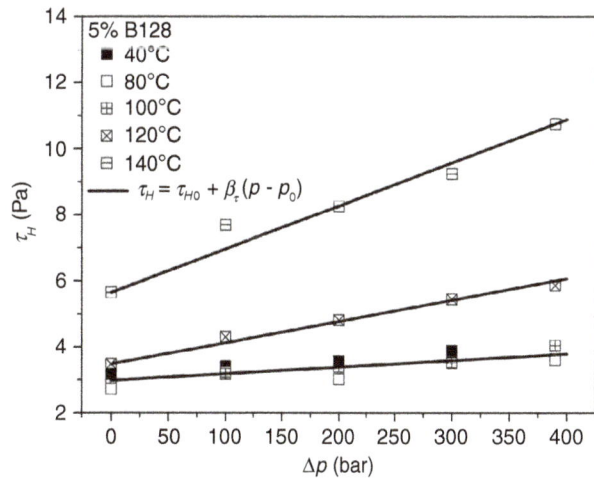

Figure 6

Effect of pressure on Herschel-Bulkley's yield stress for B128-based drilling fluid, as a function of temperature.

described by using a large variety of models, such as models based on physical properties, empirical equations or factorial equations of temperature and pressure (Tschoegl et al., 2002; Bair, 2007; Rajagopal and Saccomandi, 2009; Fuentes-Audén et al., 2005; Fan and Wang, 2011).

TABLE 5
Piezo-yield stress coefficients for the oil drilling fluids studied, at different temperatures

| | B34 | | B128 | |
| | β_τ ($\times$ 10^4, Pa/bar) | | β_τ ($\times$ 10^4, Pa/bar) | |
T (°C)	Bingham's model	Herschel-Bulkley's model	Bingham's model	Herschel-Bulkley's model
40	15.6	11.3	16.5	16.8
80	2.36	1.94	23.9	23.6
100	2.04	2.06	20.8	19.6
120	1.25	1.59	64.9	64.8
140	1.82	1.57	132	131

In this case, a factorial WLF-Barus equation has been selected:

$$\eta(p, T) = \eta_0 10^{\left(-\frac{c_1(T-T_0)}{c_2+(T-T_0)}\right)} \exp\left(\beta(T)(p-p_0)\right) \quad (4)$$

being:

$$\beta(T) = \beta_0 + \beta_1(T - T_0) \quad (5)$$

where η_0 is the plastic viscosity of the fluid at atmospheric pressure, p_0, and temperature of reference, T_0; c_1 and c_2 are empirical constants; and $\beta(T)$ is the piezo-viscous coefficient, which has been linearized with temperature to generalize the model using two parameters, β_0 and β_1.

Figure 7 shows the experimental plastic viscosity values as a function of pressure, and the fitting of WLF-Barus' model, for B34 and B128-based drilling fluids, in a temperature range comprised between 40 and 140°C. In addition, the Newtonian viscosities, at 40 and 140°C, of SR-10 base oil have been included for the sake of comparison. As can be seen in these figures, the factorial WLF-Barus model fits the pressure (in the range of 1-390 bar) and temperature (between 40 and 140°C) dependence of plastic viscosity, for both drilling fluids, fairly well. Nevertheless, it must be mentioned that, in both cases, the fitting shows a slightly higher deviation at 140°C. WLF-Barus' parameters are listed in Table 6. At 40°C, the evolution of drilling fluid viscosity with pressure is quite similar to that found for the base oil (Hermoso et al., 2014). Likewise, the decrease observed in the piezo-viscous coefficient with temperature is progressively less marked as temperature increases (Tab. 6), as reported elsewhere (Chaudemanche et al., 2009) Hence, the influence of pressure is mainly dependent on base oil flow properties,

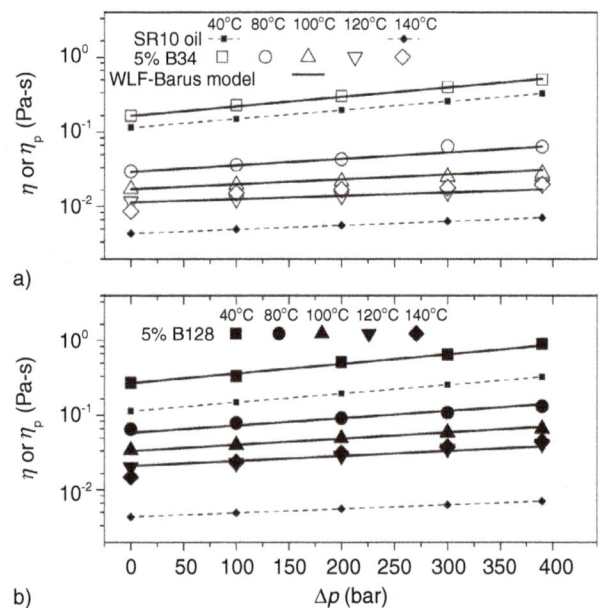

a)

b)

Figure 7

Experimental viscosities (SR-10 oil) and plastic viscosities (drilling fluids) as a function of pressure, and WLF-Barus' model fitting, in a temperature range of 40-140°C. a) B34-based drilling fluid; and b) B128-based drilling fluid.

as has been pointed out by different authors (Houwen and Geehan, 1986; Hermoso et al., 2014). It also interesting to notice that B34-based drilling fluid piezo-viscous coefficient presents a slightly higher temperature dependence than both B128-based one and SR10 oil, as can be seen in Table 6. In addition, B34-based drilling fluid exhibits lower viscosity values than B128-based one in the whole range of temperature and pressure tested. The slight differences between both drilling fluids at high shear rates should be related to changes in organoclay suspension microstructure as temperature increases.

TABLE 6

WLF-Barus' model parameters for the base oil and drilling fluids studied

Sample	η_0 (Pa·s)	c_1	c_2 (°C)	β_0 (1/bar)	β_1 (1/bar·°C)
SR-10 oil	0.114	2.54	80.65	2.62×10^{-3}	-1.43×10^{-5}
B34-based fluid	0.164	2.58	98.03	2.92×10^{-3}	-2.34×10^{-5}
B128-based fluid	0.270	3.31	159.08	2.95×10^{-3}	-1.76×10^{-5}

2.2.3 Influence of Pressure and Temperature on Herschel-Buckley's Flow Index

The effect of temperature and pressure on the Herschel-Buckley flow index of the oil drilling fluids studied can be analysed from the data shown in Tables 1 and 2. As can be seen, the flow indexes are always lower than 1 for the suspension formulated with B34 organoclay, showing a moderate shear-thinning behaviour (Park and Song, 2010). In addition, the influence of pressure on this parameter is not significant between 1 and 390 bar. It is important to remark that the flow index slightly decreases as temperature increases up to 120°C and then suddenly drops at 140°C.

On the other hand, changes in B128-based oil drilling fluid flow index with both pressure and temperature are very weak. In this sense, at 40°C, this fluid shows very similar flow index values, slightly shear-thinning, in the whole pressure range tested. At temperatures above 40°C, the flow index is independent of both temperature and pressure (Tab. 2).

CONCLUSIONS

In this study, HPHT viscous flow characterization of oil drilling fluids formulated with two different organoclays was performed by using a non-conventional geometry coupled to a rheometer. The experimental results indicate that the double helical ribbon geometry is a very useful tool to characterise the complex viscous flow behaviour of these fluids under high pressure.

From the experimental results obtained at atmospheric pressure, it can be concluded that the viscous flow behaviour of these drilling fluids is strongly affected by both the organically modified surface-oil pair solvency and oil viscosity, in the temperature range studied. The low yield stress values for B34-based drilling fluid suggest a short-range attractive force between the particles, which develop a network of aggregates weakly interconnected. On the other hand, the increase in yield stress for B128-based drilling fluid, above a threshold temperature, may be associated with structural changes due to thermal gelling.

Both Bingham's and Herschel-Bulkley's models fitted the complex viscous flow behaviour of these drilling fluids, at different pressures and temperatures, fairly well. However, Herschel-Bulkley's model is statistically better to account for changes in the viscous behaviour of B34-based drilling fluids under HPHT conditions. Drilling fluids yield stresses increase with pressure, whilst their temperature dependence is related to the type of organoclay used. This behaviour may be explained on the basis of oil compression and thermally-induced changes in the effective organoclay volume fraction. A factorial WLF-Barus model satisfactorily describes the temperature/pressure evolution of the plastic viscosity for both drilling fluids, being the viscosity at high shear rate mainly influenced by the piezo-viscous properties of the continuous phase.

ACKNOWLEDGMENTS

This work has been sponsored by the "Junta de Andalucía" FEDER-Excellence Projects Programme (Research project P08-TEP-3895). The authors gratefully acknowledge its financial support.

REFERENCES

Abu-Jdayil B. (2011) Rheology of sodium and calcium bentonite-water dispersions: Effect of electrolytes and aging time, *Int. J. Miner. Process.* **98**, 208-213.

Bair S. (2007) *High Pressure Rheology for Quantitative Elastohydrodynamics*, first edition, Elsevier BV, Oxford, UK.

Barnes H.A., Nguyen Q.D. (2001) Rotating vane rheometry - a review, *J. Non-Newtonian Fluid Mech.* **98**, 1-14.

Briscoe B.J., Luckham P.F., Ren S.R. (1994) The properties of drilling fluids at high pressures and high temperatures, *Philos. T. Roy. Soc. A* **348**, 179-207.

Burgentzlé D., Duchet J., Gérard J.F., Jupin A., Fillon B. (2004) Solvent-based nanocomposite coatings I. Dispersions of organophilic montmorillonite in organic solvents, *J. Colloid Interf. Sci.* **278**, 26-39.

Caenn R., Chillingar G.V. (1996) Drillings Fluids: State of art, *J. Pet. Sci. Eng.* **14**, 221-230.

Chaudemanche C., Henaut I., Argillier J.-F. (2009) Combined Effect of Pressure and Temperature on Rheological Properties Water-in-Crude Oil Emulsions, *Appl. Rheol.* **19**, 1-8.

Darley H.C.H., Gray G.R. (1988) *Composition and Properties of Drilling Fluids and Completion Fluids*, Gulf Professional Publishing, Houston, Texas.

Fan Y., Wang K. (2011) The Viscosity of Dimethyl Silicone Oil and the Concentration of Adsorbed Air, *AiChE J.* **57**, 3299-3304.

Fuentes-Audén C., Martínez-Boza F., Navarro F.J., Partal P., Gallegos C. (2005) Viscous flow properties and phase behavior of oil-resin blends, *Fluid Phase Equilib.* **237**, 117-122.

Gandelman R.A., Leal R.A.F., Gonçalves J.T., Aragão A.F.L., Lomba R.F., Martins A.L. (2007) Study on Gelation and Freezing Phenomena of Synthetic Drilling Fluids in Ultra Deep Water Environments, *SPE/IADC Drilling Conference*, Amsterdam, Netherlands, 20-22 Feb.

Ghalambor A., Ashrafizadeh S.N., Nasiri M. (2008) Effect of basic parameters on the viscosity of synthetic-based drilling fluids, *SPE International Symposium on Formation Damage Control*, Lafayette, Louisiana, 13-15 Feb.

Harris O.O., Osisanya S.O. (2005) Evaluation of equivalent circulating density of drilling fluids under high pressure/high temperature conditions, *SPE Annual Technical and Exhibition Conference*, Dallas, Texas, 9-12 Oct.

Hermoso J., Jofore B.D., Martínez-Boza F.J., Gallegos C. (2012) High Pressure Mixing Rheology of Drilling Fluids, *Ind. Eng. Chem. Res.* **51**, 14399-14407.

Hermoso J., Martínez-Boza F.J., Gallegos C. (2014) Influence of Viscosity Modifier Nature and Concentration on the Viscous Flow Behavior of Oil-based Drilling Fluid, *Appl. Clay Sci.* **87**, 14-21.

Herzhaft B., Peysson Y., Isambourg P., Delepoulle A., Toure A. (2001) Rheological Properties of Drilling Muds in Deep Offshore Conditions, *SPE/IADC Drilling Conference*, Amsterdam, Netherlands, 27 Feb.-1 March.

Herzhaft B., Rousseau L., Neau L., Moan M., Bossard F. (2003) Influence of temperature and clay/emulsion microstructure on oil-based mud low shear rate rheology, *SPE J.* **8**, 211-217.

Houwen O.H., Geehan T. (1986) Rheology of Oil-Base Muds, *SPE Annual Technical Conference and Exhibition*, New Orleans, Louisiana, 5-8 Oct.

Joshi R.M., Pegg M.J. (2007) Flow properties of biodiesel fuel blends at low temperatures, *Fuel* **86**, 143-151.

Kelessidis V.C., Christidis G., Makri P., Hadjistamou V., Tsamantaki C., Mihalakis A., Papanicolaou C., Foscolos A. (2007) Gelation of water-bentonite suspensions at high temperature and rheological control with lignite addition, *Appl. Clay Sci.* **36**, 221-231.

Kelessidis V.C., Papanicolaou C., Foscolos A. (2009) Application of Greek lignite as an additive for controlling rheological and filtration properties of water-bentonite suspensions at high temperatures: A review, *Int. J. Coal Geol.* **77**, 394-400.

Lu G., Li X., Chen T., Shan J., Gao Y. (2008) New method for non-linear least square estimation on rheological parameter in Casson model of drilling fluid, *Shiyou Xuebao/Acta Petrolei Sinica* **29**, 470-474.

Massinga P.H., Focke W.W., de Vaal P.L., Atanasova M. (2010) Alkyl ammonium intercalation of Mozambican bentonite, *Appl. Clay Sci.* **49**, 142-148.

Minase M., Kondo M., Onikata M., Kawamura K. (2008) The viscosity of organic liquid suspensions of trimethyldococylammonium-montmorillonite complexes, *Clays Clay Miner.* **56**, 49-65.

Møller P.C.F., Mewis J., Bonn D. (2006) Yield stress and thixotropy: On the difficulty of measuring yield stress in practice, *Soft Matter* **2**, 274-283.

Moraru V.N. (2001) Structure formation of alkylammonium montmorillonites in organic media, *Appl. Clay Sci.* **19**, 11-26.

Munawar K., Jan B.M. (2012a) Herschel-Bulkley rheological parameters of a novel environmentally friendly lightweight biopolymer drilling fluid xanthan gum and starch, *J. Appl. Polym. Sci.* **124**, 595-606.

Munawar K., Jan B.M. (2012b) Viscoplastic modeling of a novel lightweight biopolymer drilling fluid for unbalanced drilling, *Ind. Eng. Chem. Res.* **51**, 4056-4068.

O'Shea J.-P., Tallon C. (2011) Yield stress behavior of concentrated silica suspensions with temperature-responsive polymers, *Colloids Surf. A* **385**, 40-46.

Paredes X., Fandiño O., Pensado A.S., Comuñas M.J.P., Fernández J. (2012) Pressure-Viscosity Coefficients for Polyalkylene Glycol Oils and Other Ester Ionic Lubricants, *Tribol. Lett.* **45**, 89-100.

Park E.-K., Song K.-W. (2010) Rheological Evaluation of Petroleum Jelly as a Base Material in Ointment and Cream Formulations: Steady Shear Flow Behavior, *Arch. Pharm. Res.* **33**, 141-150.

Politte M.D. (1985) Invert Oil Mud Rheology as a Function of Temperature and Pressure, *SPE/IADC Drilling Conference*, New Orleans, Louisiana, 5-8 March.

Quemada D., Berli C. (2002) Energy of interaction in colloids and its implications in rheological modeling, *Adv. Colloid Interface Sci.* **98**, 51-58.

Rajagopal K.R., Saccomandi G. (2009) The mechanics and mathematics of the effect of pressure on the shear modulus of elastomers, *P. Roy. Soc. Lond. A Mat.* **465**, 3859-3874.

Rossi S., Luckham P.F., Zhu S., Briscoe B.J. (1999) High-Pressure/High-Temperature Rheology of Na$^+$-Montmorillonite Clay Suspensions, *Proceedings of the 1999 SPE International Symposium on Oilfield Chemistry*, Houston, Texas, 16-19 Feb.

Rossi S., Luckham P.F., Tadros ThF (2002) Influence of nonionic polymers on the rheological behavior of Na +-montmorillonite clay suspensions–I Nonyphenol–polypropylene oxide–polyethylene oxide copolymers, *Colloids Surf. A* **201**, 85-100.

Shahbazi K., Metha S.A., Moore R.G., Ursenbanch M.G., Fraassen K.C.V. (2007) Oxidation as a Rheology Modifier and a Potential Cause of Explosions in Oil and Synthetic-Based Drilling Fluids, *International Symposium on Oilfield Chemistry*, Houston, Texas, 28 Feb.-2 March.

Tropea C., Yarin A.L., Foss J.F. (2007) *Springer Handbook of Experimental Fluid Mechanics*, Springer Verlag, Berlin, Heidelberg, pp. 705.

Tschoegl N.W., Knauss W.G., Emri I. (2002) The Effect of Temperature and Pressure on the Mechanical Properties of Thermo-and/or Piezorheologically Simple Polymeric Materials in Thermodynamic Equilibrium – A Critical Review, *Mech. Time-Depend. Mater.* **6**, 53-99.

Williams H., Khatri D., Vaughan M., Landry G., Janner L., Mutize B., Herrera M. (2011) Particle Size Distribution-Engineered Cementing Approach Reduces Need for Polymeric Extenders in Haynesville Shale Horizontal Reach Wells, *SPE Annual Technical and Exhibition Conference*, Denver, Colorado, 30 Oct.-2 Nov.

Wu X.-J., Wang Y., Wang M., Yang W., Xie B.-H., Yang M.-B. (2012) Structure of fumed silica gels in dodecane: Enhanced network by oscillatory shear, *Colloid Polym. Sci.* **290**, 151-161.

Zhao S., Yan J., Shu Y., Li H., Li L., Ding T. (2009) Prediction model for rheological parameters of oil-based drilling fluids at high temperature and high pressure, *Shiyou Xuebao/Acta Petrolei Sinica* **30**, 603-606.

10

Emissions to the Atmosphere from Amine-Based Post Combustion CO$_2$ Capture Plant – Regulatory Aspects

Merched Azzi[1]*, Dennys Angove[1], Narendra Dave[1], Stuart Day[1], Thong Do[1], Paul Feron[1], Sunil Sharma[1], Moetaz Attalla[1] and Mohammad Abu Zahra[2]

[1] CSIRO Energy Technology, 11 Julius Avenue, 2113 NSW - Australia
[2] MASDAR Institute, Abu Dhabi
e-mail: merched.azzi@csiro.au

* Corresponding author

Résumé — **Émissions atmosphériques des installations de captage de CO$_2$ en postcombustion par les amines – Aspects réglementaires** — Le captage en postcombustion (PCC, *Post Combustion Capture*) du CO$_2$ par les amines est une technologie immédiatement disponible à même d'être déployée pour réduire les émissions de CO$_2$ des centrales électriques au charbon. Cependant, il est probable que les installations de PCC rejetteront de faibles quantités d'amines et de produits de dégradation des amines dans les gaz de fumées traités. Les effets environnementaux potentiels de ces émissions ont été examinés dans diverses études partout dans le monde. En se basant sur les gaz de fumées d'une centrale électrique au charbon ultra-supercritique de 400 MW, des simulations avec Aspen-Plus du procédé PCC ont été utilisées pour prédire les émissions atmosphériques potentielles de la centrale. De nouvelles données issues de divers projets de recherche menés dans ce domaine ont réduit significativement la perception du risque de rejet d'amines et de produits de dégradation d'amines dans l'atmosphère. En plus de la réduction des émissions de CO$_2$, la technologie PCC est aussi à même d'assister la réduction des émissions de SO$_x$ et de NO$_2$. Cependant, certains autres polluants, tels que NH$_3$ et les aérosols, risquent de voir leurs émissions augmenter si des technologies de contrôle appropriées ne sont pas adoptées. Pour étudier la photo-oxydation atmosphérique des amines, des tentatives de développement de modèles des réactions chimiques permettant l'évaluation de la qualité de l'air ont été effectuées. Cependant, une recherche plus approfondie dans ce domaine reste nécessaire pour estimer la réactivité des solvants aminés en présence d'autres polluants tels que les NO$_x$ et les autres composés organiques volatiles présents dans l'air.

Les directives de qualité de l'air existantes devront peut-être être mises à jour afin d'inclure les limites des autres polluants tels que NH$_3$, les nitrosamines et les nitramines, lorsque plus d'information sur leurs émissions devient disponible. En termes de directives de qualité de l'air et autres aspects réglementaires, cet article se focalise sur la description des prédictions des concentrations des principaux polluants potentiellement rejetés par une centrale au charbon, prédictions obtenues par des simulations de procédé ASPEN-Plus PCC.

Abstract — **Emissions to the Atmosphere from Amine-Based Post Combustion CO$_2$ Capture Plant – Regulatory Aspects** — *Amine-based Post Combustion Capture (PCC) of CO$_2$ is a readily available technology that can be deployed to reduce CO$_2$ emissions from coal fired power plants. However, PCC plants will likely release small quantities of amine and amine degradation products to the* *atmosphere along with the treated flue gas. The possible environmental effects of these emissions have been examined through different studies carried out around the world. Based on flue gas from a 400 MW ultra-supercritical coal fired power plant Aspen-Plus PCC process simulations were used to predict the potential atmospheric emissions from the plant. Different research initiatives carried out in this area have produced new knowledge that has significantly reduced the risk perception for the release of amine and amine degradation products to the atmosphere.*

In addition to the reduction of the CO_2 emissions, the PCC technology will also help in reducing SO_x and NO_2 emissions. However, some other pollutants such as NH_3 and aerosols will increase if appropriate control technologies are not adopted. To study the atmospheric photo-oxidation of amines, attempts are being made to develop chemical reaction schemes that can be used for air quality assessment. However, more research is still required in this area to estimate the reactivity of amino solvents in the presence of other pollutants such as NO_x and other volatile organic compounds in the background air.

Current air quality guidelines may need to be updated to include limits for the additional pollutants such as NH_3, nitrosamines and nitramines once more information related to their emissions is available. This paper focuses on describing the predicted concentrations of major pollutants that are expected to be released from a coal fired power plant obtained by ASPEN-Plus PCC process simulations in terms of current air quality regulations and other regulatory aspects.

INTRODUCTION

The world's dependency on coal and gas for its major share of energy demand requires the development of a comprehensive approach for capturing and storing CO_2 emissions produced during the combustion of these energy sources (IPCC, 2005). It is likely that some expected greenhouse gas reduction programs are being delayed until a carbon technology can be developed and deployed. Currently carbon tax are being introduced in different countries around the world including Australia and new performance standards limiting carbon dioxide emissions from power plants are being planned or introduced (European Commission, 2013). On March 27, 2012, the USEPA (United States Environmental Protection Agency) announced new standards that limit CO_2 emissions to 1 000 lb per megawatt-hour (454 kg MWh^{-1}) (USEPA, 2012). This means that all new coal-fired power plants may be required to adopt new technologies to capture or reduce their anticipated emissions. Therefore, any proposed carbon capture technology should be developed in concert with any emerging environmental regulations.

Carbon Capture and Sequestration (CCS), has been widely recognised as a technology that can be used to mitigate the CO_2 emissions from power plants and from other energy intensive facilities. The technology is based on separating the CO_2 from a selected flue gas to produce a pure CO_2 stream that can be captured. The captured pure CO_2 will be compressed to a dense fluid that can be transported and injected in the underground for long term storage. Currently, the most effective and available technology for CO_2 capture from the flue gas

of power plants is based on the chemical interaction between the flue gas and selected organic solvents occurring in an absorber/stripper system that can selectively absorbs then releases CO_2.

A typical 400 MW pulverised coal fired power plant produces about 1 000 000 m^3 of flue gas per hour, containing CO_2, H_2O, N_2, O_2, NO_x, SO_x and fly ash. The large quantities of gases that must be managed to carry out efficient CO_2 capture presents a technical and scientific challenge for the following several reasons:

- the high volume of gas to be treated (for a 400 MW plant around 5 000 t of CO_2 per day should be treated);
- the flue gas is at atmospheric pressure and the CO_2 is diluted (under 15% CO_2);
- the flue gas contains other trace impurities and oxygen that will affect the separation efficiency.

Amine-based Post Combustion CO_2 Capture (PCC) technology that is commercially available although it has not been tested at full scale power generation plants (GCCSI, 2012). Nevertheless, PCC shows promise as a near-term solution to capture CO_2 from flue gas of fossil fuelled power plants. This technology can be easily deployed and retrofitted to the existing fossil-fuelled power plants. Amine solvents can degrade under the interactions with different compounds existing in the flue gas and under different plant operational conditions to produce different degradation products. While the most widely used and discussed solvent is the aqueous solution of 0.3 g/g MonoEthanolAmine (MEA), different solvents are being synthesised and tested in the aim to improve the overall properties and performance of CO_2 capture.

A number of PCC pilot plants from around the world using MEA (Moser et al., 2011; Knudsen et al., 2011; Strazisar et al., 2003) have reported the presence of different degradation products at different sections of the plant, resulting from the interaction between the flue gas constituents and the amines. These products reduce the efficiency of the solvent and create a challenge for the management of wastes materials. Some of these products have the potential to be directly released from the plant as gaseous compounds while others may be entrained in droplets to reach the atmosphere. In the atmosphere, amines and their degradation products can undergo further chemical reactions to produce new products, some of which may be toxic (IEAGHG, 2012; Karl et al., 2012; Pitts et al., 1978; Grosjean, 1991). For amine-based PCC plants, therefore, the identification and quantification of major degradation products of health and environmental concern are required. The degradation compounds will be reported to regulatory agencies to assess the environmental impacts of the plant emissions. It is important that any potential risks associated with these releases are identified and mitigated before PCC systems are deployed for capturing CO_2 from coal fired power plant flue gas.

The assessment of the potential health and environmental impacts of amines and amine degradation products being released to the atmosphere from PCC plants has lately attracted a lot of research around the world (Koornneef et al., 2012; Brakstad et al., 2010; Thitakamol et al., 2007). This research is intended to reduce gaps in the current knowledge and to provide a better understanding of the various issues related to the atmospheric fate of amine release to the atmosphere. In the current paper, the use of MEA to scrub the CO_2 from a coal-fired power plant forms the focus of the following discussion. Results and analysis of major degradation products will be discussed. This paper focuses on application of process simulation towards predicting the atmospheric emissions from an amino solvent based post combustion capture plant.

1 MAIN DEGRADATION PRODUCTS

The amine used to capture CO_2 has the potential to degrade in the process and in the atmosphere after being released from the plant. Solvent will degrade in the process under the influence of the plant operating conditions and by interaction with the flue gas chemical contents and in the atmosphere by interacting with the background air chemical species.

Despite considerable efforts to understand the amine degradation pathways occurring in the process, there are still many gaps in the knowledge. Solvent degradation is a challenging issue that affects the performance the CO_2 capture of amine-based PCC system and may also increase corrosion rates which reduce equipment life (Reynolds et al., 2012; Gouedard et al., 2012; Lepaumier et al., 2011). For MonoEthanolAmine (MEA), a common and widely studied amine solvent, two recognised degradation pathways are:
- oxidative degradation driven by oxygen and other constituents in the flue gas entering the absorber (Bedell, 2009; Goff, 2005; Sexton and Rochelle, 2006),
- thermal degradation driven by high temperatures in the process (Davis, 2009; Thitakamol et al., 2007).

Oxidative degradation takes mainly place in the absorber while the thermal degradation occurs mainly in the stripper process. The solvent degradation is responsible of the production of many intermediate and final species in the process. The volatility of these compounds can be used to predict the most likely compounds that are likely to leave the plant to the atmosphere. Volatile (and some less volatile species) may be emitted at different concentrations to the atmosphere while others can accumulate in the liquid and solid waste in the reclaimer. The accumulation of non-volatile contaminants in amine solutions may also create undesirable operational problems.

It has long been recognised that CO_2, O_2, SO_x, NO_x and other compounds in flue gas from coal fired power plants undergo complex chemical reactions with MEA to produce different degradation products. The presence of selected trace metals such as Fe and Cu was found to catalyse these paths (Goff and Rochelle, 2004; Sexton and Rochelle, 2009). Ammonia, aldehydes, amides and hydroxyacetaldehyde were recognised as primary products resulting from this degradation (Sexton and Rochelle, 2011; Chi and Rochelle, 2002). Reactions between aldehydes and oxygen produce carboxylic acids that may dissociate in the solution to form heat stable salts. In addition, ammonia formed by solvent degradation reacts with MEA in the presence of oxygen to form amides and alkylamines.

Because of the complexity involved in describing the appropriate chemical reaction pathways of the solvent degradation, different degradation mechanisms have been proposed but these mechanisms are still far from being complete (Angove et al., 2011a, 2011b). As a result of the complexity of properly describing the chemistry of amine degradation, the appropriateness of process modeling results would depend on the accuracy of the adopted chemical reaction scheme.

Nitrosamines and nitramines were reported as possible degradation products resulting from the interaction between organic amines and nitrosating agents such as nitrate and nitrite (Masuda et al., 2000; Attalla and Azzi, 2009, 2010; Brakstad et al., 2010; Pedersen et al., 2010).

Sun *et al.* (2011) reported that carbonyl compounds formed by oxidative degradation of amine have the potential to catalyse the nitrosation of amines. These compounds are of major concerns because their potential risks on human health. However, it is unclear how much of these substances will be formed and emitted during the capture process or formed in the atmosphere after emission. It is known that in the atmosphere and under the influence of sunlight nitrosamines can be readily photolysed while nitramines have longer lifetimes. More research is needed for the identification and quantification of nitrosamines and nitramines.

In summary, despite the substantial progress achieved lately in identifying the generation of by-products in the process, there are still many uncertainties that need to be properly addressed.

Amines and their degradation products that leave the PCC plant to the atmosphere will undergo chemical and physical transformation in the atmosphere. Chemical transformations will be driven by complex atmospheric chemical reactions while the meteorological conditions will control the transport and dispersion of these pollutants (Angove *et al.*, 2012; Attalla and Azzi, 2009). Additional degradation products could be generated depending on their interactions with the surrounding ambient air chemical compounds. The non-linear chemistry of the process of the secondary pollutants formation makes it hard to establish linear relationships to predict these anticipated species and their concentrations. It has been reported that NH_3, aldehydes, amides and particles are produced during the atmospheric photo-oxidation of amines (Angove *et al.*, 2012; Karl *et al.*, 2012).

To study the atmospheric degradation of MEA and other amine solvents, the CSIRO smog chamber has been used to carry out experiments at different conditions to develop the database needed to elucidate the major degradation pathways for amines. The smog chamber data are used to develop and validate chemical mechanisms that can be embedded in air quality models to determine the ground level concentrations of pollutants of concerns. Air quality modeling results can be communicated to stakeholders and regulators to show the environmental performance of the plant. Figure 1 shows an example of the smog chamber results (Angove *et al.*, 2012). In this example, NH_3 and O_3 concentrations are seen to increase as a function of time due to photo-oxidation of MEA in the presence of NO_x.

2 IDENTIFICATION OF EMISSIONS FROM THE PLANT

The current assessment was based on the results obtained using ASPEN-Plus PCC process simulations to determine potential emissions of MEA and its degradation products in the treated flue gas of an MEA-based CO_2 capture plant attached at the end of a 400 MW ultra-supercritical coal fired power plant (IEAGHG – report 2012/07). The PCC plant was assumed to use 30% w/w aqueous MEA solvent at the operating conditions specified in Table 1 (Do *et al.*, 2012).

The process flow sheet developed by *Fluor Daniel Ltd* and given in the IEA GHG PH4/33 report that represents the industrial process concept based on the

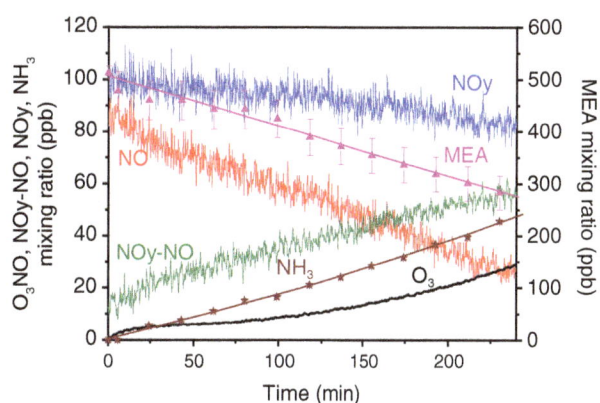

Figure 1

Smog chamber results of an MEA photo-oxidation experiment showing the formation of selected degradation products (Angove *et al.*, 2012).

TABLE 1

Feed flue gas data of the selected ultra supercritical coal-fired power plant used for the simulation (IEAGHG - Report 2012/07)

	Ultra-supercritical coal fired power plant
Flow rate (tonnes/h)	2 973
Temperature (°C)	50
Pressure (kPa)	102.3
Composition	
O_2 (mole %)	4.3
CO_2 (mole %)	12.4
H_2O (mole %)	12.2
N_2 (mole %)	71.1
SO_2 (mg/m^3)	10
NO_x (mg/m^3)	200

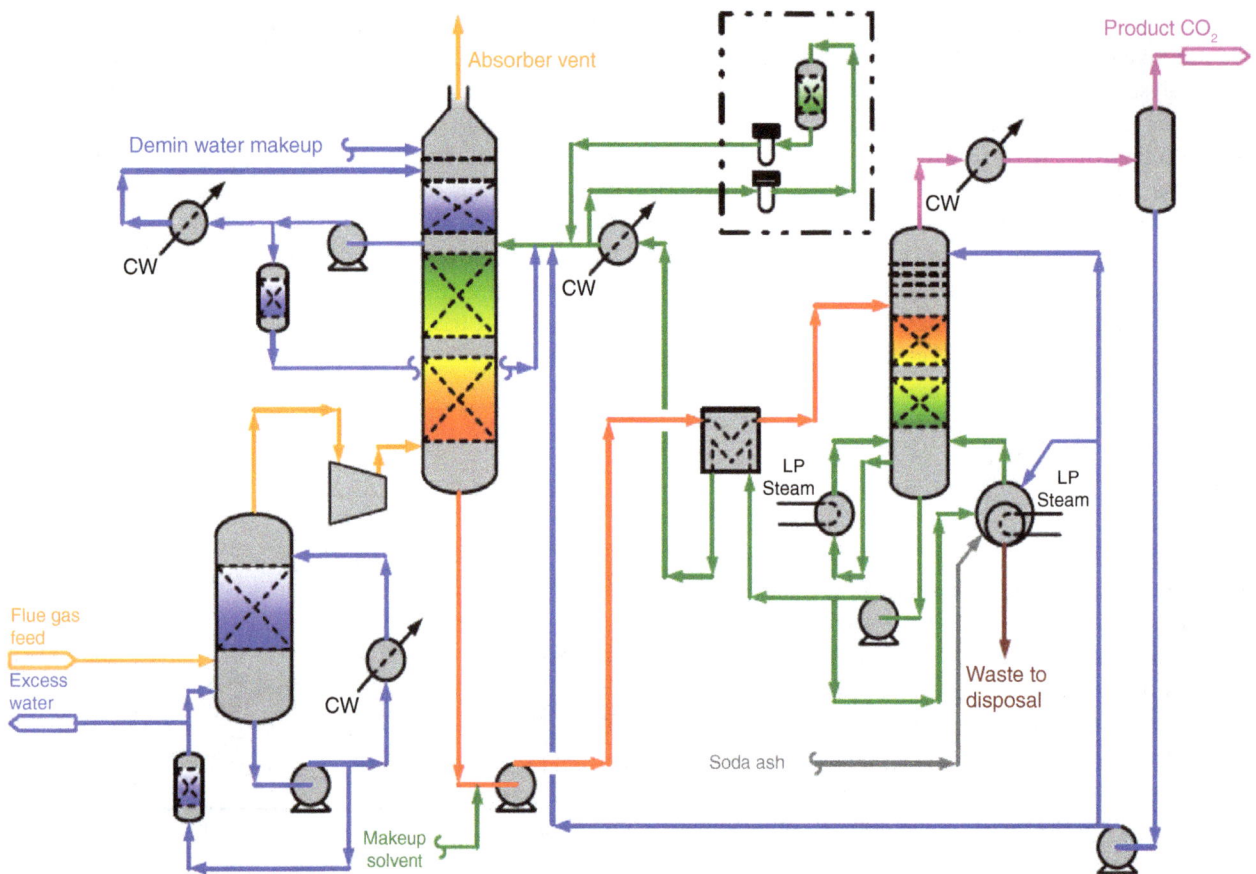

Figure 2

Standard amine-based PCC process flow sheet (Fluor, 2004).

standard amine-based PCC process was adopted for the reported ASPEN-Plus PCC process simulations given in "IEAGHG - report 2012/07". Figure 2 shows a generic process flow sheet that explains the industrial CO_2 capture process concept narrated in the same IEA report.

The CO_2 capture system consists of three main elements: a Direct Contact Cooler (DCC), where the flue gas is conditioned and prepared for the absorption process; an absorber, where CO_2 is absorbed into the amine-based solvent through a chemical reaction; and a regenerator (or stripper), where the concentrated CO_2 is released and the original solvent is recovered and recycled back through the process.

The flue gas is first cooled in the DCC then contacts a circulating sodium hydroxide (NaOH) scrubbing solution to remove up to 98% of SO_2. Removal of SO_2 would minimize the build-up of heat-stable salts in the downstream absorber-regenerator loop.

In the CO_2 stripping section, the CO_2 is liberated from the solvent using steam, producing a low pressure CO_2 product stream and regenerating the solvent for CO_2 capture use. The CO_2 rich solvent leaves the bottom of the absorber and is pumped to the solvent regeneration section of the plant. This is the section where the weakly bonded compound is broken down with the application of heat to liberate the CO_2 and leave reusable solvent behind. The rich solvent from the absorber is first heated in a heat exchanger and then enters the stripper where the rich solvent flows down the stripper through the packed beds counter current to the stripping steam which removes the CO_2 from the rich solvent. The solvent collects on the bottom chimney tray and is sent to the reboiler where low-pressure steam, which is extracted from the cogeneration plant, is used to heat the loaded solvent. The resulting vapor at the top of the stripper contains CO_2 saturated with water. In the stripper, the CO_2 is separated from most of the moisture, yielding a rich CO_2 stream that is sent to the CO_2 product compressor. The lean solvent (i.e., with CO_2 removed) is routed back to the absorber via the rich/lean heat exchanger.

A slipstream of the lean solvent is periodically treated in a solvent reclaiming section to remove contaminant by-products that can gradually build up in the circulation loop. Oxidative degradation and the presence of acid gas impurities in the flue gas lead to the formation of heat-stable salts in the sorbent stream, which cannot be dissociated even on application of heat. To avoid accumulation of these salts in the solvent stream and to recover some of this lost solvent, a slip stream of hot lean solvent from the bottom of the stripper is removed through a semi-continuous reclaiming operation and sent for proper disposal. The rate at which this reclaimer effluent waste stream is generated varies according to the capacity and running conditions of the CO_2 capture system. The reclaimer effluent comprises heat-stable salts, non-volatile solvent degradation products, and unrecovered solvent. Typical disposal measures for this type of waste are incineration and landfilling.

The SO_2 concentration in the feed flue gas from the selected power plant was set at 10 mg/m^3 in accordance with the flue gas desulphurisation plant operation upstream and 95% of 200 mg/m^3 concentration of NO_x in the flue gas was considered to be NO with rest being NO_2.

The CO_2 capture plant releases the CO_2 lean stream to the atmosphere from the top of the absorber. During the process, chemical compounds are produced as a result of complex chemical reactions occurring in the absorber. Some of these products continue to recycle in the plant while others such as amines, ammonia, aldehydes, carboxylic acid etc. may be released to the atmosphere.

The bulk composition of the treated flue gas would differ from the incoming flue gas primarily in CO_2 and water. Furthermore, the CO_2 capture system would perform trim SO_2 removal and would also remove some ammonia, HCl, and HF, plus part of the particulates and a portion of the NO_x. The incoming flue gas contains NO_x as nitrogen oxide (NO) and nitrogen dioxide (NO_2). The majority of the NO would pass directly through the CO_2 capture system.

Approximately 95% of the NO_2 would be absorbed either in the DCC or in the CO_2 capture system. The CO_2 capture system would not generate additional NO_x components. The CO_2 capture system would add trace amounts of solvent, ammonia, and other VOC into the flue gas. The incoming flue gas contains SO_x as SO_2 and sulfur trioxide (SO_3). As a result of the pre-treatment and further removal in the CO_2 absorber, the vent gas would be effectively free of SO_2. All of the incoming

SO_3 present as very fine mist would be emitted with the CO_2 capture system vent. The CO_2 capture system would not generate hydrocarbons or CO. Any hydrocarbons or CO present in the incoming flue gas would be vented out through the CO_2 capture system stack.

The operating conditions of the absorber top control the amount of the concentrated vapour that is expected to be emitted to the atmosphere. MEA and its degradation products may be released to the atmosphere in accordance with to the vapour pressures of the gas constituents which are dependent on the absorber temperature and concentrations in the solvent. For the considered case study, the used demineralised process water and cooling water inlet temperature were set at 12°C and the maximum rise in cooling water temperature was restricted to 7°C.

The concentrations of MEA and its degradation products obtained in the CO_2 lean gas stream represent the theoretical vapour phase emissions of these chemicals in the treated flue gas and not the physical liquid entrainment based emissions at the plant operating condition. For the purposes of this paper, the highest value of 0.13 m^3 per million m^3 in the range given in the Handbook of Gas Processors Suppliers' Association (2004) was used to estimate physical carryover of wash water in the treated gas that leaves the wash section. From the quantity of entrained wash water and its chemical composition, the actual physical entrainment losses of MEA solvent and its degradation products were calculated. The chemical composition of the entrained wash water was considered to be same as that of the wash water circulating inside the wash section at steady state operation.

The process simulation assumed CO_2 absorber to have 32 theoretical stages whereas the solvent regenerator (stripper) was assumed to have 38 theoretical stages. The high number of stages was used to get a smooth temperature profile and achieve a maximum likely rise in temperature along the absorber and stripper. This was used to study the effect of maximum temperature rise on the degradation of MEA. For practical purposes the columns should be further optimised. Both the absorber and the stripper were assumed to operate according to chemical rate kinetics with mass transfer considerations (ASPEN Rate-Sep Models). The water wash tower (after the absorber) was assumed to have three theoretical stages and the flue gas temperature at the outlet of the wash tower was set at 45°C. The electrolyte NRTL model from ASPEN's property data bank was used to track the ionic species generated during the CO_2 absorption/regeneration process and determine the overall physical properties of various process streams.

The major pathways leading to solvent loss during the CO_2 absorption process are linked to:

- physical liquid entrainment in the treated gas,
- vapour phase carryover in the treated gas,
- solution degradation,
- mist formation produced by nucleation of SO_3 present in the incoming flue gas.

2.1 Simulation Results

During normal operation, the CO_2 capture system vent would emit the treated flue gas. The bulk composition of the treated flue gas would differ from the incoming flue gas primarily in reduced CO_2 and water emissions. Furthermore, the CO_2 capture system would perform trim SO_2 removal, as discussed above, and would also remove some ammonia, HCl, and HF, plus part of the particulates and a portion of the NO_x. The incoming flue gas contains NO_x as nitrogen oxide (NO) and nitrogen dioxide (NO_2). The majority of the NO would pass directly through the CO_2 capture system. Approximately 95% of the NO_2 would be absorbed either in the DCC or in the CO_2 capture system. The CO_2 capture system would not generate additional NO_x components. The CO_2 capture system would add trace amounts of solvent, ammonia, and other VOC into the flue gas.

The incoming flue gas contains SO_x as SO_2 and sulfur trioxide (SO_3). Over 98% of the incoming SO_2 would be removed in the DCC. The remaining SO_2 in the flue gas would react with the amine-based solvent in the CO_2 capture system; therefore, the vent gas would be effectively free of SO_2. However, for achieving improved solvent performance, it is recommended to reduce SO_3 below 10 mg/m^3 before entering the absorber. SO_3 contributes to solvent losses due to the potential formation of aerosols that would be emitted with the CO_2 capture system vent. The CO_2 capture system would not generate hydrocarbons or CO. Any hydrocarbons or CO present in the incoming flue gas would be vented out through the CO_2 capture system stack.

The CO_2 capture system was designed to achieve a 90% CO_2 capture efficiency during steady-state operations. While the CO_2 capture system may offer the additional benefit of reducing some other residual emissions as described above, these other reductions would be ancillary benefits and not the focus of the CO_2 capture process.

With the exception of VOC emissions (*i.e.*, amine solvent and acetaldehyde, as discussed below), the CO_2 capture system is not expected to increase the emission rates of any regulated emissions in the flue gas stream exiting the facility.

The chemical reactions used in the current simulations were based on information available in the open literature. However, because of the gaps in the available information not all reactions could be modelled and not all compounds were available in the Aspen-Plus database. To overcome these difficulties, we have utilised the following power-law kinetic model developed by Uyanga and Idem (2007) to predict the rate of MEA degradation:

$$-R_{MEA} = 0.00745^* e^{-45258/RT} * [MEA]^{1.9} * [CO_2]^{-0.3}$$
$$\{[SO_2]^{3.4} + [O_2]^{2.8}\}$$

where, $-R_{MEA}$ is the rate of O_2- and SO_2-induced degradation of MEA (mol/(L.h)) and R the gas constant (8.314 J/(mol.K)). [MEA], $[SO_2]$, $[O_2]$, and $[CO_2]$ are the respective concentrations of MEA, SO_2, O_2, and CO_2 (presented in units of mol/L).

The simulation results show that NH_3, aldehydes, amides will be emitted from the plant at different rates. However, the three principle heavier degradation products oxazolidone and other low volatility products shown in Table 2 were found to have extremely low emission levels in the current simulation. The selected single stage water wash appeared to be very effective in removing these low volatility compounds. In addition, carboxylic acids (HCOOH) that are produced during the

TABLE 2

Likely maximum atmospheric emissions of MEA and its degradation products from the water wash tower (IEAGHG, 2012)

Emissions	Total (mg/Nm^3(dry))
MEA	1.38E-02
Ammonia	1.14E+00
DEA	4.43E-05
Formaldehyde	2.99E-01
Acetaldehyde	3.16E-01
Acetone	3.84E-01
Methylamine	2.44E-01
Acetamide	1.94E-04
HEEDA	6.72E-13
Oxazolidone	1.55E-09
HEIA	2.98E-12
Trimer	0.00E+00
Cyclic urea	0.00E+00
Polymer	0.00E+00

degradation of MEA are trapped as stable salts in the solvent.

Emissions of SO_2 from the flue gas slipstream would be effectively eliminated and emissions of HCl, HF would be substantially reduced when the capture system is operational. Emissions of amine solvent NH_3, and acetaldehyde, which may be formed by reaction of the amine-based solvent with oxygen present in the flue gas, would be increased.

As mentioned earlier, it has been recognised that the use of amine-based solvents to capture CO_2 may result in emissions of a group of nitrosamines and nitramines that are potentially harmful for human health and the environment. These substances that are formed in the PCC process during interactions between the amines and selected flue gas constituents may exist at extremely low concentration levels. It is anticipated that small amounts of these compounds may be emitted with the treated flue gas to air or captured in other waste streams. The exact reaction mechanisms describing the formation of these substances are very complex and more knowledge is needed before proceeding with any predictions of these substances. The modeling results of nitrosamines and nitramines were not included in the present paper.

Table 2 summarizes the estimated concentrations of the treated flue gas exiting the CO_2 capture facility. It shows the likely atmospheric emission compounds and their calculated concentrations in the vapour and droplet phases for the selected case study of a post combustion CO_2 capture plant that processes coal-fired power plant flue gas using 30% w/w aqueous MEA solution. The shaded section of Table 2 highlights the predicted extremely low concentration values of some of these compounds.

The list of the identified compounds that are expected to be released from a PCC plant includes reactive species that are likely to affect the levels of air pollution present over downwind areas depending on the type of the prevailing ambient conditions. Each of these compounds would generate different types of pollution ranging from the production of ozone, NH_3, aerosols, NO_2, etc. Amines degradation in the atmosphere has attracted a lot of attention where different research programs were initiated to improve our understanding on the topic.

3 EMISSION STANDARDS AND LEGISLATION

Emissions from amine-based PCC plant would be examined in terms of their types and amounts. The ambient concentration of each pollutant is determined by the amount of emissions released from the plant and the prevailing meteorological conditions that are able to transport and dilute such emissions. Analysis of potential short-term and long-term air quality impacts of the proposed plant and the use of applicable regulations are the basis for qualifying future operation of a PCC plant. Mitigation measures will be recommended, as necessary, to reduce significant air quality impacts.

Environmental setting and environmental impacts could be used to assess emissions from an industrial plant:

- environmental setting consists of the description of existing regional and local environmental conditions relevant to the plant location;
- environmental impacts describe the significance of the anticipated pollutant emissions relatively to the existing environmental thresholds above which the anticipated emissions would have significant effects on the environment.

Regulatory agencies specify maximum acceptable levels of air pollutants for a selected region. Emissions of carbon monoxide (CO), NO_x, SO_2, Hydrogen Fluoride (HF), Hydrogen Chloride (HCl), and Particulate Matter (PM) from the flue gas are currently authorized under existing air permits issued by regulators. Permit alterations are used to update a permit in cases where there is no increase in emission limits.

Future deployment of amine-based PCC capture plant of CO_2 will be associated with emissions of different pollutants in the gaseous and droplet phases to the atmosphere. These emissions can be reduced by deploying the appropriate control technology such water wash systems and demisters. If nitrosamines were found to be emitted at concentrations that may raise health concerns, UV destruction techniques (IEA GHG, 2012) or other techniques can used to resolve this issue.

Aspen-Plus PCC process simulations were used to provide a reasonable assessment for the emissions expected from the plant. The results obtained would depend on the accuracy of the inputs used to execute the simulations. The outputs of the Aspen Plus modeling scenarios provide the knowledge needed to determine which of these compounds should be targeted for further air quality modeling to determine its fate in the environment. The air quality assessment of emissions to air will need to address all expected plant operations including the business as usual operations. The air quality modeling results will be used to provide the required information for regulators, stakeholders, and the industry to assess the plant operation in terms of meeting the current air quality regulations. If emission levels are exceeding the implemented guidelines other control technology may be implemented to reduce selected compounds.

Emissions from PCC plants could contain chemical substances related to the used amines and their degradation products that may raise different environmental concerns. Consequently, industrial-scale PCC systems will be subject to pollution mitigation regulations. Currently, little is known about the health risks related to amines that are expected to be emitted from PCC. Most countries have legislations that regulate emissions of common pollutant species, especially for the following "criteria" pollutants, which are considered as the measure of pollution in air (Azzi et al., 2012). These are ozone, carbon monoxide, sulphur dioxide, nitrogen dioxide, lead and particulate matter as these are commonly found in the air. These pollutants have been regulated by developing human health-based and/or environmentally-based criteria for setting permissible levels. The set of limits based on human health is called primary standards. Another set of limits intended to prevent environmental and property damage is called secondary standards.

Ozone is not likely to be produced directly in the PCC process but it will be produced by secondary atmospheric chemical reactions (see *Fig. 1*). In addition to the criteria pollutants there is a large number of other toxic materials that are of concern globally. In the United States, more than 180 substances have been classified as Hazardous Air Pollutants (HAP) and similar lists have been adopted in many other countries. Regulatory agencies throughout the world either have already developed, e.g. the US, or are now actively developing rules for industry and other sources to minimise emissions of these materials. Some of the compounds that have been identified as potential atmospheric emissions from PCC units are classified as Toxic Air Contaminants (TAC) such as the identified formaldehyde and acetaldehyde. Table 3 shows the potential contribution of each pollutants to atmospheric pollution where air quality regulations should be met.

TAC are defined as air pollutants that may cause or contribute to an increase in mortality or in serious illness, or may pose a hazard to human health. For these compounds there is no threshold level below which adverse health impacts may not be expected to occur even at low concentrations. Best available control technology for toxics is used to reduce these pollutants to the lowest possible level. A toxic-emission inventory would be required by regulators. This would also apply to the groups of nitrosamines and nitramines.

The deployment of PCC units will affect the emission inventories for compounds that are anticipated to be emitted from the plant. It is expected that SO_x and NO_2 will be reduced during the PCC operations. However, volatile organic compounds and NH_3 will increase

TABLE 3

Potential contribution of each pollutant to atmospheric pollution

Pollutants	Potential contribution to atmospheric pollution
MEA	Potential of photochemical smog oxidants formation
Ammonia	Secondary aerosols
DEA	Potential of photochemical smog oxidants formation
Formaldehyde	Potential of photochemical smog oxidants formation
Acetaldehyde	Potential of photochemical smog oxidants formation
Acetone	Potential of photochemical smog oxidants formation
Methylamine	Potential of photochemical smog oxidants formation
Amides	Secondary aerosols

along with some other compounds such as amides and nitrosamines. There is also clear indication that aerosols would also increase as primary emissions from the plant and as secondary products produced by chemical reactions in the atmosphere. Appropriate air quality assessment of the potential impacts of the plant on human health and the environment would require the following:

– an updated emissions inventory that includes all major compounds that are expected to be emitted from the plant;
– atmospheric chemical reactions mechanisms describing the degradation of reactive species in the atmosphere.

Air emissions standards are also needed to evaluate exposure and health impacts resulting from different plant operating scenarios. All these requirements are to be understood before the deployment of commercial plants.

CONCLUSION

Amine-based post combustion capture of CO_2 technology is a readily available technology that can be used to reduce CO_2 emissions from coal fired power plants but it is important that emissions from the deployment of PCC meet the implemented air quality regulations.

The deployment of PCC will reduce SO_x and NO_2 emissions from the normal flue gas emissions. However, some other pollutants will be produced if appropriate control emission technologies were not adopted. Of these pollutants, NH_3 and aerosols emissions can be assessed with current air quality models to determine their air quality impacts. Attempts are being made to develop chemical reaction schemes to describe the photo-oxidation of MEA emissions and other amines. However, this topic is far from being complete and more research is still required to improve this area.

Current air quality guidelines may require to be updated to include limits for NH_3, nitrosamines and nitramines once more information related to their emissions is conclusive and being made available.

ACKNOWLEDGMENTS

The authors would like to acknowledge the financial support from the IEA GHG R&D and CSIRO Advanced Coal Technology Portfolio for completion of this work.

REFERENCES

Angove D., Azzi M., Tibbett A., Campbell I. (2012) An Investigation into the Photochemistry of Monoethanolamine (MEA) in NO_x, *ACS Symposium Series* **1097**, 265-273.

Angove D., Jackson P., Lambropoulos N., Azzi M., Attalla M. (2011a) CO_2 Capture Mongstad-Theoritical evaluation of the potential to form and emit harmful compounds-Atmospheric Chemistry.

Angove D., Jackson P., Lambropoulos N., Azzi M., Attalla M. (2011b) CO_2 Capture Mongstad-Theoritical evaluation of the potential to form and emit harmful compounds-Atmospheric Chemistry.

Attalla M., Azzi M. (2009) Environmental impact of emissions from post combustion capture, IEA Greenhouse Gas R&D Programme, Regina Canada.

Attalla M., Azzi M. (2010) Environmental impact of emissions from post combustion capture, IEA Greenhouse Gas R&D Programme. Oslo, Norway.

Azzi M., Day S., French D., Fry R., Lavrencic S. (2012) Analysis of environmental legislative and regulatory requirements for the use of amine-based PCC in Australia. CSIRO Report prepared for the Australian National Low Emissions Coal R&D.

Azzi M. (2011) Environmental impacts of amine-based PCC plant, Presented at: *CLIMIT-Workshop: Amine emission from post combustion CO_2 Capture*, Oslo Norway.

Bedell S. (2009) Oxidative degradation mechanisms for amines in flue gas capture, *Energy Procedia* **1**, 771-778.

Brackstad O.G., da Silva E.F., Syversen T. (2010) TCM Amine Project: Support on input to environmental discharges. Evaluation of degradation components. Version 3, SINTEF F16202, Trondheim: SINTEF.

Chi S., Rochelle G. (2002) Oxidative degradation of MEA, *Ind. Eng. Chem. Res.* **41**, 17, 945-969.

Davis J. (2009) Thermal degradation of aqueous amines used for carbon dioxide capture, *Thesis*, University of Texas Austin.

Do T., Narendra D., Feron P., Azzi M. (2012) Process Modelling for Amine-based Post-Combustion Capture plant. Deliverable 3.2 Report prepared by CSIRO for the "Australian National Low Emissions Coal Research and Development".

European Commission (2013) Climate Action.

Fluor (2004) Improvement in power generation with post combustion capture of CO_2. IEA GHG R&D Programme, Nov., Report No: PH4/33, 272.

GCCSI (2012) CO_2 capture technologies, Post Combustion Capture.

Goff G., Rochelle G. (2004) MEA degradation: O_2 mass transfer effects under CO_2 capture conditions, *Ind. Eng. Chem. Res.* **43**, 20, 6400-6408.

Goff G. (2005) Oxidative degradation of aqueous MEA in CO_2 capture processes, *PhD Dissertation*, University of Texas Austin, TX.

Gouedard C., Picq D., Launay F., Carrette P.-L. (2012) Amine degradation in CO_2 capture. I. A review, *Int. J. Greenhouse Gas Control* **10**, 244-270.

Grosjean D. (1991) Atmospheric chemistry of toxic contaminants: 6. Nitrosamines: Dialkylamine and N-nitrosomorpholine, *J. Air Waste Manage. Assoc.* **41**, 306-311.

Handbook of Gas Processors Suppliers Association (2004) 12th edition, Chap. 7, p. 5, GPSA.

IEAGHG R&D Program (2012) Gaseous Emissions from amine based PCC processes and their deep removal, Report 2012/07.

IPCC (Intergovernmental Panel on Climate Change) (2005) Special Report. Carbon Dioxide Capture and Storage, Summary for Policymakers. Intergovernmental Panel on Climate Change, http://www.ipcc.ch/pdf/special-reports/srccs/srccs_ summaryforpolicymakers.pdf.

Karl M., Dye C., Schmidbauer N., Wisthaler A., Mikoviny T., D'Anna B., Muller M., Borras E., Clemente E., Muñoz A., Porras R., Ródenas M., Vázquez M., Brauers T. (2012) Study of OH-initiated degradation of 2-aminoethanol, *Atmos. Chem. Phys.* **12**, 1881-1901.

Knudsen J., Andersen J., Jensen J. (2011) Results from test campaigns at the 1 t/h CO_2 PCC pilot plant in Esberg under the EU FP7 CESAR Project, *IEAGHG PCC1 Conference*, 17-19 May 2011, Abu Dhabi, UAE.

Koornneef J., Ramirez A., Turkenburg W., Faaij A. (2012) The environmental impact and risk assessment of CO_2 capture, transport and storage – An evaluation of the knowledge base, *Progr. Energ. Combust. Sci.* **38**, 62-86.

Lepaumier H., Da Silva E., Einbu A., Grimstvedt A., Knudsen J., Zahlen K., Svendsen H. (2011) Comparison of MEA degradation in pilot-scale with lab-scale experiments, *Energy Procedia* **4**, 1652-1659.

Masuda M., Mower H., Pignatelli B., Celan I., Friesen M. (2000) Formation of N-nitrosamines and N-nitramines by the reaction of secondar amines with peroxynitrite and other reactive nitrogen species, *Chem. Res. Toxicol.* **13**, 4, 301-308.

Moser P., Schmidt S., Stahl K. (2011) Performance of MEA in a long-term test at the PCC plant in Niederaussem, *Int. J. GHG Control* **5**, 620-627.

Pedersen S., Sjovoll M., Foastas B. (2010) Flue gas degradation of amines, *IEA GHG Workshop*, Oslo, 16 Fev.

Pitts J., Grosjean D., Vanmcauwenberghe K., Schmidt J., Fitz D. (1978) Photooxidation of aliphatic amines under simulated atmospheric conditions: Formation of nitrosamines, nitramines, amides, and photochemical oxidant, *Environ. Sci. Technol.* **12**, 946-953.

Reynolds A., Verheyen V., Adeloju S., Meuleman E., Feron P. (2012) Towards commercial scale postcombustion capture of CO_2 with MEA solvent-Key considerations for solvent management and environmental impacts, *Environ. Sci. Technol.* **46**, 3643-3654.

Sexton A., Rochelle G. (2006) Oxidation products of amines in CO_2 capture, Eight International Conference on Greenhouse Gas Control Technologies, GHGT-8, Trondheim, Norway, 19-2 June.

Sexton A., Rochelle G. (2009) Catalysts and inhibitors for oxidative degradation of MEA, *Int. J. Greenhouse Gas Control* **3**, 6, 704-711.

Sexton A., Rochelle G. (2011) Reaction products from the oxidative degradation of MEA, *Ind. Eng. Chem. Res.* **50**, 2, 667-673.

Strazisar B., Anderson R., White C. (2003) Degradation pathways for MEA in a CO_2 capture facility, *Energy Fuels* **17**, 1034-1039.

Sun Z., Liu Y., Zhong R. (2011) Carbon dioxide in the nitrosation of amine: catalyst or inhibitor? *J. Phy. Chem.* **115**, 26, 7753-7764.

Thitakamol B., Veawab A., Aroonwilas A. (2007) Environmental impacts of absorption-based CO_2 capture unit for post-combustion treatment of flue gas from coal-fired power plant, *Int. J. Greenhouse Gas Control* **1**, 318-342.

USEPA (2012) epa.gov/carbonpollutionstandard/pdfs/20120327 proposal.pdf.

Uyanga I.J., Idem R. (2007) Studies of SO_2 and O_2 induced degradation of aqueous MEA during CO_2 capture from power plant flue gas streams, *Ind. Eng. Chem. Res.* **46**, 2558-2566.

11

Assessing the Permeability in Anisotropic and Weakly Permeable Porous Rocks Using Radial Pulse Tests

Richard Giot*, Albert Giraud and Christophe Auvray

Laboratoire Géoressources, ENSG-Université de Lorraine, UMR 7359, 2 rue du Doyen Marcel Roubault, TSA 70605, 54518 Vandœuvre-Lès-Nancy - France
e-mail: richard.giot@univ-lorraine.fr - albert.giraud@univ-lorraine.fr - christophe.auvray@univ-lorraine.fr

* Corresponding author

Résumé — **Estimation de la perméabilité d'une roche anisotrope très faiblement perméable par pulse tests radiaux** — Le pulse test est généralement considéré comme un essai adapté pour la mesure de la perméabilité des roches poreuses très faiblement perméables. Classiquement, l'essai consiste à imposer un saut de pression à la base d'un échantillon cylindrique et à mesurer les variations de pression dans les réservoirs amont et aval. Dans les présents travaux, nous proposons un nouveau dispositif et une nouvelle procédure sur échantillon de type cylindre creux : le saut de pression est imposé dans un trou axial et la pression mesurée dans le trou et à la circonférence de l'échantillon. Cette configuration génère un écoulement à la fois axial et transversal, et non seulement axial comme dans le cas de l'essai classique. Pour les roches isotropes transverses, cela permet la détermination de la perméabilité dans les directions parallèle et perpendiculaire aux plans d'isotropie, à partir d'un seul échantillon, judicieusement orienté. L'essai est couplé hydro-mécaniquement, et en conséquence aucune solution analytique ne peut être considérée. L'essai est alors interprété en ayant recours à des modélisations numériques couplées HM 3D, prenant en compte l'anisotropie des échantillons. Dans les travaux précédents, une loi poro-élastique isotrope transverse, ainsi qu'une méthode inverse couplée aux modélisations numériques pour identifier des paramètres, ont été développées et implantées dans le code aux éléments finis Code_Aster (Edf). La méthode est adaptée pour l'interprétation de l'essai de pulse radial et appliquée à des échantillons cylindriques creux d'argilites de Meuse/Haute-Marne, avec axe soit perpendiculaire soit parallèle aux plans d'isotropie. Alors que pour l'essai de pulse axial classique, la méthode requiert deux échantillons pour déterminer les perméabilités intrinsèques dans les directions parallèle et perpendiculaire aux plans d'isotropie, la méthode appliquée à l'essai de pulse radial permet la détermination des perméabilités intrinsèques dans ces deux directions sur un seul échantillon. Ceci permet de s'affranchir d'effets d'hétérogénéités des échantillons.

Abstract — Assessing the Permeability in Anisotropic and Weakly Permeable Porous Rocks Using Radial Pulse Tests — The pulse test is usually considered to be an efficient method for measuring the permeability of weakly permeable porous rocks. Classically, the test consists of imposing a pressure drop on the base of a cylindrical sample and measuring the pressure variations in the upstream and downstream reservoirs. In the present work, we propose a new apparatus and procedure for hollow cylindrical samples in which the pressure drop is imposed in an axial hole and the pressure is measured both in the hole and on the circumference of the sample. Unlike the classical axial pulse test, this configuration results in a flow in both the axial and transversal directions rather than only in the axial direction. For transverse isotropic rocks, this configuration allows the assessment of the permeability in the isotropy planes and normal to the isotropy planes in a single sample when the samples are appropriately oriented. The test is fully hydro-mechanically coupled; therefore, no analytical solution exists. The test is then interpreted through fully coupled numerical modeling in 3D, considering the anisotropy of the samples. In previous works, we developed and implemented a transverse isotropic poroelastic constitutive law in the finite element code Code_Aster (Edf), as well as an inverse method coupled to the numerical modeling for parameter identification. The method is adapted to the radial pulse test and then applied to hollow cylindrical samples of Meuse/Haute-Marne argillite with the axis either parallel or perpendicular to the isotropy planes. Although this method requires 2 samples for the assessment of permeability in the isotropy planes and normal to the isotropy planes in the axial pulse test, the method applied to the radial pulse test allows the assessment of intrinsic permeability in both directions on a single sample, which allows freeing ourselves from a heterogeneity effect.

NOMENCLATURE

σ	Second-order stress tensor (Pa)
ε	Second-order strain tensor (-)
σ_m	Mean stress (Pa)
s	Deviatoric stress tensor (Pa)
ε_v	Volumetric strain (-)
e	Deviatoric strain tensor (-)
F^m	Resultant of mass forces (N)
m_{lq}	Liquid mass supply (kg.m^{-3})
ρ_{lq}	Liquid density (kg.m^{-3})
p_{lq}	Liquid pressure (Pa)
K_o	Drained bulk modulus (Pa)
K_{un}	Undrained bulk modulus (Pa)
b	Biot coefficient in the isotropic case (-)
b_1	Biot coefficient in the isotropy plane (-)
b_3	Biot coefficient in the direction normal to the isotropy plane (-)
M	Biot modulus (Pa)
λ_{lq}	Darcy's conductivity for liquid (Pa^{-1}.m^2.s^{-1})
k_{lq}	Liquid conductivity (m.s^{-1})
k_{int}	Intrinsic permeability in the isotropic case (m^2)
K_1^{int}	Intrinsic permeability in the isotropy plane (m^2)
K_3^{int}	Intrinsic permeability in the direction normal to the isotropy plane (m^2)
μ_{lq}	Liquid dynamic viscosity (Pa.s)
M_{lq}	Hydraulic flow of water (kg.s^{-1}.m^{-2})
p_{re}^u	Pressure in the upstream reservoir (Pa)
p_{re}^d	Pressure in the downstream reservoir (Pa)
M_{lq}^u	Mass of liquid contained in the upstream reservoir (kg)
M_{lq}^d	Mass of liquid contained in the downstream reservoir (kg)

ζ^u	Volume of liquid contained in the upstream reservoir (m^3)
ζ^d	Volume of liquid contained in the downstream reservoir (m^3)
C_{re}^u	Stiffness of the upstream reservoir (Pa.m^{-3})
C_{re}^d	Stiffness of the downstream reservoir (Pa.m^{-3})
S_{re}^u	Upstream reservoir storage coefficient (kg.Pa^{-1})
S_{re}^d	Downstream reservoir storage coefficient (kg.Pa^{-1})
γ_{lq}	Volumetric weight of liquid (kg.m^{-3})
D	Liquid diffusivity (m^2.s^{-1})
S_s	Specific storage coefficient (kg.Pa^{-1}.m^{-3})
K_{sq}	Matrix bulk modulus (Pa)
K_{lq}	Liquid bulk modulus (Pa)
f_0	Initial Lagrangian porosity (-)
E_o	Drained Young's modulus (Pa)
E_1	Young's modulus in the isotropy plane (Pa)
E_3	Young's modulus in the direction normal to the isotropy plane (Pa)
v_o	Drained Poisson's ratio (-)
$\mathbb{C}$	Fourth-order elasticity tensor of the porous medium
$\mathbb{C}^s$	Fourth-order elasticity tensor of the matrix
B	Second order tensor of Biot coefficients (-)
M_ϕ	Pore compressibility (Pa)
K^{int}	Second-order tensor of intrinsic permeability (m^2)
λ^{lq}	Second-order tensor of Darcy's conductivity (Pa^{-1}.m^2.s^{-1})
ϕ, θ, ψ	Euler angles (°)
g	Gravity acceleration (m.s^{-2})

INTRODUCTION

In weakly permeable porous rocks, such as argillite, liquid and gas transfers are governed by permeability. As a consequence, permeability is a key parameter in most of the engineering applications involving these natural geomaterials, for example, nuclear waste or gas storage (for which they can be used as formation for geological sequestration of CO_2). Consequently, permeability is a key issue and has been widely studied for years. For the range of intrinsic permeability values of weakly permeable porous media, that is 10^{-22}-10^{-20} m^2, the classic measurement techniques based on a permanent flow are not adapted. Indeed, for this range of permeability, it is nearly impossible, in practice, to conduct drained tests. As an alternative, one can use transient flow methods, such as the so-called pulse test proposed by Brace et al. [1]. The pulse test allows the assessment of both the intrinsic permeability and specific storage, considering samples fully saturated with one fluid. The specific storage is defined as the volume of water, per unit volume of saturated rock, injected into the pores when exposed to a unit increase of pore fluid pressure. During the pulse test, the intrinsic permeability and the specific storage govern the transient evolution and the final equilibrium of pore pressure distribution, respectively. Classically, the pulse test is assumed to be a one-dimensional pore pressure diffusion problem in the axial direction considering simplifying hypotheses, such as constant mean stress, constant strain or uniaxial strain. The matching solutions are then based on the resolution of a hydraulic diffusion-type equation assuming a decoupling between hydraulic and mechanical behaviours. Under the assumption of constant mean stress, Hsieh et al. [2] solved the system of equations and proposed an analytical solution for the pulse test. Neuzil et al. [3] utilized this solution to offer a graphical method for identifying both the permeability and the specific storage. The analytical solution for the constant mean stress was also applied by Homand et al. [4] for the interpretation of pulse tests on argillites, but with an inverse method rather than a graphical method. Sevaldurai et al. [5] also considered the classical 1D solutions of the pulse test on saturated samples for the interpretation of pulse tests on Lindsay limestone. The measured pressure curves were compared to calculated curves obtained with the analytical solution, and they assessed permeability on the order of 10^{-23}-10^{-22} m^2.

Even though solutions considering an uncoupling between hydraulic and mechanical behaviours are widespread, Wang [6] and Adachi and Detournay [7] demonstrated that the pulse test is a fully coupled problem with poroelastic strains coupled with the pore pressure field.

Walder and Nur [8] showed that neglecting the poromechanical couplings could result in errors in the assessment of permeability. They showed that this simplification leads to a sample size effect that could result in an error on the order of 50% in the estimation of permeability in tight rocks. The coupling between hydraulic and mechanical responses leads to a three-dimensional or two-dimensional axisymmetrical problem, and an accurate solution can only be determined by coupled hydro-mechanical modeling, for example, with a finite element code.

Giot et al. [9] also showed the importance of considering poromechanical couplings through 2D axisymmetrical coupled analysis of the pulse test. They developed an inverse method for the interpretation of the pulse test, associating a fully coupled analysis of the pulse test by finite element modeling to an inversion algorithm for the identification of poromechanical parameters. They focused on intrinsic permeability and parameters influencing the specific storage, more precisely, the Biot coefficient, drained Young's modulus and the reservoirs' stiffnesses. The pulse test can then be viewed as a parameter identification problem. The inverse method consists in minimizing, with a gradient-based optimization algorithm, a cost-function that measures the differences between measured and calculated reservoir pressures (the latter being functions of the parameters to be identified). The method was applied on 4 tests on Meuse/Haute-Marne claystone, and a 2D back analysis accounting for couplings gave more accurate results than a single 1D analysis neglecting these couplings. The quality of the fitting given by both analyses is quite similar, but the values of the estimated parameters are different. Concerning the intrinsic permeability, the difference between parameters estimated in 1D and 2D is approximately 5 to 15%, whereas for the specific storage, it can reach 80%. The results of the 4 tests revealed a transverse isotropy of the argillite.

The present work addresses Meuse/Haute-Marne argillites, which are a potential host for radioactive waste disposal. Numerous experimental studies have been conducted on this claystone and have demonstrated its transverse isotropy. Amongst others, Zhang and Rothfuchs [10] showed a gas permeability ratio of 10 and a Young's moduli ratio of 1.5 between directions parallel and normal to the isotropy planes. For partially saturated samples, Cariou et al. [11] showed that the anisotropy ratio's effect on argillite stiffnesses increases when water content decreases, whereas the Biot coefficient appears to be isotropic. It is thus essential to consider this anisotropy when conducting permeability measurements on Meuse/Haute-Marne argillites. More generally, when dealing with claystone, Selvadurai et al. [5]

underlined the importance of accounting for the anisotropy of rock based on the interpretation of the axial pulse tests they conducted on Lindsay limestone. Marschall *et al.* [12] also reported the intrinsic permeability anisotropy of Opalinus Clay, which was determined through long lasting steady state tests. The order of permeability of this material is 10^{-21}-10^{-20} m^2, with an anisotropy ratio less than 10. Thus, Giot *et al.* [13] adapted their interpretation method of the axial pulse test to account for this anisotropy in the analysis of the test. While studying anisotropy, the 2D fully coupled inversion of the pulse test was replaced by a 3D fully coupled inversion, and a fully coupled transverse isotropic poromechanical model was developed and implemented in the finite element code and then used for the modeling of the pulse test. Nevertheless, they showed that the axial pulse test only allows for the assessment of the permeability in the axial direction of the sample, the flow being only axial. 2 tests on 2 different samples are then required to determine permeability in the isotropy planes and normal to the isotropy planes, which can result in a bias due to heterogeneity effects. Thus, in the present study, we propose a new test, the radial pulse test. This test is performed on hollow cylinders of rock and aims to characterise the whole transverse isotropic permeability, *i.e.*, the intrinsic permeability in the directions parallel and normal to isotropy planes, on one single sample. This method allows freeing ourselves from the heterogeneity of the rock when dealing with different samples and to limit the number of tests. The same approach used for the axial pulse test is used for interpreting the radial pulse test. We make use of the same fully coupled transverse isotropic poromechanical model and inversion procedure to interpret the radial pulse test. Such an inverse problem may have a non-unique solution. Nevertheless, the developed inversion procedure has proven to be efficient in the field of geomechanics, for example, on the axial pulse test (Giot *et al.* [9, 13]) or overcoring (Giot *et al.* [14]), particularly concerning the issues of local minima and the consideration of uncertainties on measurements.

Permeability measurements with hollow cylinders were previously proposed by Alarcon-Ruiz *et al.* [15] on concrete samples of 265 mm height and 350 mm external diameter. The pressure was applied by nitrogen gas rather than with water. The measurements were combined with numerical modeling, without consideration of any mechanical processes, to assess intrinsic permeability. They found permeability on the order of 10^{-16} m^2, which is a rather high permeability and justifies their method for samples of great dimensions. Part of the concrete samples used by Alarcon-Ruiz *et al.* were extracted from a larger sized

hollow cylinder that had previously been used for evaluating the intrinsic permeability when submitted to temperature variations. This work was due to Dal Pont *et al.* [16], with the concrete hollow cylinder being 1.5 m high, 0.55 m of external radius with an internal cylinder of 0.25 m. The initial permeability of the concrete is on the order of 10^{-17} m^2, which is also greater than the argillaceous material considered in our case. The gas pressure was measured in the hollow cylinder through 4 cylindrical sensors placed at different distances from the heated surface. This experimental set-up has been coupled to a numerical analysis with a finite element method, and both experimental and numerical results (temperature and gas pressure) have been compared to identify the values of the parameters of the relationship describing the evolution of permeability with temperature.

Even if hollow cylinders have previously been used for relative permeability measurements, none of the previous analyses were aimed at studying the anisotropy of the materials considered, and the sizes of the samples were greater than the one considered in the present study. Moreover, those tests were aimed at identifying gas relative permeability, which is significantly different from water relative permeability in the materials we are concerned with in the present work. Davy *et al.* [17] indicated that water permeability is 2 to 8 orders of magnitude smaller than gas permeability on the same claystone we study in the present work. In terms of physical processes, the radial pulse test is conducted on water-saturated samples and aims to identify intrinsic permeability, which is independent from fluid permeability. There is no gas implied in this test, and thus, there is no need to consider gas diffusion (Fick's law) or even mechanics of partially saturated media. In the present paper, we focus on the pore-pressure diffusion process (in other words, advection, described by Darcy's law) in fully water-saturated medium under isothermal conditions (climate control). Gas diffusion, as described by Fick's law, is not concerned with the present research, or the mechanics of partially saturated media.

1 TRANSVERSE ISOTROPIC COUPLED POROELASTIC CONSTITUTIVE LAW

In this section, we present the main constitutive equations of the transverse isotropic poroelastic model for the saturated case. This model is based on Biot's mechanics for fluid-saturated porous media [6, 27-32]. The corresponding isotropic poroelastic model is available in the finite element code Code_Aster (Edf) [32-33] and has been detailed in Giot et al. [9]. For the Meuse/Haute-Marne (MHM) argillites, we focused on transverse isotropy, with the axis of revolution denoted as O_{x3} and the supposed isotropic behaviour in the plane O_{x1x2}. We assume that the porous medium and the matrix (skeleton) are transverse isotropic elastic with the same symmetry axis O_{x3}.

The full transverse isotropic model is detailed in Giot et al. [13]. This model accounts for microscopic parameters and relations between macroscopic and microscopic properties of the constituents. It is an improved version of the model presented in Noiret et al. [18]. It is inspired by the anisotropic poroelastic model proposed by Cheng [19] that takes into account micromechanical considerations. Cheng provided the relations between macromechanical and micromechanical parameters in the case of transverse isotropy. Abousleiman et al. [20], Abousleiman and Cui [21] and Cui et al. [22, 23] used this type of model in geomechanics and developed analytical solutions for wellbore and consolidation problems. Numerical applications of anisotropic poroelasticity can be found, among others, in Cui et al. [24], Kanj and Abousleiman [25] and Ekbote and Abousleiman [26].

In the following, we consider a porous medium that is composed of a deformable matrix and fully saturated by a liquid (subscript lq). The advection of the liquid is accounted for through Darcy's law. Isothermal conditions are assumed (due to climate control) such that the linear porous elastic model consists of one balance equation for liquid mass and linear momentum of the media.

1.1 Constitutive Equations for the Fully Saturated Medium

The non-linear poroelastic constitutive equations for the fully saturated media [28] are written incrementally. In the case of anisotropy, these constitutive equations have to be written in tensorial form as follows:

$$d\boldsymbol{\sigma} = \mathbb{C} : d\boldsymbol{\varepsilon} - \mathbf{B} dp_{lq} \tag{1}$$

$$\frac{dm_{lq}}{\rho_{lq}} = \mathbf{B} : d\boldsymbol{\varepsilon} + \left(\frac{1}{M_\phi} + \frac{\phi}{K_{lq}}\right) dp_{lq} \tag{2}$$

where $\boldsymbol{\sigma}$ and $\boldsymbol{\varepsilon}$ represent the second-order stress and strain tensors, respectively. p_{lq} refers to the liquid pressure. $\mathbb{C}$ is the fourth-order elasticity tensor, characterized by 2 Young's moduli (E_1, E_3), 2 Poisson coefficients (v_{12}, v_{13}) and a shear coefficient (G_{13}).

The components of the drained elastic stiffness tensor $\mathbb{C}$ of the porous medium can be found in Giot et al. [13].

$\mathbf{B}$ refers to the second-order tensor of the Biot coefficients, which can be written as follows:

$$\mathbf{B} = b_1(\mathbf{e_1} \otimes \mathbf{e_1} + \mathbf{e_2} \otimes \mathbf{e_2}) + b_3 \mathbf{e_3} \otimes \mathbf{e_3} \tag{3}$$

M_ϕ is the solid Biot modulus and ϕ is the Lagrangian porosity. m_{lq}, ρ_{lq} and K_{lq} represent the liquid mass supply, the liquid density and the liquid compressibility, respectively. M_ϕ can be linked to the microscopic poroelastic properties of the matrix, the Biot coefficient and the porosity (Giot et al. [13] for details).

Concerning the microscopic poroelastic properties of the matrix, the components of the elastic tensor of the matrix $\mathbb{C}^s$ are given in Giot et al. [13]. The constitutive equations for the transverse isotropic poroelastic behaviour of the matrix can be written as follows:

$$d\boldsymbol{\sigma} = \mathbb{C}^s : d\boldsymbol{\varepsilon} - \mathbf{B} dp_{lq} \tag{4}$$

$$d\phi = \mathbf{B} : d\boldsymbol{\varepsilon} + \frac{dp_{lq}}{M_\phi} \tag{5}$$

Micro-macro relations can then be established for the Biot tensor and are detailed in Giot et al. [13]. It appears that the components of the Biot tensor are functions of 4 elastic moduli of the transverse isotropic elastic matrix tensor and on the corresponding 4 moduli of the transverse isotropic drained elastic tensor of the porous medium, which justifies that the same Biot tensor is considered for the fully saturated porous media and the matrix.

1.2 Conduction Equations

By applying Darcy's generalised law in a fully saturated medium and neglecting gravity, the velocity of liquid is ruled by:

$$\frac{M_{lq}}{\rho_{lq}} = \boldsymbol{\lambda}^{lq} \cdot \left(-\nabla p_{lq} + \rho_{lq} \mathbf{F^m}\right) \tag{6}$$

where $\boldsymbol{\lambda}^{lq}$ designates Darcy's conductivity for liquid and is linked, for the fully saturated porous medium, to the intrinsic permeability $\mathbf{K^{int}}$ through:

$$\boldsymbol{\lambda}^{lq} = \mathbf{K^{int}}(\phi) \frac{1}{\mu_{lq}} \tag{7a}$$

μ_{lq} represents the liquid dynamic viscosity. Darcy's conductivity for a liquid λ_{lq} is linked to liquid conductivity k_{lq}, which is homogeneous to a velocity (m.s^{-1}), through:

$$\lambda_{lq} = \frac{k_{lq}}{\rho_{lq}g} \tag{7b}$$

Considering the transverse isotropy, the intrinsic permeability, and thus Darcy's conductivity, must be written as second-order tensors in the conduction equations as follows:

$$\mathbf{K}^{int} = K_1^{int}(\mathbf{e_1} \otimes \mathbf{e_1} + \mathbf{e_2} \otimes \mathbf{e_2}) + K_3^{int}\mathbf{e_3} \otimes \mathbf{e_3} \tag{8a}$$

$$\boldsymbol{\lambda}^{lq} = \lambda_1^{lq}(\mathbf{e_1} \otimes \mathbf{e_1} + \mathbf{e_2} \otimes \mathbf{e_2}) + \lambda_3^{lq}\mathbf{e_3} \otimes \mathbf{e_3} \tag{8b}$$

1.3 Momentum and Mass Conservation Equations

If we omit gravity, the linear momentum equation is written as follows:

$$\nabla.\sigma = 0 \tag{9}$$

The mass conservation equation for the liquid is expressed as follows:

$$\frac{\partial m_{lq}}{\partial t} = -\nabla.M_{lq} \tag{10}$$

1.4 Numerical Implementation and Model Parameters

The transverse isotropic poroelastic model was implemented in the finite element code Code_Aster (Edf), taking advantage of the isotropic poroelastic model available in Code_Aster [32-33]. The model was written for the general partially saturated case, and it also accounts for the couplings with thermal phenomena. Nevertheless, in this paper, we focus on the poromechanical couplings because the pulse test is assumed to be an isothermal process in which the temperature is fixed

during the test. The parameters of the model, which were introduced in addition to the isotropic model, are:
- the elastic parameters of the skeleton: E_1^s, E_3^s, v_{13}^s, v_{12}^s and G_{13}^s, which can be assessed from measurements of the macromechanical parameters and micro-macro relations;
- the elastic parameters of the porous medium: E_1, E_3, v_{13}, v_{12} and G_{13};
- the hydromechanical coupling parameters of the Biot tensors: b_1 and b_3;
- the conduction parameters K_1^{int} and K_3^{int} for Darcy's law;
- the geometrical parameters (the 3 Euler angles, ϕ, θ, and ψ) which allow us to define the orientation of the isotropy planes (*Fig. 1*).

Moreover, as in the case of the axial pulse test, the model was simplified assuming micro-isotropy (Giot *et al.* [13]):

$$E_1^s = E_3^s = E^s; v_{12}^s = v_{13}^s = v^s \text{ and } G_{13}^s = G^s = \frac{E^s}{2(1 + v^s)} \tag{11}$$

This assumption means that the solid constituent of the matrix is isotropic at the microscopic level. The macroscopic anisotropy comes from a structural origin and is a consequence of directional pore or fissure arrangement (Cheng [19]). This assumption is consistent with the actual knowledge of the microstructural elastic parameters of the studied rock type material. It reduces the number of elastic parameters to be assessed for the grain constituents. In the case of micro-isotropy assumption for the solid phase, the Biot tensor only depends on 4 elastic constants of the porous medium (E_1, E_3, v_{13}, v_{12}) and on the bulk modulus of the solid

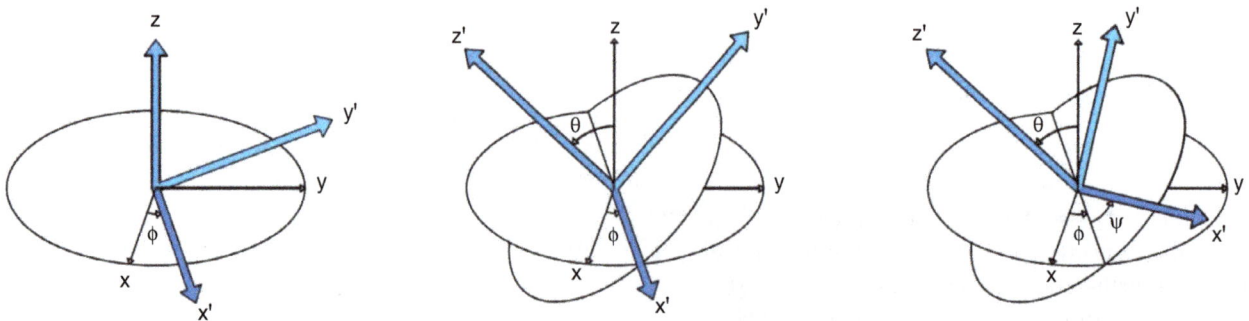

Figure 1

Euler angles convention.

phase k^s. The Biot tensor does not depend on the shear coefficients G_{13} and G^s.

2 RADIAL PULSE TEST: PRINCIPLE AND MODELING

2.1 Description of the Test

The experimental apparatus is presented in Figure 2. A hollow cylindrical rock sample, with an inner radius of 5 mm, an outer radius of 25 mm and a height of 50 mm, is placed in a load cell. Both the top and base of the sample are insulated to prevent water and air flow. The inner hole is connected to an upstream reservoir, while the outer boundary of the sample is connected to a downstream reservoir. The height of the inner chamber, which is connected to the upstream reservoir, is 30 mm and is less than the height of the sample. The diameter of the sample is less than the diameter of the cell. The remaining space between the sample and the body is 5 mm thick and is filled with silica glass balls of 1 and 2 mm diameters, which reproduces a porous network, allowing for the drainage of water, a homogeneous application of the confinement stress and a possible deformation of the sample. During the test, both the inner (upstream) and outer (downstream) pressures are measured using pressure sensors. Concerning the inner (upstream) pressure, it is measured using 2 different types of sensors. An internal sensor is placed in the cell, just under the draining wedge. This sensor consists of a little tablet that is 5 mm in diameter and 2 mm thick, with a resolution of 0.01 MPa. An external sensor, which also has a resolution of 0.01 MPa, is placed in the water circuit outside of the cell. Concerning the outer (downstream) pressure, it is not possible to place an internal sensor in the cell at the circumference of the sample (as for the upstream pressure); therefore, the pressure is only measured using an external pressure sensor placed in the water circuit. During the test, the temperature is maintained constant at 22°C thanks to a climate control. This climate control functions with successive cycles of emission of hot and cold air, lasting 5 minutes each. The external pressure sensors, for both the upstream and downstream chambers, are directly connected to those cycles of climate control; therefore, they exhibit pressure oscillations. The effects of the climate control are damped for the internal sensor placed in the cell, and as a consequence, the oscillations are softened.

The measurement system was calibrated based on a long time experiment (more than 15 years) on permeability measurements on very low permeability claystones, and particularly with the axial pulse test.

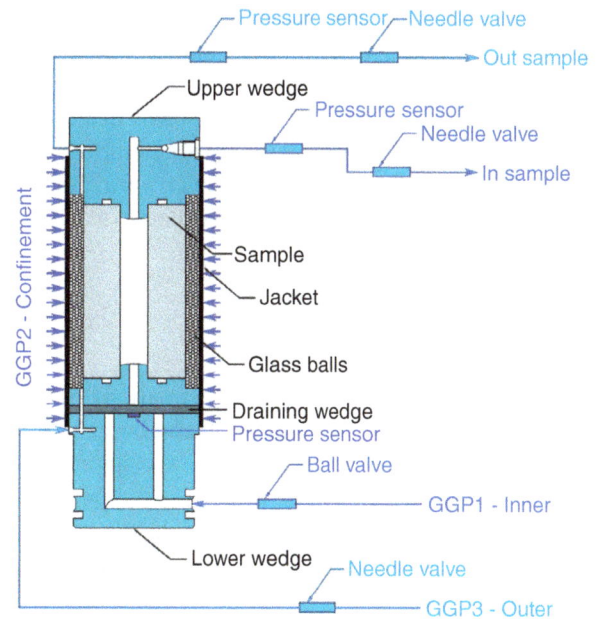

Figure 2

Experimental apparatus for the radial pulse test.

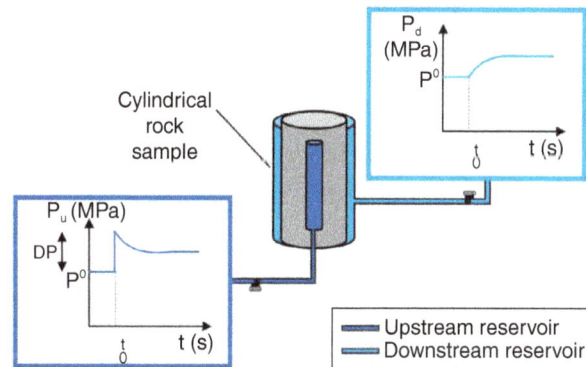

Figure 3

Principle of the radial pulse test.

Moreover, numerical modeling of the test was achieved considering fictitious isotropic samples for the design of the test.

The principle of the test is illustrated in Figure 3. The test is similar to the axial pulse test. After sample saturation and pore pressure homogenisation, the pressure is suddenly increased in the upstream reservoir (connected to the inner hole of the sample). The evolution of the pressures in the upstream (connected to the inner hole) and downstream (connected to the external circumference of the sample) reservoirs are then measured. Because the length of the inner chamber is shorter than the length of the sample, the flow is bi-dimensional (both axial and radial); previous works have demonstrated that the flow is purely axial in the case of the axial pulse test. As a consequence, as will be shown *infra*, both axial and radial permeabilities are stimulated in the radial pulse test.

2.2 Initial and Boundary Conditions

In addition to the field and constitutive equations presented in Section 2, initial and boundary conditions for the radial pulse test must be expressed. The cylindrical sample of inner radius R_{in}, outer radius R_{ou}, and height L is submitted to a hydrostatic stress state. The sample is initially fully saturated, and both the liquid pressure and mean stress inside the sample are homogeneous:

$$p_{lq}(\mathbf{x}, t = 0) = p_{lq}^0 \tag{12}$$

$$\sigma_m(\mathbf{x}, t = 0) = \sigma_m^0 \tag{13}$$

In these equations, $\mathbf{x}$ denotes the position vector. The base Γ_0 and top Γ_L of the sample are insulated:

$$\mathrm{M}_{lq}.\mathbf{n} = 0 \quad \text{on } \Gamma_0 \text{ and } \Gamma_L \tag{14}$$

The liquid pressure is assumed to be homogeneous in both the upstream ($r = R_{in}$) and downstream ($r = R_{ou}$) reservoirs, where r represents the radial coordinate:

$$p_{lq}(z, t) = p_{re}^u(t) \quad \text{on } \Gamma_{Rin} \tag{15a}$$

$$p_{lq}(z, t) = p_{re}^d(t) \quad \text{on } \Gamma_{Rou} \tag{15b}$$

Γ_{Rin} is the circumference corresponding to the inner hole minus the first centimetres in the base and on the top of this hole. The height of the inner chamber for liquid injection is 30 mm, compared with the 50 mm height of the sample. The rest of the inner surface is insulated, like the top and base of the sample.

The liquid pressure is suddenly increased in the upstream reservoir:

$$p_{re}^u(0^+) = p_{lq}^0 + \Delta p \tag{16}$$

The conservation of liquid mass provides the 2 boundary conditions between the sample and the reservoirs:

$$\frac{\partial \xi^u}{\partial t} = \int_{\Gamma_{Rin}} -\lambda_{lq} \nabla p_{lq}(z, t).\mathbf{n} da \tag{17a}$$

$$\frac{\partial \xi^d}{\partial t} = \int_{\Gamma_{Rou}} -\lambda_{lq} \nabla p_{lq}(z, t).\mathbf{n} da \tag{17b}$$

$$\xi^u = \frac{M_{lq}^u}{\rho_{lq}} \quad \xi^d = \frac{M_{lq}^d}{\rho_{lq}} \tag{17c}$$

M_{lq}^u, M_{lq}^d, ξ^u and ξ^d represent the masses and volumes of liquid contained in the upstream and downstream reservoirs, respectively. Design tests on the experimental apparatus with a steel sample allowed for the assumption of a linear relationship between the volume of liquid content in the reservoirs and the liquid pressure. This linear relationship can be accepted in the tested range of pore pressures and for the increase of pressure considered in the tests. This relationship permits us to write the following equations:

$$\xi^u = \frac{p_{re}^u}{C_{re}^u} \quad \xi^d = \frac{p_{re}^d}{C_{re}^d} \tag{18}$$

The 2 coefficients C_{re}^u and C_{re}^d denote the stiffnesses of the reservoirs that can be bound to the reservoir storage coefficients S_{re}^u and S_{re}^d through:

$$C_{re}^u = \frac{\gamma_{lq}}{S_{re}^u} \quad C_{re}^d = \frac{\gamma_{lq}}{S_{re}^d} \tag{19}$$

γ_{lq} represents the volumetric weight of liquid. The boundary conditions between the reservoirs and the rock sample can then be expressed as:

$$\int_{R_{in}} -\lambda_{lq} \nabla \mathbf{p}_{lq}(\mathbf{x}, t).\mathbf{n} da = \frac{1}{C_{re}^u} \frac{\partial p_{lq}(\mathbf{x}, t)}{\partial t}$$
$$p_{lq}(\mathbf{x}, t) p_{re}^u(t) \quad \text{on } \Gamma_{R_{in}} \tag{20a}$$

$$\int_{R_{ou}} -\lambda_{lq} \nabla \mathbf{p}_{lq}(\mathbf{x}, t).\mathbf{n} da = \frac{1}{C_{re}^d} \frac{\partial p_{lq}(\mathbf{x}, t)}{\partial t}$$
$$p_{lq}(\mathbf{x}, t) p_{re}^d(t) \quad \text{on } \Gamma_{R_{ou}} \tag{20b}$$

The boundary conditions were implemented within the finite element code Code_Aster (Edf) through the adaptation of Fortran routines developed in the framework of the 3D numerical modeling of the axial pulse test (Giot *et al.* [13]). The integration was carried out with a Gaussian quadrature method, and an explicit time integration scheme was considered for implementing the boundary conditions (20):

$$p_{re}^u(t_{n+1}) = p_{re}^u(t_n) + C_{re}^u \Delta t \int_{R_{in}} -\lambda_{lq} \nabla \mathbf{p}_{lq}(\mathbf{z}, t_n).\mathbf{n} da \tag{21a}$$

$$p_{re}^d(t_{n+1}) = p_{re}^d(t_n) + C_{re}^d \Delta t \int_{R_{ou}} -\lambda_{lq} \nabla \mathbf{p}_{lq}(\mathbf{z}, t_n).\mathbf{n} da \tag{21b}$$

2.3 Geometry and Mesh

The geometry of the radial pulse allows us to focus on a 2D-axisymmetrical mesh. Thus, only half of the sample could be represented with a vertical symmetry axis. Nevertheless, the present work aims at investigating the effects of anisotropy. Except for the case of isotropy planes normal to the axis of the sample, the whole sample has to be considered and the geometrical model must be fully 3D. Figure 4 shows the geometry and mesh considered for the modeling. Several meshes, of different fineness or coarseness, were considered for the modeling. The mesh presented in Figure 4 is a simplified one. For the accuracy of the calculation, the mesh was refined on the inner and outer circumferences of the hollow cylinder. This refinement improves the calculation of the fluxes on these boundaries.

2.4 Main Results

Based on the results of the axial pulse test (Giot *et al.* [9, 13]), a sensitivity analysis was conducted on 2 types of anisotropic parameters, transfer parameters, more precisely intrinsic permeabilities K_1^{int} and K_3^{int}, and deformability parameters, specifically Young's moduli E_1 and E_3. Concerning the coupling parameters, Biot coefficients b_1 and b_3, they are linked to the Young's moduli through the specific storage coefficient, and thus we decided to focus on œdometric tests for the assessment of these parameters. Moreover, the previous works on the axial pulse test showed strong correlations between

the Biot coefficients and the Young's moduli. Finally, the selected parameters control both the transient evolution (liquid diffusivity, D) and the final equilibrium (specific storage coefficient, S_s). In the isotropic case, the liquid diffusivity is defined as the ratio between the liquid conductivity k_{lq} and the specific storage coefficient:

$$D = \frac{k_{lq}}{S_S}$$

The reservoirs stiffnesses are intrinsic parameters of the experimental apparatus and are not impacted by anisotropy; thus, the stiffnesses were not considered to be of paramount importance when focusing on anisotropy, as they were for the axial pulse test.

In the following modeling, we considered both a transverse sample and a parallel sample to better understand the effects of the parameters on the results of the radial pulse test. The parameters with an index 1 represent the direction in the isotropy plane, whereas parameters with an index 3 represent the direction perpendicular to the isotropy plane. Each parameter will be either normal or axial, depending on the orientation of the sample.

2.4.1 Intrinsic Permeabilities

We first consider a transverse sample with the axis perpendicular to the isotropy planes. On such a sample, the axis of the sample coincides with the normal to the isotropy plane, denoted by $\mathbf{e}_3$. Figure 5 shows the curves

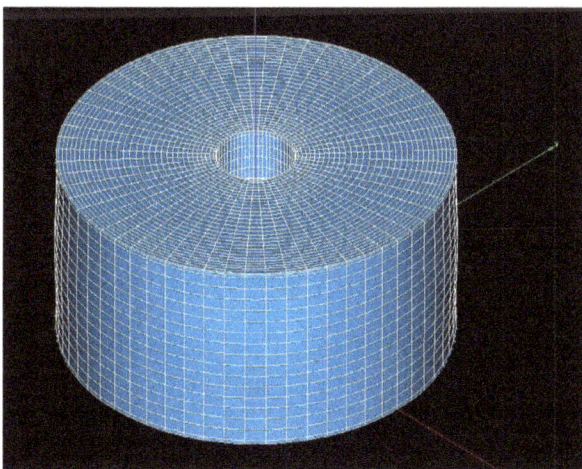

Figure 4

Example of a (simplified) 3D mesh used for numerical modeling of the radial pulse test.

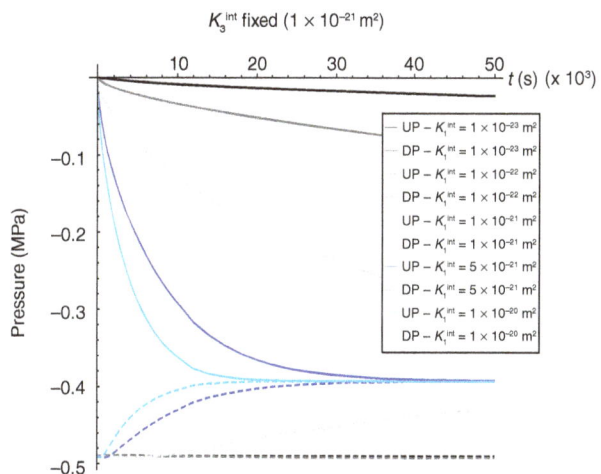

Figure 5

Evolution of reservoir pressure during the pulse test – influence of permeability K_1^{int} on the calculated pressure curves on the transverse sample (UP: Upstream pressure; DP: Downstream pressure).

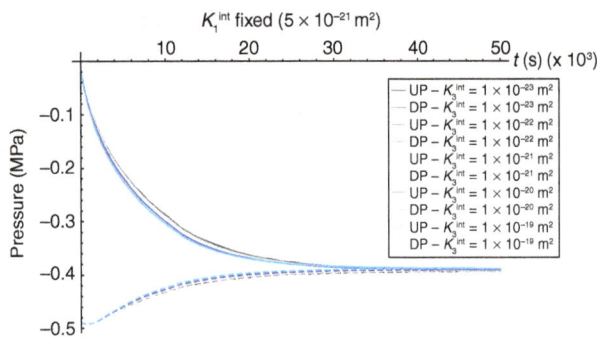

Figure 6

Evolution of reservoir pressure during the pulse test – influence of permeability K_3^{int} on the calculated pressure curves on the transverse sample (UP: Upstream pressure; DP: Downstream pressure).

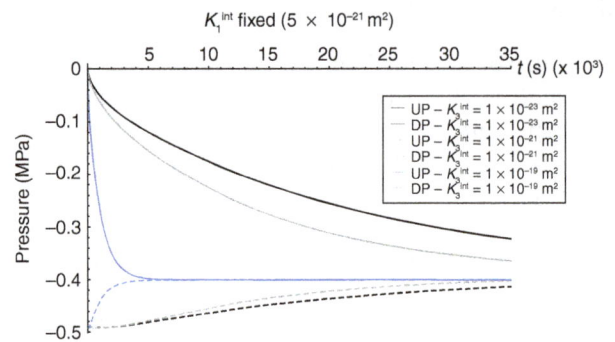

Figure 7

Evolution of reservoir pressure during the pulse test – influence of permeability K_3^{int} on the calculated pressure curves on the parallel sample (UP: Upstream pressure; DP: Downstream pressure).

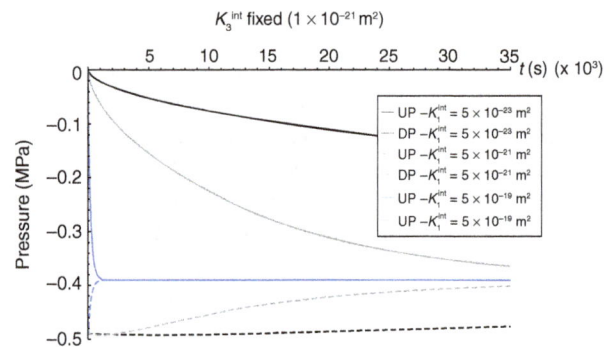

Figure 8

Evolution of reservoir pressure during the pulse test – influence of permeability K_1^{int} on the calculated pressure curves on the parallel sample (UP: Upstream pressure; DP: Downstream pressure).

of the calculated upstream and downstream pressures for such a transverse sample, with constant axial permeability K_3^{int} and varying radial permeability K_1^{int}. This means that in this sample, we vary the permeability in the direction perpendicular to the axis of the sample, thus in the direction of the isotropy plane. As expected, increasing K_1^{int} results in reaching equilibrium more quickly because intrinsic permeability controls the liquid diffusivity (transient evolution) and intrinsic permeability has no impact on the specific storage (final pressure equilibrium). Figure 6 presents the curves of the calculated upstream and downstream pressures for the same transverse sample but with constant radial permeability K_1^{int} and varying axial permeability K_3^{int}. The effect of changing K_3^{int} is less obvious than changing K_1^{int}. Once again increasing K_3^{int} results in reaching equilibrium more quickly, but this is less clear than for K_1^{int}. In fact, it appears that on such a transverse sample, the flow is mainly imposed in the direction of the isotropy plane that corresponds to the radial direction, and the hydraulic solicitation is mainly radial. As a consequence, the convergence of hydraulic solicitation and orientation of the isotropy planes favours a mainly radial flow and solicits the radial permeability K_1^{int}, while the permeability in the axial direction has little influence.

We can then consider the case of a parallel sample with the axis of the sample parallel to the isotropy plane. In such a case, the radial direction of the flow does not only coincide with the isotropy planes. Figure 7 presents the curves of the calculated upstream and downstream pressures for a parallel sample with constant axial permeability K_1^{int} and varying permeability K_3^{int}, with the direction $\mathbf{e}_3$ being a radial direction perpendicular to

the axis of the sample. Once again, we can see that increasing K_3^{int} results in reaching equilibrium more quickly. Figure 8 presents the curves of the calculated upstream and downstream pressures for the same parallel sample, but with constant permeability K_3^{int} and varying permeability K_1^{int}. In this case, contrary to the transverse sample, the permeability K_1^{int} has a considerable influence. The influence is the same as for the other cases; increasing K_1^{int} results in reaching equilibrium more quickly. Therefore, in the case of the parallel sample, both intrinsic permeabilities influence the results of the radial pulse test. Indeed, in such a configuration, the boundary conditions that would favour a radial flow do not coincide with the isotropy planes, and thus boundary conditions (hydraulic solicitation) and geometrical configuration do not act jointly in only stimulating the direction of the isotropy planes.

Based on these results, we can see that for transverse samples, the permeability K_1^{int} primarily controls the pressure curves obtained during the radial pulse test, whereas K_3^{int} has little influence. On the contrary, for parallel samples, both K_1^{int} and K_3^{int} control the pressure evolution. As a consequence, to identify both intrinsic permeabilities K_1^{int} and K_3^{int}, it is more convenient to consider parallel samples than transverse samples. Generally, the intrinsic permeability controls the liquid diffusivity (transient evolution) but not the specific storage (final pressure equilibrium).

2.4.2 Young's Moduli

We first consider a transverse sample with the axis perpendicular to the isotropy planes. Figure 9 shows the effects of Young's modulus E_1 (the Young's modulus in the isotropy plane) on the upstream and downstream pressure curves obtained during the radial pulse test. One can see that E_1 slightly influences the specific storage coefficient (final equilibrium of the pressure) and has little impact on the liquid diffusivity (transient evolution). Concerning E_3, Figure 10 presents the upstream and downstream pressure curves obtained on a transverse sample for different values of E_3, with E_1 being fixed. This figure clearly demonstrates that E_3 has no influence on the pressure curves, neither on the liquid diffusivity nor on the specific storage. This response is similar to the axial pulse test. Indeed, Giot et al. [13] showed that for the axial pulse test on a transverse sample, the axial Young's modulus E_3 has no influence on the final equilibrium and a very slight influence on the liquid diffusivity, whereas E_1 influences both the liquid diffusivity and the specific storage coefficient.

Figures 11 and 12 present the upstream and downstream pressure curves for a parallel obtained during the pulse test for different values of E_3 (E_1 fixed) and E_1 (E_3 fixed). One can see that both Young's moduli influence the specific storage and have little influence on the liquid diffusivity. The influence of E_3 is a little more obvious than the influence of E_1. Thus, for the intrinsic permeability, the parallel sample appears better adapted than the transverse sample for assessing the Young's moduli anisotropy. Nevertheless, both Young's moduli act the same way, with a decrease of the Young's moduli resulting in a decrease of the final pressure, whichever the modulus considered. This could hinder the simultaneous identification of both moduli in the same test.

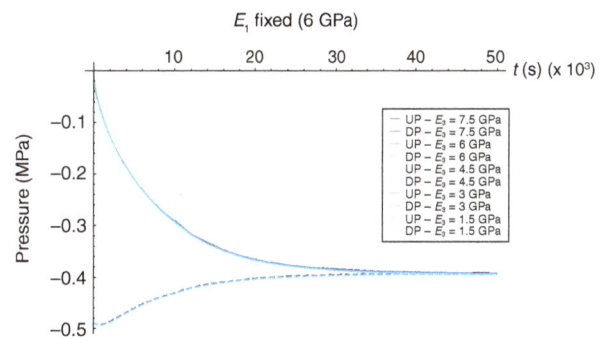

Figure 10

Evolution of reservoir pressure during the pulse test – influence of Young's modulus E_3 on the calculated pressure curves on the transverse sample (UP: Upstream pressure; DP: Downstream pressure).

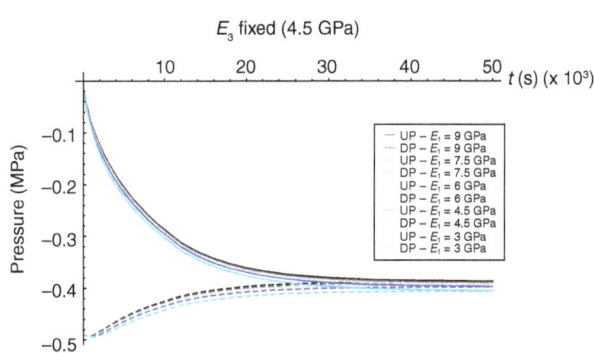

Figure 9

Evolution of reservoir pressure during the pulse test – influence of Young's modulus E_1 on the calculated pressure curves on the transverse sample (UP: Upstream pressure; DP: Downstream pressure).

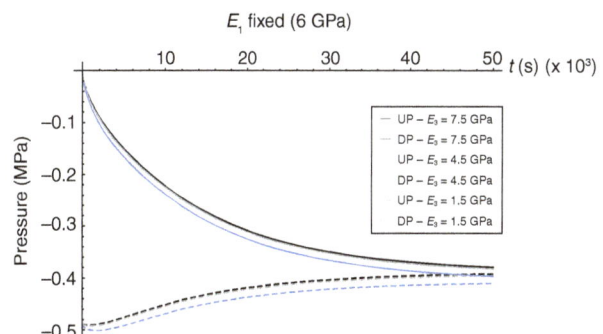

Figure 11

Evolution of reservoir pressure during the pulse test – influence of Young's modulus E_3 on the calculated pressure curves of the parallel sample (UP: Upstream pressure; DP: Downstream pressure).

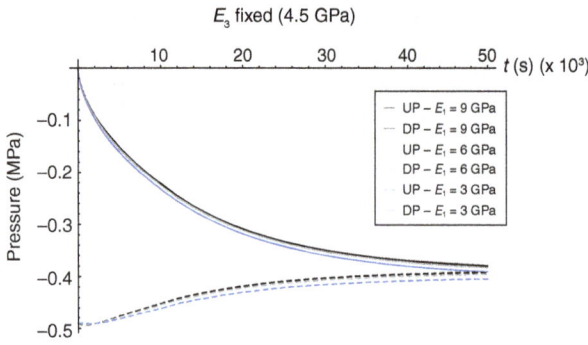

Figure 12

Evolution of reservoir pressure during the pulse test – Influence of Young's modulus E_1 on the calculated pressure curves of the parallel sample (UP: Upstream pressure; DP: Downstream pressure).

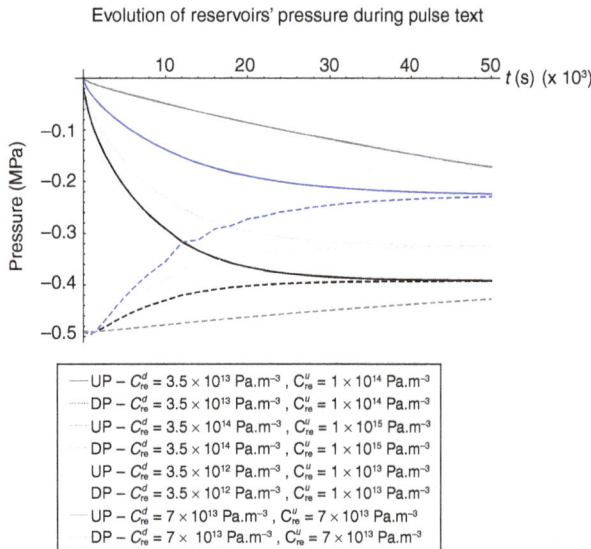

Figure 13

Evolution of reservoir pressure during the pulse test – influence of reservoir stiffnesses on the calculated pressure curves of the transverse sample (UP: Upstream pressure; DP: Downstream pressure).

In general, one can note that the tendency is the same for all of the modeling: the greater the Young's modulus, the greater the value of the pressure at final equilibrium.

2.4.3 Reservoir Stiffnesses

Figure 13 presents the upstream and downstream curves obtained during a radial pulse test on a transverse sample and for different values of the reservoir stiffnesses.

The final equilibrium is achieved more quickly with an increase of the reservoir stiffness. This figure also shows that the reservoir stiffness influences both the kinetics of the test and the final equilibrium. In that way, the reservoir stiffnesses can influence the identification of both the liquid diffusivity and the specific storage. It is then required to have knowledge of the reservoir stiffness before the test or incorporate it in the set of parameter to identify. Nevertheless, the last solution is less efficient because the identification of the values of the stiffnesses could interfere in the identification of the other parameters.

3 INVERSE PROBLEM: INTERPRETATION OF LABORATORY PULSE TESTS

3.1 Inverse Method

Interpretation of the radial pulse test couples the numerical modeling (direct problem) with an inversion algorithm (inverse problem) to identify the uppermost liquid transport parameter and more specifically, the intrinsic permeability in the normal and parallel directions and possibly the poromechanical coupling parameters impacting the specific storage, such as the Biot coefficients and Young's moduli. This parameter identification problem is expressed as an optimisation problem aiming at minimising a cost-functional that quantifies the differences between experimentally measured and numerically calculated upstream and downstream pressures, which corresponds to an inverse problem that has been detailed in Giot *et al.* [9, 13]. The cost-functional is of the least-squares type and incorporates a term accounting for previous knowledge of the parameters, inferred, for example, from the interpretation of the classical axial pulse test performed on the same material:

$$\chi[\mathbf{c}] = \frac{1}{2}\sum_{i=1}^{Nmes} \omega_i(\gamma[\mathbf{c}, t_i] - \gamma_{mes}[t_i])^2 + \frac{1}{2}\sum_{j=1}^{Npar} v_i\left(c_j - c_j^{prior}\right)^2$$

(22a)

$$\gamma_{mes}[t_i] = p_{re}^{mes}[t_i]$$ (22b)

$$\gamma[\mathbf{c}, t_i] = p_{re}^u[t_i] \quad \text{or} \quad \gamma[\mathbf{c}, t_i] = p_{re}^d[t_i]$$ (22c)

In Equations (22), γ_{mes}, γ, *Nmes*, ω_i and v_j represent the measured and calculated pressure levels in the upstream and downstream reservoirs (that is, the pore pressures in the inner hole and at the circumference of the sample, respectively), the total number of measurements, and weighting coefficients on measurements and italic > a priori information, respectively. γ is a function of the parameters to be identified that make up the vector **c**.

c^{prior} consists of prior values of the parameters to be identified. p_{re}^{u} and p_{re}^{d} are calculated by finite element resolution of the previous system of Equations (1) to (21) expressing the direct problem of the radial pulse test.

Finally, the inverse problem of the pulse test may be formulated as follows:

$$c^{identified} = \arg\min[\chi[c]]$$

The inverse method is of the probabilistic type. The minimisation algorithm is of the gradient type, the Levenberg-Marquardt algorithm to be more specific. This algorithm requires the assessment of the derivatives of the calculated pressures with respect to the parameters to be identified. This process corresponds to a sensitivity analysis that is conducted using a finite differences method.

From a numerical point of view, the second term of the cost-functional (22a), taking into account the prior knowledge of the parameters, smooths the cost-functional and decreases the probability of finding a local minimum, which is a common issue in inverse problems such as the one considered for the radial pulse test. The weighting coefficients w_i and u_j can be used to incorporate simple statistical information on both measurements and prior knowledge, and these terms stabilise the cost-functional. Moreover, these terms permit us to give relative importance to the prior knowledge in comparison with measurements, and consequently must be used with great care to avoid giving too much weight to this information. The weighting coefficients also allow us to render the cost-functional dimensionless, which is of paramount importance when several types of measurements of different orders are available (for example, pore pressures and deformations) as well as when several parameters are to be identified (also of different orders). This is the case of the radial pulse test because the gap between the intrinsic permeability and Young's moduli is of 30 orders of magnitude. The weighting coefficients on the measurements, ω_i, permit the consideration of the measurements' accuracy, and thus reduce the bias in parameter identification due to errors on pressure measurements. Errors on the measurements, which are on the order of the sensor's accuracy, should not result in a different solution to the inverse problem.

3.2 Correlations Between Parameters

Figures 14 and 15 provide the correlations between the sensitivities of both intrinsic permeabilities for a transverse and a parallel sample, respectively. The sensitivity of vector p_{re} of the reservoirs' pressures with respect to parameter c_i is generally given by:

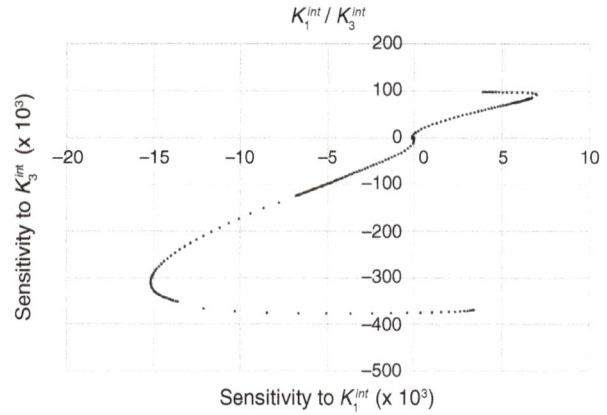

Figure 14

Correlations between the parameters' sensitivities for a transverse sample, example of intrinsic permeabilities.

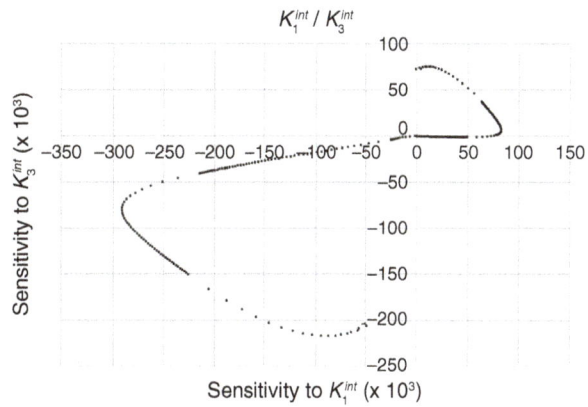

Figure 15

Correlations between parameters' sensitivities for a parallel sample, example of intrinsic permeabilities.

$$\delta_{c_i}p_{re} = \frac{\partial p_{re}}{\partial c_i} \tag{23}$$

For application to the radial pulse test, the sensitivity with respect to parameter c_i is assessed through a finite difference approximation:

$$\delta_{c_i}p_{re} = \frac{p_{re}[c_i + \Delta c_i] - p_{re}[c_i]}{\Delta c_i} \tag{24}$$

Although the sensitivities between all possible couples of parameters (intrinsic permeabilities and Young's moduli) for both transverse and parallel samples were studied, we confine ourselves to presenting the correlations between the sensitivities of intrinsic permeabilities (because the others are similar and the analysis is still valid). The linear correlation coefficient $R(i, j)$ between the sensitivities with respect to parameters c_i and c_j is given by:

$$R(i,j) = \frac{cov(i,j)}{\sqrt{var(i)}\sqrt{var(j)}} \quad (25)$$

where $cov(i,j)$ denotes the covariance between parameters c_i and c_j, and $var(i)$ is the variance of parameter c_i.

$R(i, j)$ was calculated for all the couples of parameters and was significantly less than 1, confirming that no linear correlation could be identified between parameters. Thus, from these correlations, it appears that there is no clear linear relation between the parameters on which we decided to focus for the interpretation of the radial pulse test. Nevertheless, on each curve, one must distinguish 2 parts, one corresponding to the correlations concerning the upstream reservoir pressures, and the other concerning the downstream reservoir pressure. Due to the effects of climate control on the external pressure sensors, we decided to primarily focus on the upstream reservoir pressure for parameter identification. Then, it seems wise to confine the correlations to the upstream reservoir pressures. In such a case, the correlation coefficients are still low, always less than 0.7, indicating that no linear correlation can be identified between the parameters, except for the Young's moduli for a parallel sample (*Fig. 16*, $R^2 = 0.99$). Indeed, for such a sample, both moduli are solicited during the radial pulse test and, as shown in Figures 11 and 12, both parameters have the same influence on the pressure curves. This

result confirms the previous observations and makes it difficult to identify the Young's moduli on a parallel sample. Nevertheless, the identification of these parameters is still interesting because we already have knowledge of these parameters and only want to identify as accurate values as possible. The main parameter being estimated through the pulse test is the intrinsic permeability, and the asset of the radial pulse test is the simultaneous identification of the intrinsic permeabilities in directions both parallel and normal to the isotropy planes. It is interesting to see that no clear correlations can be identified for these parameters, especially for the measurements of the pressures in the upstream reservoir on which we focused (for experimental reasons). More specifically, no correlation can be found between both permeabilities on both types of samples. As a consequence, there is no limitation on the simultaneous identification of both permeabilities, *i.e.*, the assessment of one permeability does not influence the assessment of the other permeability.

3.3 Parameters to be Identified

The previous work on the classical axial pulse test showed that anisotropy does not concern the reservoirs' stiffnesses, which are intrinsic parameters of the experimental apparatus. Moreover, strong correlations were found between those stiffnesses and the rheological parameters (permeability, Young's modulus and Biot coefficient); the values of the reservoirs' stiffnesses influence the transient evolution and the final equilibrium, which are, respectively, linked to the permeability and storage coefficient, the latter being a function of Young's modulus and Biot coefficient. As a consequence, we focused on the identification of rheological parameters and left the stiffnesses aside for the inverse problem. Additionally, strong correlations between the Young's modulus and Biot coefficient were observed in the isotropic case. Moreover, in the anisotropic case, it can be shown that the Biot coefficients can be expressed as functions of Young's moduli (Giot *et al.* [13]). Therefore, to limit the number of parameters to be identified, the identification was limited to intrinsic permeabilities, which influences the transient behaviour, and the Young's moduli, which influences the specific storage (the final equilibrium). When considering transverse isotropy for the classical axial pulse test (Giot *et al.* [13]) and based on a sensitivity analysis, the identification process focused on 4 parameters: the "axial" permeability, that is to say, K_1^{int} for the parallel samples and K_3^{int} for the transverse samples; and the "normal" Young's moduli, that is to say, E_1 for the transverse samples and E_3 for the axial samples. Thus, the identification focused on

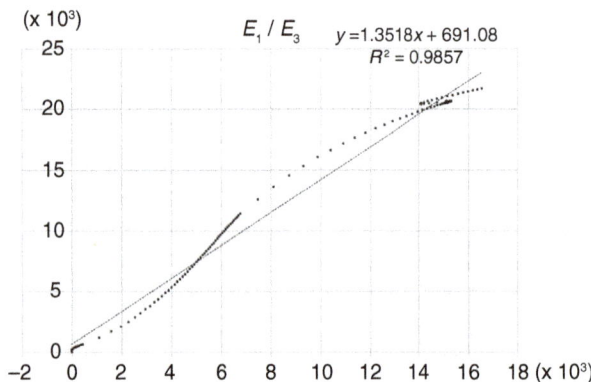

Figure 16

Correlations between parameters' sensitivities for a parallel sample, example of Young's moduli for upstream pressure.

2 parameters for each sample, which were the "axial" intrinsic permeability and the "normal" Young's modulus.

In comparison to the axial pulse test, the aim of the radial pulse test is to simultaneously identify both the axial and normal intrinsic permeabilities and Young's moduli on the same sample. The sensitivity analysis and the assessment of the correlations between these 4 parameters show that such identification is possible and that the only correlation that can be found is between both Young's moduli for a parallel sample (for upstream pressure). Nevertheless, the sensitivity analysis showed that only K_1^{int} could be identified for the transverse samples, whereas both K_1^{int} and K_3^{int} could be identified when considering parallel samples. The intrinsic permeabilities are the paramount parameters for the pulse test; therefore, we decided to focus on parallel samples for the identification of both intrinsic permeabilities. The price to pay is that the identification of both Young's moduli is more delicate for the parallel samples due to the strong correlation that exists between both parameters for such samples, more so because the identification favours the upstream pressure measurements (because of effects of climate control on downstream pressure) and the strong correlation concerns these upstream measurements. As a consequence, this correlation prevents the gradient-based inverse method being used for a wide range of values of the Young's moduli. In such a case, as for the inversion of the axial pulse test, the identification method presented previously must be coupled to a preliminary direct search method. The results of the direct search method and the previous knowledge on the Young's moduli are introduced in the gradient-based inverse method to define the *a priori* information as well as the initial values and tight boundaries on these parameters. Regarding the Young's moduli, the gradient-based inversion is then an optimisation or correction measure aiming at identifying the most accurate values in the range of values assessed by the direct search method.

Finally, the parameters to be identified can be noted as follows:

$$\mathbf{c}^{\textbf{identified}} = \left\{ K_1^{int\,opt}, K_3^{int\,opt}, E_1^{opt}, E_3^{opt} \right\}$$

3.4 Application

The inverse method combined with numerical modeling of the radial pulse test was applied to the interpretation of laboratory tests on 2 hollow cylindrical samples on Meuse/Haute-Marne claystones, a soft rock exhibiting transverse isotropy. Both samples were parallel samples

to allow the identification of both permeabilities in the directions parallel and normal to isotropy planes. The physical data, dimensions and tests conditions are provided in Table 1.

Tables 2 and 3 give the values of the identified parameters in each case as well as the initial and *a priori* information values used to initiate the inverse analysis. Figures 17 and 18 present the experimental and fitted numerical curves obtained after the identification for both samples. One can observe the oscillations due to climate control in the experimental downstream pressure curves. The downstream pressure is measured using an external pressure sensor, while the upstream pressure is measured using an internal pressure sensor placed in the cell, which is less subjected to the climate control effects. As a consequence, the fitting is quite fair for the upstream pressure and better than for the downstream pressure. Indeed, we gave priority in the inversion method to the upstream pressure because its measurement is easier, more accurate and more reliable than the measurement of downstream pressure. This priority was achieved through the weighting parameters ω_i on measurements in the cost-functional (22a). For the upstream pressure, both the transient evolution and the final level are well reproduced numerically, indicating that both the intrinsic permeabilities and the Young's moduli were fairly identified, respectively.

For both samples, a meaningful set of parameters was identified. The values of permeabilities and Young's moduli obtained after inversion are consistent with the intrinsic permeability values, and tend to lie in the upper

TABLE 1

Physical data, dimensions and test conditions

Sample	44368-2	44368-3
Inner diameter (m)	0.0095	0.0095
Outer diameter (m)	0.04935	0.04969
Height (m)	0.05106	0.05005
Depth (m) NGF	−123.62	−123.62
Porosity	0.18	0.18
Orientation	Parallel to bedding	Parallel to bedding
Confinement (MPa)	4	4
Initial pore pressure (MPa)	1.00	1.02
Pore pressure increment (MPa)	0.41	0.40

range of values for the considered claystone. This result could be explained by the development of micro-cracks during drilling of the inner hole, which generates a damaged zone around this inner hole, increasing the permeability in that zone. For both samples, it appears that the anisotropy of transfer coefficients (permeabilities) is quite low, with a K_1^{int}/K_3^{int} ratio of 1.10 and 1.18 for samples 44368-2 and 44368-3, respectively. On the contrary, the mechanical anisotropy is much more marked, with a E_1/E_3 ratio of 2.14 for sample 44368-2 and a ratio of 1.28 for sample 44368-3. This result is similar to the one obtained for axial pulse tests (Giot *et al.* [13]). One can note that the anisotropy is less expressed for sample 44368-3 than for sample 44368-2 (*Tab. 4*).

TABLE 2

Results of the inversion on pulse test 44368-2

	Initial value	*A priori* value	Boundaries	Identified value
K_1^{int} (m^2)	8.0×10^{-20}	2×10^{-20}	1×10^{-23} / 1×10^{-18}	1.67×10^{-19}
K_3^{int} (m^2)	8.0×10^{-20}	2×10^{-20}	1×10^{-23} / 1×10^{-18}	1.51×10^{-19}
E_1 (Pa)	4.5×10^9	4.5×10^9	1×10^9 / 12×10^9	3.25×10^9
E_3 (Pa)	3.0×10^9	3×10^9	1×10^9 / 12×10^9	1.52×10^9
Nb iterations	11			

TABLE 3

Results of the inversion on pulse test 44368-3

	Initial value	*A priori* value	Boundaries	Identified value
K_1^{int} (m^2)	1.67×10^{-20}	2×10^{-20}	1×10^{-23} / 1×10^{-18}	1.44×10^{-19}
K_3^{int} (m^2)	1.51×10^{-20}	2×10^{-20}	1×10^{-23} / 1×10^{-18}	1.22×10^{-19}
E_1 (Pa)	3.25×10^9	4.5×10^9	1×10^9 / 12×10^9	6.77×10^9
E_3 (Pa)	1.52×10^9	3×10^9	1×10^9 / 12×10^9	5.27×10^9
Nb iterations	10			

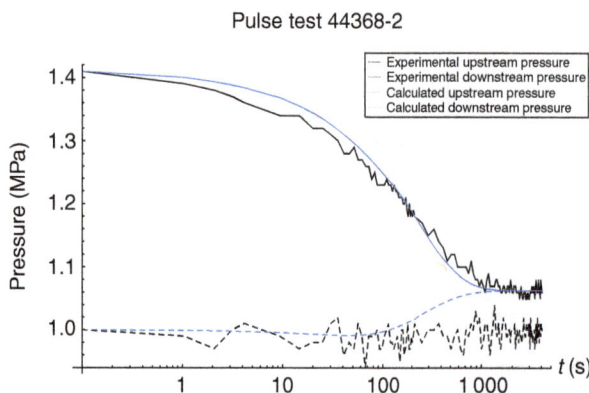

Figure 17

Pulse test 44368-2: comparison between experimental and fitted reservoir pressures.

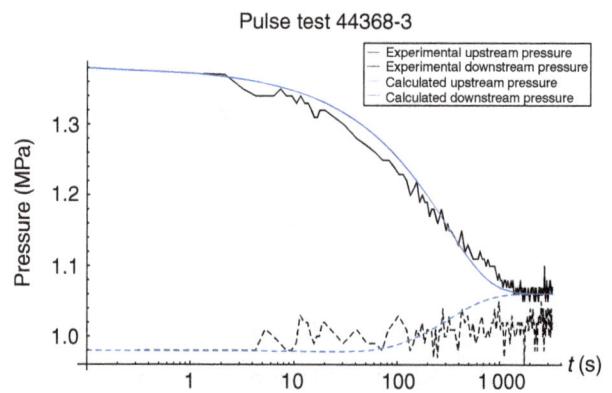

Figure 18

Pulse test 44368-3: comparison between experimental and fitted reservoir pressures.

TABLE 4

Comparison of the parameters inferred for both tests

Sample	44368-2	44368-3
Orientation e_y, e_z, e_x		
K_1^{int} (m^2)	1.67×10^{-19}	1.44×10^{-19}
K_3^{int} (m^2)	1.52×10^{-19}	1.22×10^{-19}
E_1 (GPa)	3.25	6.77
E_3 (GPa)	1.52	5.27

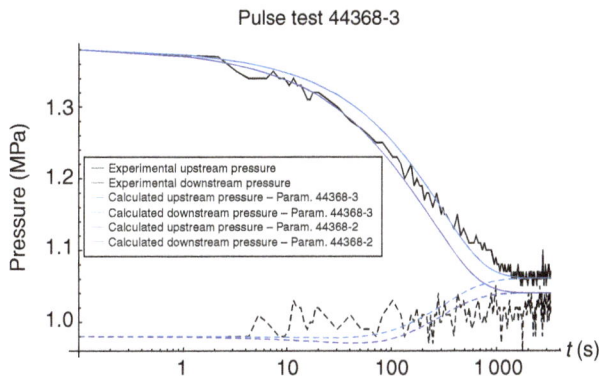

Figure 19

Pulse test 44368-3: comparison between experimental, fitted reservoir pressures and reservoir pressures calculated with the set of parameter identified on sample 44368-2.

The difference between the values identified for the intrinsic permeabilities and the Young's moduli between both samples may be explained by the heterogeneity of the rock mass, even if the samples are taken from the same borehole. This difference is not a consequence of the inversion method, as shown in Figure 19. This figure presents a comparison between the experimental curves for sample 44368-3, the calculated curves with the set of parameters identified on sample 44368-3 and the calculated curves with the set of parameters identified on sample 44368-2. This figure shows that the inverse procedure on sample 44368-3 gives a better fitting; the difference on the values of parameters identified is due to experimental data rather than to the identification process.

Let us note that for sample 44368-2, the inversion was conducted in 2 steps as follows: first, identification of both intrinsic permeabilities, with fixed values of the

Young's moduli; then a second identification with the 4 parameters, considering for the permeabilities the values identified through the first identification. For sample 44368-3, we chose the parameter set identified on sample 44368-2 as the initial iterate because both samples were extracted from the same cell, and were thus geographically quite similar. In both cases, the value of the cost-functional was significantly reduced, by one order of magnitude for sample 44368-2, and half an order of magnitude for sample 44368-3 (for which the initial set was closer to the minimum because it integrated the results of the identification on sample 44368-2). The convergence was obtained on a condition on the gradient of the cost-functional, where the values of the parameters did not change significantly (less than 1% of the value) from one iteration to the next. To ensure the uniqueness of the solution, we tried different initial sets, but always in the same range of values, and verified that the same minimum was given by the identification process. This is mainly allowed by the first direct search step of the identification process, which always provides sets of parameters rather similar to the final set (corresponding to the minimum of the cost-functional (22a) used for the gradient based inversion algorithm).

Finally, let us add that a slight modification of the data (on the order of 10%) does not modify the values of the identified parameters, which means that the inversion method is robust. This is due to the consideration of the weighting parameters ω_i on the measurements in the cost-functional. Indeed, these terms smooth the cost-functional, taking into account the uncertainties or inaccuracies on the measurements. This process has always proven to be efficient on the axial pulse test and overcoring test, among others.

CONCLUSION

The interpretation of the axial pulse test on claystones, considering poromechanical couplings, underlined the anisotropy of this rock and the necessity to take anisotropy into account for the mechanical and transfer properties. Incorporating the transverse isotropy of the claystone in the poromechanical back-analysis of the test showed that 3D transverse isotropic modeling provides much more meaningful values of the identified parameters than the 2D isotropic modeling, particularly for the mechanical parameters. Nevertheless, these interpretations suffered from a lack a consistency on some results, which was attributed to the heterogeneity of the rock varying from one sample to the other. To overcome this problem, it was required to measure the 2 Young's moduli and 2 intrinsic permeabilities on the same sample, and the radial pulse test was developed to achieve this aim.

The present paper addresses the radial pulse test, which consists in generating a sudden increase of pressure in the central hole of a hollow cylinder and measuring the evolution of pressure in the upstream (inner) reservoir and downstream (outer) reservoir. The interpretation is based on poromechanical numerical modeling, taking into account the transverse anisotropy's effects on the mechanical and transfer properties. The poromechanical modeling is coupled with an inverse method, as well as the poromechanical transverse isotropic constitutive law, with both being implemented in the finite element code Code_Aster (Edf).

The sensitive analysis performed on some parameters, including the intrinsic permeabilities and Young's moduli in directions both parallel and normal to the isotropy planes, showed that both intrinsic permeabilities could be identified during a single test only if parallel samples (cylindrical samples with the axis parallel to the isotropy planes) were considered. The identification of Young's moduli is more delicate due to a strong correlation between these 2 parameters and requires an adaptation of the inversion method. Indeed, the gradient-based inverse method must be coupled to a preliminary direct search method to reduce the range of variation in the values of the Young's moduli.

The identification method was used on laboratory tests on Meuse/Haute-Marne argillites, on parallel cylindrical samples with the axis parallel to the isotropy planes. The results are quite fair and encouraging, showing a clearer anisotropy on the mechanical parameters than on the transfer parameters.

In future works, the radial pulse test could be coupled to the classical axial pulse test on samples of the same origin, to characterize the effects of damage on rock permeability. Indeed, the configuration of the radial pulse test, i.e., hollow cylinder, is close to the configuration of gallery excavation and can thus generate some type of excavation damaged zone at small scales. Comparing the values of the permeabilities identified on both pulse tests could allow identification of the effects of this damaged zone on permeability. This methodology could be used on isotropic samples, at first.

The next step of the present work is to identify some other poromechanical parameters to obtain a complete set of parameters for the claystone. To achieve this aim, other types of tests have to be considered and back-analysed using the inverse method presented in this paper, coupled with numerical modeling and accounting for the poromechanical transverse isotropic model. Drying tests allow identification of transfer parameters in partially saturated conditions (relative permeability and sorption isotherm, to be more specific), while the œdometric test allows the identification of Biot coefficients.

ACKNOWLEDGMENTS

This work was supported by ANDRA, Scientific Division, 1/7 rue Jean Monnet, 92290 Châtenay-Malabry, France. The author expresses sincere gratitude to the organization.

REFERENCES

1 Brace W.F., Walsh J.B., Frangos W.T. (1968) Permeability of granite under high pressure, *J. Geophys. Res.* **73**, 2, 2225-2236.

2 Hsieh P.A., Tracy J.V., Neuzil C.E., Bredehoeft J.D., Silliman S.E. (1981) A transient laboratory method for determining the hydraulic properties of tight rocks. I. Theory, *Int. J. Rock Mech. Min. Sci. Geomech. Abstr.* **18**, 245-252.

3 Neuzil C.E., Cooley C., Silliman S.E., Bredehoeft J.D., Hsieh P.A. (1981) A transient laboratory method for determining the hydraulic properties of tight rocks. II. Application, *Int. J. Roch Mech. Min. Sci. Geomech. Abstr.* **18**, 253-258.

4 Homand F., Giraud A., Escoffier S., Koriche A., Hoxha D. (2004) Permeability determination of a deep argillite in saturated and partially saturated conditions, *Int. J. Heat Mass Trans.* **47**, 3517-3531.

5 Selvadurai A.P., Letendre A., Hekimi B. (2001) Axial flow hydraulic pulse testing of an argillaceous limestone, *Environ. Earth Sci.* **64**, 2047-2058.

6 Wang H.F. (2003) *Theory of linear poroelasticity*, Princeton University Press, Princeton and Oxford.

7 Adachi J.I., Detournay E. (1997) A poroelastic solution of the oscillating pore pressure method to measure permeabilities of tight rocks, *Int. J. Rock Mech. Min. Sci. Geomech. Abstr.* **34**, 3-4, paper 062.

8 Walder J., Nur J. (1986) Permeability measurement by the pulse decay method: effect of poroelastic phenomena and non linear pore pressure diffusion, *Int. J. Rock Mech. Min. Sci. Geomech. Abstr.* **23**, 3, 225-232.

9 Giot R., Giraud A., Auvray C., Homand F., Guillon T. (2011) Fully coupled prormechanical back analysis of the pulse test by inverse method, *Int. J. Numer. Anal. Methods Geomech.* **35**, 3, 329-359.

10 Zhang C., Rothfuchs T. (2004) Experimental study of the hydro-mechanical behavior of the Callovo-Oxfordian argillite, *Appl. Clay Sci.* **26**, 325-336.

11 Cariou S., Duan Z., Davy C., Skoczylas F., Dormieux L. (2012) Poromechanics of partially saturated Cox argillite, *Appl. Clay Sci.* **56**, 36-47.

12 Marschall P., Horseman S., Gimmi T. (2005) Characterisation of Gas Transport Properties of the Opalinus Clay, a Potential Host Rock Formation for Radioactive Waste Disposal, *Oil Gas Sci. – Technol. Rev. IFP* **60**, 1, 121-139.

13 Giot R., Giraud A., Guillon T., Auvray C. (2012) Three-dimensional poromechanical back analysis of the pulse test accounting for transverse isotropy, *Acta Geotechnica* **7**, 151-165.

14 Giot R., Giraud A., Homand F. (2005) Three Dimensional modelling of stress relaxation tests with finite element in anisotropic clayey medium: direct problem and back analysis, *Geotech. Geol. Eng.* **24**, 919-947.

15 Alarcon-Ruiz L., Brocato M., Dal Pont S., Feraille A. (2010) Size Effect in Concrete Intrinsic Permeability Measurements, *Transport Porous Med.* **85**, 541-564.

16 Dal Pont S., Schrefler B.A., Ehrlacher A. (2005) Intrinsic Permeability Evolution in High Temperature Concrete: An Experimental and Numerical Analysis, *Transport Porous Med.* **60**, 43-74.

17 Davy C., Skoczylas F., Barnichon J.-D., Lebon P. (2007) Permeability of macro-cracked argillite under confinement: Gas and water testing, *Phys. Chem. Earth Parts A/B/C* **32**, 667-680.

18 Noiret A., Giot R., Bemer E., Giraud A., Homand F. (2011) Hydromechanical behavior of Tournemire argillites: measurement of the poroelastic parameters and estimation of the intrinsic permeability by œdometric tests, *Int. J. Numer. Anal. Methods Geomech.* **35**, 4, 496-518.

19 Cheng A.H.-D. (1997) Material Coefficient of Anisotropic Poroelasticity, *Int. J. Rock Mech. Min. Sci. Geomech. Abstr.* **34**, 2, 199-205.

20 Abousleiman Y., Cheng A.H.-D., Cui L., Detournay E., Roegiers J.-C. (1996) Mandel's problem revisited, *Geotechnique* **46**, 2, 187-195.

21 Abousleiman Y., Cui L. (1998) Poroelastic solutions in transversely isotropic media for wellbore and cylinder, *Int. J. Solids Structures* **35**, 34-35, 4905-4929.

22 Cui L., Abousleiman Y., Cheng A.H.-D., Roegiers J.-C. (1996) Anisotropy effect on one-dimensional consolidation, *ASCE, EM 11th conference*, FT Lauderdale, FL, May 19-22, pp. 471-474.

23 Cui L., Abousleiman Y., Roegiers J.-C. (1998) Solutions for hollow cylinders in transversely isotropic porous materials, *Int. J. Rock Mech. Min. Sci.* **35**, 4-5, 635-636.

24 Cui L., Cheng A.H.D., Kaliakin V.N., Abousleiman Y., Roegiers J.-C. (1996) Finite element analysis of anisotropic poroelasticity: A generalized Mandel's problem and an inclined borehole problem, *Int. J. Numer. Anal. Methods Geomech.* **20**, 6, 381-401.

25 Kanj M., Abousleiman Y. (2005) Porothermoelastic analyses of anisotropic hollow cylinders with applications, *Int. J. Numer. Anal. Methods Geomech.* **29**, 2, 103-126.

26 Ekbote S., Abousleiman Y. (2006) Porochemoelastic solution for an inclined borehole in a transversely isotropic formation, *J. Eng. Mech.* **137**, 7, 754-763.

27 Lewis R.W., Schrefler B.A., Simoni L. (1991) Coupling *versus* uncoupling in soil consolidation, *Int. J. Numer. Anal. Methods Geomech.* **15**, 533-548.

28 Coussy O. (2004) *Poromechanics*, John Wiley and Sons, Paris, ISBN 0-470-84920-7.

29 Detournay E., Cheng A.H.-D. (1993) Fundamentals of poroelasticity, in *Comprehensive Rock Engineering: Principles, Practice and Projects, Vol. II, Analysis and Design Method*, Chapter 5, Fairhurst C. (ed.), Pergamon Press, Oxford, pp. 113-171.

30 Schrefler B.A., Gawin D. (1996) The effective stress principle: incremental or finite form? *Int. J. Numer. Anal. Methods Geomech.* **20**, 785-814.

31 Cheng A.H.D. (2008) Abousleiman Y. Intrisic poroelasticity constants and a semilinear model, *Int. J. Numer. Anal. Methods Geomech.* **32**, 7, 803-831.

32 Chavant C. (2001) Modélisations THHM, généralités et algorithmes, *Official Documentation of Code_Aster*, R7.01.10a, www.code-aster.org.

33 Chavant C., Granet S. (2005) Modèles de comportement THHM, *Official Documentation of Code_Aster*, R7.01.11b, www.code-aster.org.

12

Validation of a Liquid Chromatography Tandem Mass Spectrometry Method for Targeted Degradation Compounds of Ethanolamine Used in CO_2 Capture: Application to Real Samples

Vincent Cuzuel[1], Julien Brunet[1], Aurélien Rey[1], José Dugay[1]*, Jérôme Vial[1], Valérie Pichon[1] and Pierre-Louis Carrette[2]

[1] LSABM, UMR CBI 8231, ESPCI – CNRS, 10 rue Vauquelin, 75005 Paris - France
[2] IFP Energies nouvelles, Rond-point de l'échangeur de Solaize, BP 3, 69360 Solaize - France
e-mail: vincent.cuzuel@espci.fr - julien.brunet@espci.fr - aurelien.rey@espci.fr - jose.dugay@espci.fr - jerome.vial@espci.fr
valerie.pichon@espci.fr - p-louis.carrette@ifpen.fr

* Corresponding author

Résumé — **Validation d'une méthode de chromatographie en phase liquide couplée à la spectrométrie de masse en tandem pour des composés de dégradation ciblés de l'éthanolamine utilisée dans le captage du CO_2 : application à des échantillons réels** — Dans le domaine des émissions de gaz à effet de serre, une approche prometteuse consiste à capter et stocker le CO_2. Cependant la plupart des procédés mis en œuvre sont basés sur l'utilisation de solutions d'amines qui sont susceptibles de se dégrader et produire des composés potentiellement dangereux pour l'homme et l'environnement. Il y a donc un véritable besoin de méthodes d'analyse pour identifier et quantifier ces produits. La monoéthanolamine est choisie comme composé modèle pour les amines utilisées lors du captage du CO_2.

Une méthode de chromatographie en phase liquide couplée à la spectrométrie de masse en tandem a été développée et validée pour la quantification de six produits de dégradation de la monoéthanolamine (Glycine, N-(2-hydroxyéthyle)glycine, N-glycylglycine, bicine, N,N'-bis-(2-hydroxyéthyle) urée et diéthanolamine) qui ont été systématiquement retrouvés avec une méthode LC-MS en mode « scan » dans des échantillons réels issus de procédés de captage du CO_2 en vue de son stockage ultérieur. La principale difficulté de cette étude et son originalité se situent dans la stratégie développée pour surmonter les difficultés liées à la complexité de la matrice qui est un mélange d'eau et d'amine (70/30) : l'utilisation combinée de composés deutérés comme étalons internes et d'une approche chimiométrique récente pour valider la méthode, *i.e.* le profil d'exactitude. Pour cinq composés, il a été possible de valider la méthode avec une limite d'acceptabilité de 20 %. Cette méthode a ensuite été appliquée avec succès à l'analyse d'échantillons réels issus de pilotes et d'expériences de laboratoire.

Abstract — *Validation of a Liquid Chromatography Tandem Mass Spectrometry Method for Targeted Degradation Compounds of Ethanolamine Used in CO_2 Capture: Application to Real Samples* — *In the field of greenhouse gas emission, a promising approach consists in CO_2 storage and capture. However most of the processes are based on amine solutions which are likely to* *degrade and produce potentially harmful compounds. So there is a need for analytical methods to identify and quantify these products. Monoethanolamine was used as a model compound for the amines used for CO₂ capture.*

A liquid chromatography tandem mass spectrometry method was developed and validated for the quantification of six products of degradation of monoethanolamine (Glycine, N-(2-hydroxyethyl) glycine, N-glycylglycine, bicine, N,N'-bis-(2-hydroxyethyl) urea (BHE Urea), and diethanolamine) that were systematically detected with a LC-MS Scan method in real samples from CO₂ capture and storage processes. The main difficulty of this study and its originality ly in the strategy developed to overcome the complexity of the matrix which is a mix of water and amine (70/30): the combined use of deuterated internal standards and a recent chemiometric approach to validate the method, i.e. the accuracy profile. For five compounds, it was possible to validate the method with acceptance limits of 20%. This method was then successfully applied to real samples from pilot plant and lab-scale experiments.

HIGHLIGHTS

- An analytical method based on LC/MS-MS was developed and validated using the accuracy profile;
- 6 priority compounds issued from MEA degradation were quantified in pilot plant and lab-scale experiments samples;
- Use of deuterated internal standards was found to be relevant to overcome the complexity of the matrix.

INTRODUCTION

CO₂ capture and storage is one of the promising technol ogies to reduce greenhouse gas emissions. To be used, this technology needs economic but also environmental acceptance. In some processes, amines are known to react with flue gas components (O_2, CO_2, NO_x, SO_x, etc.) to form degradation products, and some of them could be potentially dangerous to humans or environment according to their toxicity and their concentration. These products could be discharged to the atmosphere essentially with treated flue gas. Such amine degradation causes also amine loss, therefore additional costs, and can lead to corrosion [1], solid deposit [2] and foaming. Therefore it is necessary to list all the degradation products of amines used in CO_2 capture, to understand their formation and to study their toxicity. Alkanolamines are the most studied molecules. The benchmark molecule is MonoEthanolAmine (MEA) [3-8], but some other amines were studied: mainly DiEthanolAmine (DEA) [9-11], MethylDiEthanolAmine (MDEA) [12-14], PiperaZine (PZ) [15] and 2-Amino-2-MethylPropan-1-ol (AMP) [16]. Some alkyl amines and polyamines were studied [17-20]. The identification of amine degradation products and their mechanisms of formation were recently reviewed [21].

Amine degradation in post-combustion CO_2 capture is a main problem because of its consequences on process units and the potential impact of degradation products on environment. Therefore, amine degradation study is a key point for CO_2 capture acceptance. This is the reason why methods are required to detect, identify and quantify degradation products. DALMATIEN (Degradation of Amines in Liquid Matrix and Analysis: Toxicity or Innocuousness for ENvironment?) is an industrial research project dedicated to post-combustion. The goal of this project is to list all the degradation products of amines used in CO_2 capture, to understand their formation and to study their toxicity. A recent article [22] showed the presence of ten new degradation products (pyrazine and nine alkyl derivatives) using an analytical method based on Head Space Solid Phase Micro Extraction (HS-SPME) and Gaz Chromatograhy Mass Spectrometry (GC–MS). To go further into the analysis of degradation products, the study focused on six other compounds (Glycine, N-(2-hydroxyethyl)glycine, N-glycylglycine, bicine, N,N'-bis-(2-hydroxyethyl) urea, and diethanolamine) which were systematically detected with a LC-MS Scan method in real samples from *IFP Energies nouvelles* (IFPEN) pilot plant and lab-scale experiments. As those six compounds were considered as priority compounds by people in charge of the CO_2 capture process (IFPEN) and as most of them were not compatible with GC-MS analysis, a liquid chromatography approach had to be developed and validated to determine if they can be quantified in such a complex matrix. The use of a porous graphitic carbon column was found to be relevant in this study according to the complexity of the matrix and the high range of polarity of compounds. Validation was carried out using the total error concept and the accuracy profile which will be detailed in Section 1.4.

TABLE 1
Molecular weight, formula, retention time and MRM parameters of compounds of interest

Compound	M (g/mol)	Formula	Retention time (min)	Parent ion (m/z)	Transition (m/z)	Collision energy (eV)
MEA	61.08	C_2H_7NO	1.8	46.3	30.4*	17
DEA	105.14	$C_4H_{11}NO_2$	1.8	106.0	88.2*	11
					70.4	13
DEA-d8	113.19	$C_4H_3D_8NO_2$	1.8	114.0	96.2*	12
					78.3	15
GlyGly	132.12	$C_4H_8N_2O_3$	2.3	133.1	76.4*	8
					115.2	5
Gly	75.07	$C_2H_5NO_2$	2.3	76.2	30.6*	10
					31.6	26
Gly-d5	80.10	$C_2D_5NO_2$	2.3	78.2	32.4*	14
					33.6	32
HEGly	119.12	$C_4H_9NO_3$	2.7	120.2	74.4*	12
					56.4	19
Bicine	163.17	$C_6H_{13}NO_4$	3.7	164.1	118.2*	14
					146.2	12
BHE Urea	148.16	$C_5H_{12}N_2O_3$	17.8	149.2	62.4*	11
					44.6	19

* Transition used for quantification.

1 MATERIAL AND METHODS

1.1 Chemicals and Reagents

MonoEthanolAmine (MEA), Glycine (Gly), N-(2-HydroxyEthyl) Glycine (HEGly), N-Glycylglycine (Glygly), Bicine, Oxazolidine, Piperazine, PyraZine (PZ), N,N′-Bis-(2-HydroxyEthyl) Urea (BHE Urea), N-(2-HydroxyEthyl) EthyleneDiamine (HEEDA), N, N′-Bis-(2-HydroxyEthyl) EthyleneDiAmine (BHEE-DA), DiMethylAmine (DMA), N-(2-HydroxyEthyl) ImidAzolidinone (HEIA), DiEthanolAmine (DEA) were purchased from *Sigma-Aldrich* (Saint-Quentin-Fallavier, France). DiEthanolAmine-d8 and Glycine-d5 were bought from *Eurisotop* (Saint-Aubin, France). Methanol and formic acid were purchased from *Carlo Erba Reagents* (Fontenay-sous-bois, France). Ultra pure water was produced using a Direct-Q UV 3 system (18.2 MΩ/cm) from *Millipore* (Molsheim, France).

1.2 LC-MS-MS Instrumentation and Conditions

Analyses were performed on a LC Thermo Scientific Dionex Ultimate 3000 (Analytical Autosampler WPS-3000SL, Quaternary Analytical Pump LPG-3400SD) coupled with a MS Thermo Scientific TSQ Quantum Access MAX (HESI-II source) (*Thermo Scientific*, Illkirch, France). It was used in positive mode, probe in position C, electrospray voltage of 2 500 V and capillary temperature of 200°C. The sheath gas was at a flow rate of 40 mL/min and the auxiliary gas at 8 mL/min. Data were acquired in MRM (Multiple Reaction Monitoring) mode with Xcalibur (Thermo software). Transitions and collision energy were optimized by infusion of each individual product (*Tab. 1*).

A Thermo HyperCarb column (PGC) 150 mm × 3 mm, 5 μm-particles (*Thermo Scientific*, Illkirch, France), an Agilent Polaris 3 Amide-C18 column 100 mm × 3 mm, 3 μm-particles (*Agilent Technologies*, Massy, France) and a Waters Symmetry shield RP18

TABLE 2

Composition of synthetic samples (concentrations given in mg/L)

Calibration	Gly	DEA	HEGly	GlyGly	BHE Urea	Bicine	DEA d8	Gly d5	MEA
Mix 1	0.2	0.2	0.5	0.01	0.5	0.02	1	1	-
Mix 2	0.5	0.5	1	0.05	1	0.05	1	1	-
Mix 3	1	1	10	0.1	5	0.1	1	1	-
Mix 4	5	5	25	0.2	10	1	1	1	-
Validation	Gly	DEA	HEGly	GlyGly	BHE Urea	Bicine	DEA d8	Gly d5	MEA
Level A	0.1	0.1	0.75	0.02	0.25	0.01	1	1	300
Level B	0.3	0.3	5	0.075	0.75	0.03	1	1	300
Level C	0.75	0.75	15	0.15	2	0.075	1	1	300
Level D	4	4	30	0.25	7	0.5	1	1	300

column 150 mm × 2.1 mm, 3.5 µm-particules (*Waters*, Saint-Quentin-en-Yvelines, France) were studied to determine the most relevant stationary phase to conduct the chromatographic separation. For both Polaris and Symmetry shield columns, the mobile phase was water with 0.1% formic acid at a flow rate of 350 µL/min. As for the PGC column, the mobile phase was a mixture of (A) water with 0.1% formic acid and (B) methanol with 0.1% formic acid at a flow rate of 350 µL/min. A prerun rinse of 100% A for 8 min was performed, then the solvent gradient started at 100% A for 0 to 10 min before being changed to 80:20 (A:B v:v) in 8 min and held for 12 min (total duration of the gradient: 30 min). 5 µL of each sample are injected. The column was at room temperature, *i.e.* 22°C maintained by the lab air conditioning system.

1.3 Sample Preparation

The range of concentration for the validation was chosen from rough estimations of their concentration in real samples using external calibration. Matrix of real samples is made of a mix water and MEA (70:30 v:v), so they are 1 000-fold diluted before injection to avoid irreversible contamination of the mass spectrometer.

Two types of samples were prepared: mix for calibration and levels for validation. The four mixes for calibration were prepared in pure water with various concentrations of the six priority compounds and deuterated Internal Standard (ISTD). Four synthetic samples, with 0.3 g/L of MEA to mimic real samples 1 000-fold diluted, were prepared with the compounds and deuterated ISTD. Table 2 provides the concentrations used for compounds and ISTD for all the samples used. Five sets of samples were prepared independently on five different days. For each one of them, preparations of calibration mix and of levels for validation were performed independently of each other.

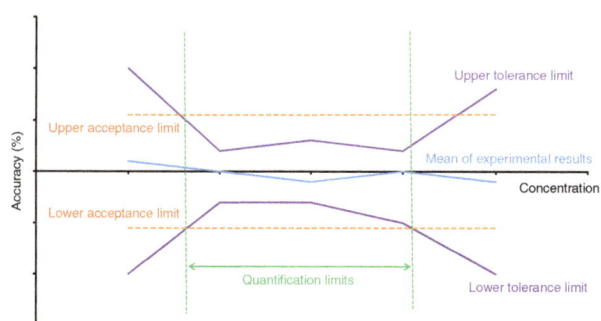

Figure 1

Example of an accuracy profile.

1.4 Validation Strategy

Validation was performed using the total error concept and the accuracy profile [23-29]. This original statistical approach was successfully applied in various contexts [30-32]. For example, methods for neurotoxic 1-2-amino-3-methylaminopropionic acid (BMAA) detection and quantification in complex matrix were validated

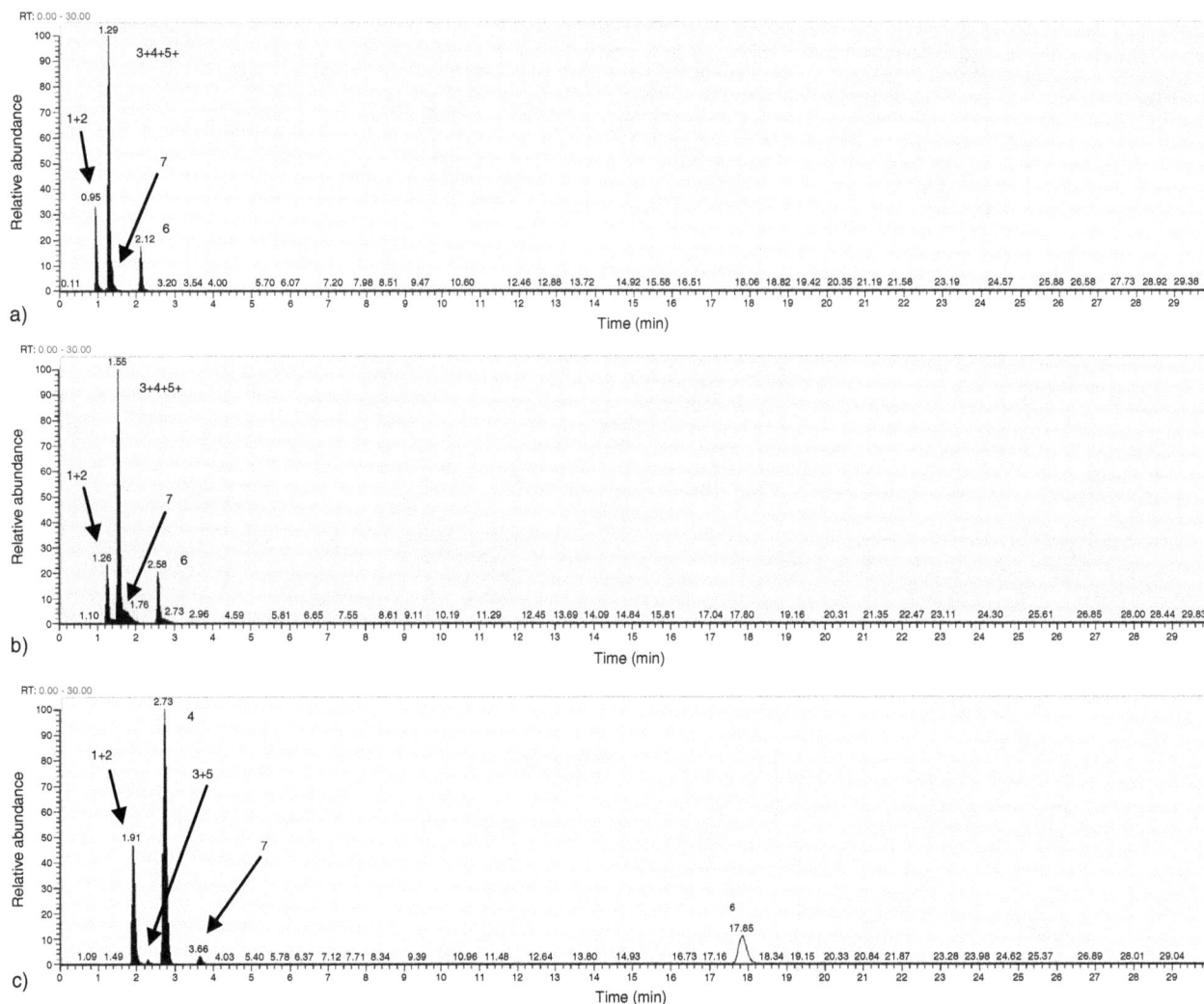

Figure 2

Comparison of chromatographic separation on a) polaris C18, b) symmetry shield RP 18, and c) PGC columns of a synthetic sample: MEA 300 mg/L, DEA 5 mg/L, Gly 5 mg/L, HEGly 25 mg/L, Glygly 0.2 mg/L, BHE Urea 10 mg/L, Bicine 1 mg/L.

using this kind of profile [33]. Those promising results on complex matrix led us to apply it to the LC/MS-MS analysis of degradation products on water/MEA matrices used for CO_2 capture and storage processes.

Validation aims at establishing, based on experimental results, if the performances of the method are compliant with its requirement. So it requires the evaluation of both the trueness and of the intermediate precision, *i.e.* the precision in the same laboratory, under different conditions (*e.g.* different days or different solvents or different apparatus or different operators). The combination of trueness and precision is called total error. In our study, the validation experiments were carried out on five series (on five different days by two different operators) and in conditions as close as possible to those

that will be met during the routine analysis. Every day, new samples were prepared according to the procedure described in Section 1.3. Each sample was analyzed in triplicates using material and methods previously described. For each set of samples and each replicate, the concentration of target compounds is determined using the calibration and internal standards. Those calculated concentrations are used to know trueness and precision of measures.

The accuracy profile shows the β expectation tolerance interval of the analytical method. The upper β tolerance and the lower β tolerance calculated are both plotted at each concentration level of the validation standards and take into account their estimated intermediate standard deviation and their bias [24, 25].

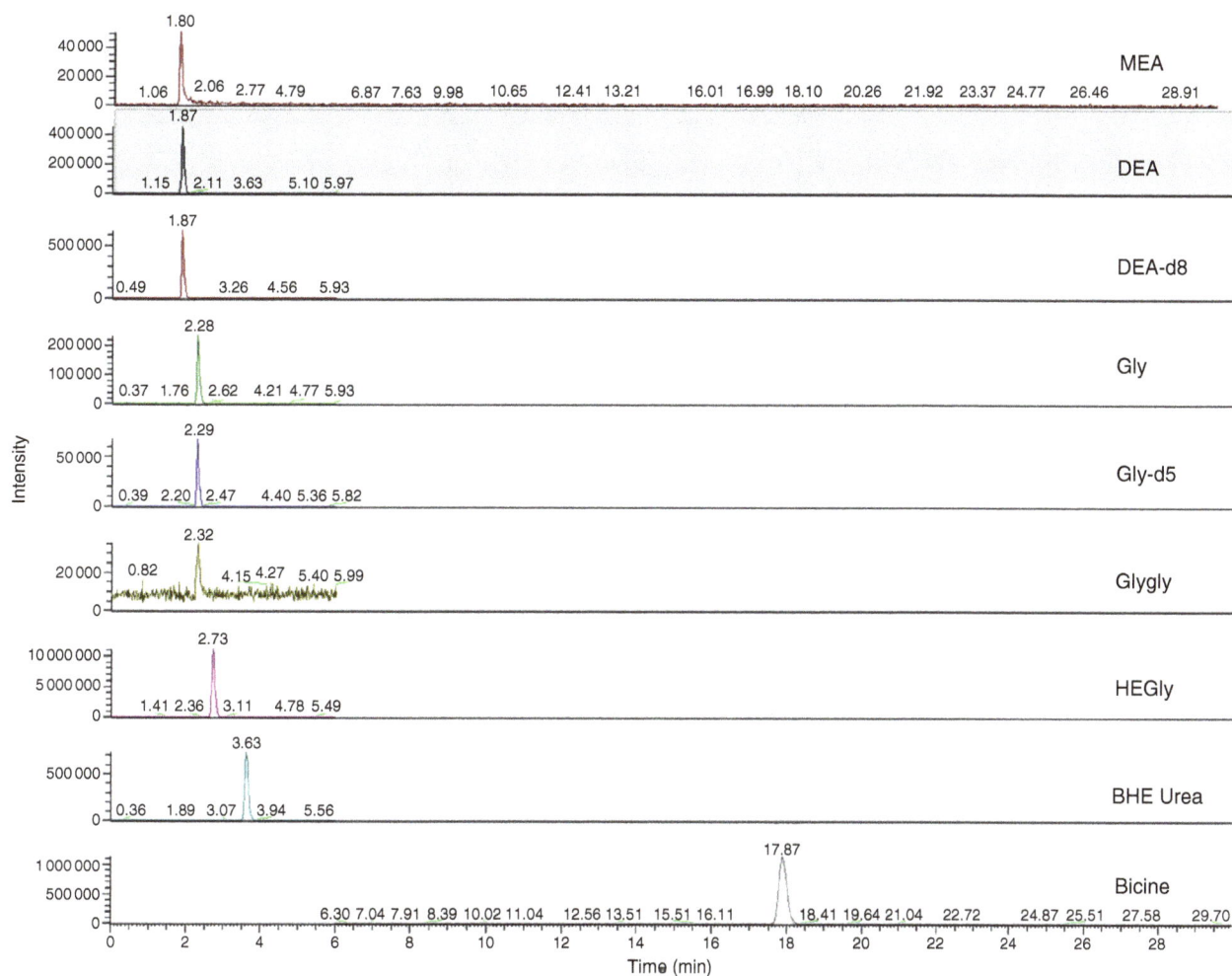

Figure 3

Chromatogram LS-ESI-MS of a synthetic sample (mix 4): MEA 10 mg/L, DEA 5 mg/L, DEA-d8 1 mg/L, Gly 5 mg/L, Gly-d5 1 mg/L, HEGly 25 mg/L, Glygly 0.2 mg/L, BHE Urea 10 mg/L, Bicine 1 mg/L – MRM transition of target compounds.

The β expectation tolerance limits were set at 80% probability level, which corresponds to an interval which will contain a future result 4 times out of 5. This graphical tool helps to define a region where each future result generated by the analytical procedure has 80% chance to fall, *i.e.* the region within four out of five future results will fall. In a word, the accuracy profile reflects directly the analytical procedure potential, and makes possible to appreciate the adequacy of different practices and to make decisions. So, the analytical method is said to be valid as long as the β interval is included between the lines representing the acceptability limits. Those limits depend on the complexity of the matrix and the study needs, in this case, and in the absence of regulatory requirements in this field, they were settled at 20%

according to previous studies about quantification by LC/MS–MS [28]. Relative error (%) is plotted *versus* the concentration. The mean of experimental results give us informations about a potential bias and limits of quantification of the analytical procedure are given by the intersection of the acceptability and tolerance limits. An example of an accuracy profile is given in Figure 1.

2 RESULTS

2.1 Chromatographic Separation

As it was previously described, real liquid samples are made of a mix of water and MEA (70:30 v:v) and were

TABLE 3

Figures of merit of the validation of the LC-MS/MS method

	Concentration (mg/L)	Truness	Precision	
		Relative bias (%)	Repeatability (RSD %)*	Intermediate precision (RSD %)
DEA	0.1	0.12	2.42	15.04
	0.3	0.04	1.96	6.87
	0.75	0.02	1.22	4.29
	4	0.03	1.27	4.42
Gly	0.1	-0.07	12.23	35.49
	0.3	-0.08	8.15	11.55
	0.75	-0.09	8.00	8.09
	4	-0.06	4.49	8.20
Glygly	0.02	-0.01	24.79	27.41
	0.075	0.03	12.76	12.76
	0.15	0.05	6.11	6.11
	0.25	-0.04	4.94	10.94
HEGly	0.75	0.35	2.94	10.98
	5	0.01	1.60	4.62
	15	-0.06	2.94	7.11
	30	0.03	2.83	7.21
BHE Urea	0.25	0.63	6.35	16.06
	0.75	0.08	5.82	10.40
	2	-0.09	5.07	5.28
	7	-0.01	4.32	9.64
Bicine	0.01	0.09	17.35	47.21
	0.03	-0.28	14.21	20.57
	0.075	-0.35	9.55	9.55
	0.5	-0.24	6.06	15.20

* RSD = Relative Standard Deviation

obtained from lab-scale experiments (sample A) and pilot plant (samples B and C) at IFPEN.

Various degradation products were listed within the framework of DALMATIEN. To perform an efficient separation despite the huge diversity of physico-chemical properties of those products, three columns were tested. The Porous Graphitic Carbon (PGC) column was chosen. Indeed, its original retention mode allowed to perform a reverse phase mode analysis, keeping a higher affinity for polar compounds than frequently-used C_{18} columns where compounds are not retained enough even with 100% water (Fig. 2) which correspond to the smallest elution strength possible in reverse phase. Separation window is six times wider with the PGC column than with the two C_{18} ones and allows us to conduct a better separation of complex real samples. Retention of the last compound is even so strong that a gradient was necessary to maintain a reasonable analysis time with this column. Although the method could seem not optimized

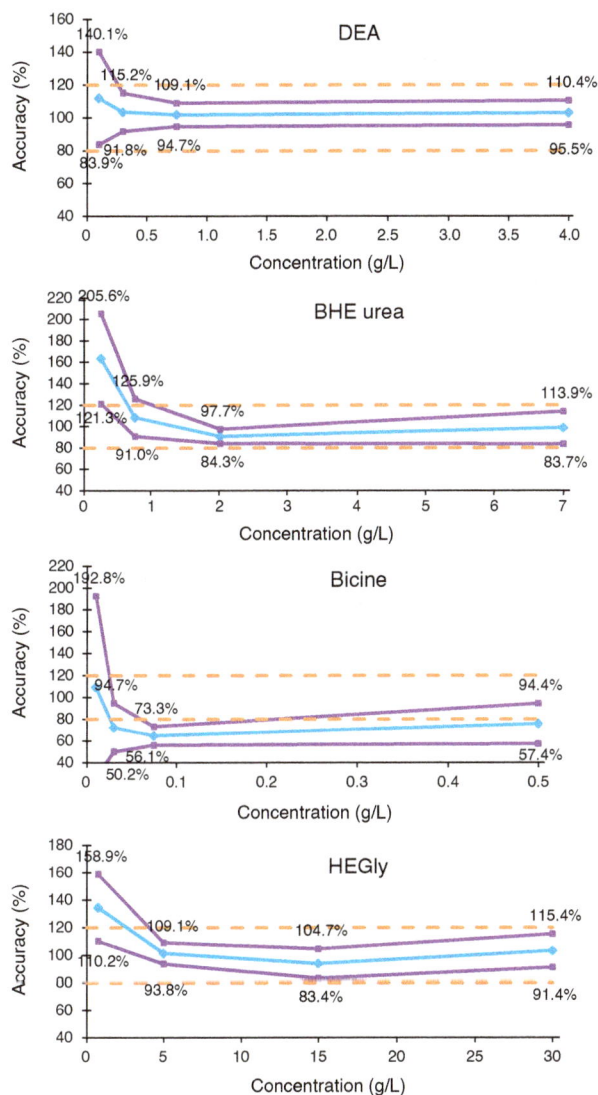

Figure 4

Accuracy profiles realized with the internal standard DEA-d8.

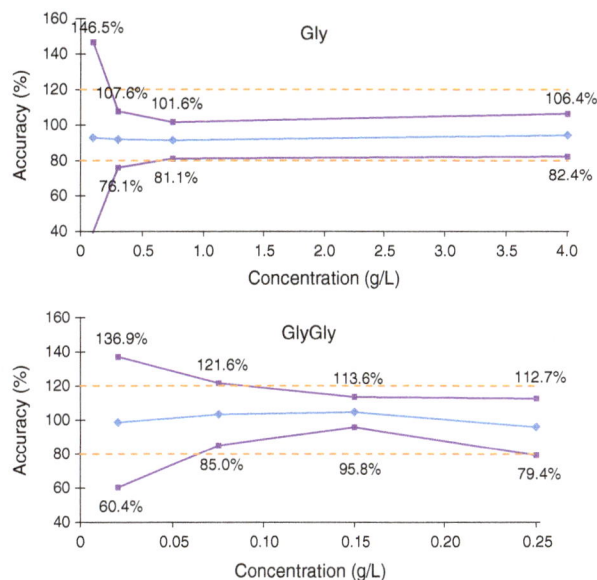

Figure 5

Accuracy profiles realized with the internal standard Gly-d5.

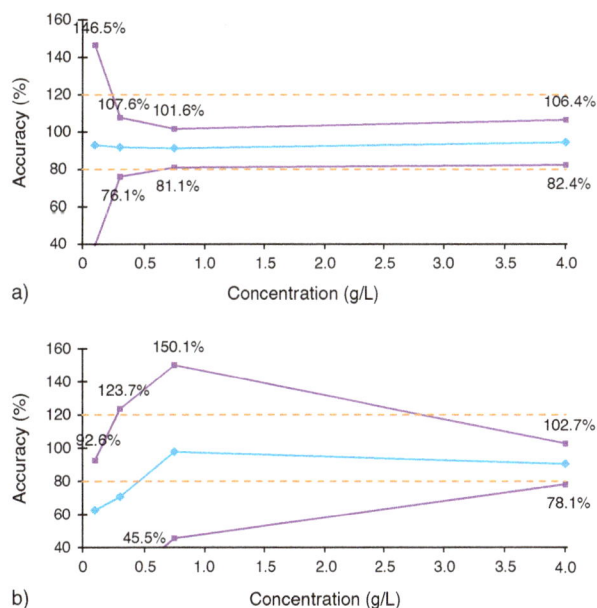

Figure 6

Comparison of accuracy profiles with a) ISTD calibration and b) external calibration for Gly.

looking at the gap between the two last eluted compounds, in fact it is. The running conditions were defined to avoid interfering compounds, present in real samples but not visible with MS-MS detection, could coelute with compound of interest and affect their ionization. Moreover, analyses can be conducted in water, at a pH between 0 and 14. This is important considering that pH of real samples is around 10. The high concentration of amine solvents raised many problems to perform analytical procedures. Samples had to be at least 1 000-fold diluted before injection to prevent the mass spectrometer from being polluted by MEA. Its

retention time is 1.8 min but MEA can be detected during the whole analysis, impacting on the MS ionization recovery. And, if the concentration is too high, adducts can be even formed and pollute the device. Moreover, even with an optimized separation, most

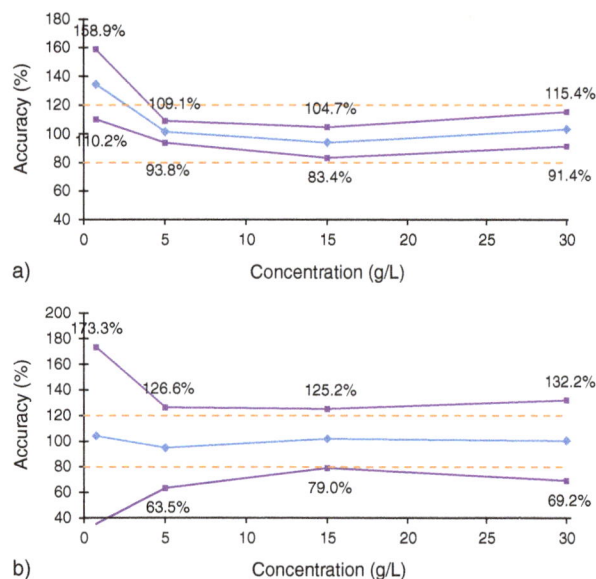

a)

b)

Figure 7

Comparison of accuracy profiles with a) ISTD calibration and b) external calibration for HEGly.

target compounds had a retention of less than 4 min except for BHE Urea which eluted around 18 min (*Fig. 3*). Diluting the sample allowed us to conduct analysis, but the limits of quantification are increased by 3 orders of magnitude. A compromise had to be found to get LOQ as low as possible.

2.2 Validation of the Quantification of the Six Target Compounds

Accuracy profiles were performed with concentrations calculated using deuterated ISTD DEA-d8 and Gly-d5. For each compound, two profiles were plotted, for each ISTD. Only the best one (shape, lower tolerance) was kept. Table 3 presents a recap chart of figures of merit for the validation of this LC-MS/MS method.

DEA, BHE Urea, bicine and HEGly profiles were based on DEA-d8 as illustrated by Figure 4, Gly and Glygly on Gly-d5 as illustrated by Figure 5. First thing to notice is that the method could be validated with acceptance limits set at 20% whereas, with external calibration only (without ISTD), it is impossible to reach

TABLE 4

Results of validation using accuracy profiles (concentration given in real samples before dilution)

Compounds	Validation	ISTD	Order of magnitude of limits of protection (g/L)	Valid range of concentrations (g/L)
DEA	☑ ± 20%	DEA d8	0.02	from 0.2 to 4
Glygly	☑ ± 20%	Gly d5	0.03	from 0.08 to 0.25
Gly	☑ ± 20%	Gly d5	0.2	from 0.6 to 4
HEGly	☑ ± 20%	DEA d8	0.02	from 4 to 30
Bicine	☒ bias	DEA d8	0.02	☒
BHE Urea	☑ ± 20%	DEA d8	0.05	from 1 to 7

TABLE 5

Concentration of the six target compounds in real samples from IFPEN pilot plant and lab-scale experiments

Real samples	Tolerance (%)	A (g/L)	B (g/L)	C (g/L)
DEA	± 20	0.17	0.13*	0.12*
Glygly	± 20	NF	NF	NF
HEGly	± 20	13.1	0.77*	0.8*
Gly	± 20	0.46	NQ	NF
BHE Urea	± 20	5.5	3.9	2.8

* Values out of the range of the valid concentrations, NF: not found, NQ: not quantified.

such values and the acceptance limits should be raised to 40%, which is not compliant with the present requirements! The case of Gly (*Fig. 6*) is truly relevant to demonstrate the necessity of the use of deuterated ISTD as the accuracy profile got with external calibration without ISTD is totally improper for any quantification, both the shape and the acceptance limits are not correct. DEA profile is also, not surprisingly, perfectly corrected by the use of the corresponding deuterated ISTD. Moreover, BHE Urea, HEGly (*Fig. 7*) and Glygly compounds can be validated with those ISTD, even if it is not exactly the same molecule. Results are not as good as it could have been with the correct ISTD, but this is interesting for routine analyses if only a limited number of deuterated ISTD is available to quantify various compounds.

However, for bicine profile, even if the shape is improved and the confidence interval is narrower, a bias around 30% still remains and prevents from validating this compound. A corrective factor could have been applied, but this solution may not be relevant as the matrix complexity may have some unpredictable effects. Moreover the use of any corrective factor always leads to an increase of the interval. The use of deuterated bicine seems to be the only solution to quantify properly this degradation product. Table 4 presents a recap chart for the limits of detection and the valid range of concentrations where this method can be applied for quantification. LOQ values correspond to the lower limit of the concentration range provided.

Despite a robustness study would have been beyond the scope of the present paper, it can be noticed that the success of the method validation justify *a posteriori* the good robustness of the method. Effectively the condition taken for intermediate precision, different days and operator, are quite representative of the small changes likely to occur when the method is implemented.

2.3 Application on Real Samples

This method was applied on real samples from IFPEN lab-scale experiments (sample A) and pilot plant (samples B and C). Results shown in Table 5 reveal that Glygly is under limit of detection for samples A. Otherwise, DEA, Gly, HEGly and BHE Urea concentrations can be estimated in those two real samples. Accuracy profiles were found to be relevant to determine what range of concentration was valid and to check easily if those priority compounds can be quantified in real samples. Those informations are useful to study more precisely ageing of amine solvents and dynamic of formation of degradation products. Some of them could

be potentially dangerous to humans or environment according to their toxicity and their concentration.

CONCLUSION

To our knowledge, this study is the first one which presents the development and validation, based on the total error approach and the accuracy profile of an analytical method for degradation products occurring in amine based CO_2 capture process. This LC/MS method is based on a chromatographic separation conducted with a PGC column which allowed a large screening of degradation compounds. It also enabled quantification of priority compounds which were found to be relevant by specialists in charge of the CO_2 capture process. This study showed that the complexity of this kind of matrix can be partially overcome with deuterated ISTD. Acceptation limits of ±20% can be reached. For some compounds, it is not indispensable to use an ISTD which is the exact corresponding deuterated molecule, as it was shown for HEGly, BHE Urea and Glygly. But only validation data enabled to check if the ISTD used was relevant, and results must be cautiously interpreted as matrix effects still remain and might be sources of a non predicted increase in variability of results. Considering those promising results, this study could be pursued with other degradation products for which this LC/MS-MS method is relevant. If not, the same approach is going to be developed in GC-MS for other degradation products defined as priority compounds.

ACKNOWLEDGMENTS

We would like to acknowledge financial support from French ANR (Agence Nationale de la Recherche) (Research Project DALMATIEN: degradation of liquid amines and methods of analysis: toxicity or innocuousness for the environment).

REFERENCES

1 Martin S., Lepaumier H., Picq D., Kittel J., De Bruin T., Faraj A., Carrette P.-L. (2012) New Amines for CO_2 Capture. IV. Degradation, Corrosion, and Quantitative Structure Property Relationship Model, *Ind. Eng. Chem. Res.* **51**, 6283-6289.

2 Chakma A., Meisen A. (1987) Degradation of aqueous DEA solutions in a heat transfer tube, *Can. J. Chem. Eng.* **65**, 264-273.

3 Davis J., Rochelle G. (2009) Thermal degradation of monoethanolamine at stripper conditions, *Energy Procedia* **1**, 327-333.

4 Fostås B., Gangstad A., Nenseter B., Pedersen S., Sjøvoll M., Sørensen A.L. (2011) Effects of NO$_x$ in the flue gas degradation of MEA, *Energy Procedia* **4**, 1566-1573.

5 Lepaumier H., Grimstvedt A., Vernstad K., Zahlsen K., Svendsen H.F. (2011) Degradation of MMEA at absorber and stripper conditions, *Chem. Eng. Sci.* **66**, 3491-3498.

6 Strazisar B.R., Anderson R.R., White C.M. (2003) Degradation Pathways for Monoethanolamine in a CO$_2$ Capture Facility, *Energy and Fuels* **17**, 1034-1039.

7 Supap T., Idem R., Tontiwachwuthikul P., Saiwan C. (2006) Analysis of Monoethanolamine and Its Oxidative Degradation Products during CO$_2$ Absorption from Flue Gases: A Comparative Study of GC-MS, HPLC-RID, and CE-DAD Analytical Techniques and Possible Optimum Combinations, *Ind. Eng. Chem. Res.* **45**, 2437-2451.

8 Vevelstad S.J., Eide-Haugmo I., da Silva E.F., Svendsen H.F. (2011) Degradation of MEA; a theoretical study, *Energy Procedia* **4**, 1608-1615.

9 Chakma A., Meisen A. (1986) Corrosivity of Diethanolamine Solutions and their Degradation Products, *Ind. Eng. Chem. Res.* **25**, 627-630.

10 Choy E.T., Meisen A. (1980) Gas chromatographic detection of diethanolamine and its degradation products, *J. Chromatogr. A.* **187**, 145-152.

11 Holub P.E., Critchfield J.E., Su W.-Y. (1998) Amine degradation chemistry in CO$_2$ service, *48th Annu. Laurance Reid Gas Cond. Conf.*, pp. 146-160.

12 Lawal O., Bello A., Idem R. (2005) The Role of Methyl Diethanolamine (MDEA) in Preventing the Oxidative Degradation of CO$_2$ Loaded and Concentrated Aqueous Monoethanolamine (MEA) - MDEA Blends during CO$_2$ Absorption from Flue Gases, *Ind. Eng. Chem. Res.* **44**, 1874-1896.

13 Chakma A., Meisen A. (1988) Identification of methyl diethanolamine degradation products by gas chromatography and gas chromatography-mass spectrometry, *J. Chromatogr. A.* **457**, 287-297.

14 Chakma A., Meisen A. (1997) Methyl-diethanolamine degradation — Mechanism and kinetics, *Can. J. Chem. Eng.* **75**, 861-871.

15 Freeman S. a., Davis J., Rochelle G.T. (2010) Degradation of aqueous piperazine in carbon dioxide capture, *Int. J. Greenh. Gas Control.* **4**, 756-761.

16 Wang T., Jens K.-J. (2012) Oxidative Degradation of Aqueous 2-Amino-2-methyl-1-propanol Solvent for Postcombustion CO$_2$ Capture, *Ind. Eng. Chem. Res.* **51**, 6529-6536.

17 Lepaumier H., Picq D., Carrette P.L. (2009) Degradation study of new solvents for CO$_2$ capture in post-combustion, *Energy Procedia* **1**, 893-900.

18 Lepaumier H., Picq D., Carrette P.L. (2009) New Amines for CO$_2$ Capture. I. Mechanisms of Amine Degradation in the Presence of CO$_2$, *Ind. Eng. Chem. Res.* **48**, 9061-9067.

19 Lepaumier H., Picq D., Carrette P.L. (2009) New Amines for CO$_2$ Capture. II. Oxidative Degradation Mechanisms, *Ind. Eng. Chem. Res.* **48**, 9068-9075.

20 Lepaumier H., Martin S., Picq D., Delfort B. (2010) New Amines for CO$_2$ Capture. III. Effect of Alkyl Chain Length between Amine Functions on Polyamines Degradation, *Ind. Eng. Chem. Res.* **49**, 4553-4560.

21 Gouedard C., Picq D., Launay F., Carrette P.L. (2012) Amine degradation in CO$_2$ capture. I. A review, *Int. J. Greenh. Gas Control.* **10**, 244-270.

22 Rey A., Gouedard C., Ledirac N., Cohen M., Dugay J., Vial J., Pichon V., Bertomeu L., Picq D., Bontemps D., Chopin F. Carrette P.-L. (2013) Amine degradation in CO$_2$ capture. 2. New degradation products of MEA. Pyrazine and alkylpyrazines: Analysis, mechanism of formation and toxicity, *Int. J. Greenh. Gas Control* **19**, 576-583.

23 Boulanger B., Chapuzet E., Cohen N., Compagnon P.A., Dewe W., Feinberg M., Laurentie M., Mercier N., Muzard G., Valat L. (2006) Validation des procédures analytiques quantitatives: harmonisation des démarches - Partie II: Statistiques, *Commision SFSTP. STP Pharma Pratiques* **16**, 1-31.

24 Hubert Ph., Nguyen-Huu J.J., Boulanger B., Chapuzet E., Chiap P., Cohen N., Compagnon P.A., Dewe W., Feinberg M., Lallier M., Laurentie M., Mercier N., Muzard G., Nivet C., Valat L. (2003) Validation des procédures analytiques quantitatives harmonisation des démarches, *Commision SFSTP. STP Pharma Pratiques* **13**, 101-138.

25 Feinberg M. (2012) *Guide de validation des méthodes d'analyse - LaboStat*, Lavoisier, Paris.

26 Hubert Ph., Nguyen-Huu J.-J., Boulanger B., Chapuzet E., Chiap P., Cohen N., Compagnon P.-A., Dewé W., Feinberg M., Lallier M., Laurentie M., Mercier N., Muzard G., Nivet C., Valat L. (2004) Harmonization of strategies for the validation of quantitative analytical procedures: A SFSTP proposal - Part I, *J. Pharma. Biomedical Anal.* **36**, 3, 579-586.

27 Hubert Ph., Nguyen-Huu J.-J., Boulanger B., Chapuzet E., Chiap P., Cohen N., Compagnon P.-A., Dewé W., Feinberg M., Lallier M., Laurentie M., Mercier N., Muzard G., Nivet C., Valat L., Rozet E. (2007) Harmonization of strategies for the validation of quantitative analytical procedures: A SFSTP proposal - Part II, *J. Pharm. Biomed. Anal.* **45**, 1, 70-81.

28 Hubert Ph., Nguyen-Huu J.-J., Boulanger B., Chapuzet E., Cohen N., Compagnon P.-A., Dewé W., Feinberg M., Laurentie M., Mercier N., Muzard G., Valat L., Rozet E. (2007) Harmonization of strategies for the validation of quantitative analytical procedures: A SFSTP proposal - Part III, *J. Pharm. Biomed. Anal.* **45**, 1, 82-96.

29 Hubert P., Nguyen-Huu J.J., Boulanger B., Chapuzet E., Cohen N., Compagnon P.A., Dewé W., Feinberg M., Laurentie M., Mercier N., Muzard G., Valat L., Rozet E. (2008) Harmonization of strategies for the validation of quantitative analytical procedures: a SFSTP proposal - Part IV. Examples of application, *J. Pharm. Biomed. Anal.* **48**, 3, 760-771.

30 Esters V., Angenot L., Brandt V., Frédérich M., Tits M., Van Nerum C., Wauters J.N., Hubert P. (2006) Validation of a high-performance thin-layer chromatography/densitometry method for the quantitative determination of glucosamine in a herbal dietary supplement, *J. Chromatogr. A.* **1112**, 1-2, 156-164.

31 Marini R.D., Servais A.-C., Rozet E., Chiap P., Boulanger B., Rudaz S., Crommen J., Hubert P., Fillet M. (2006) Nonaqueous capillary electrophoresis method for the enantiomeric purity determination of S-timolol using heptakis (2,3-di-O-methyl-6-O-sulfo)-beta-cyclodextrin: validation using the accuracy profile strategy and estimation of uncertainty, *J. Chromatogr. A.* **1120**, 1-2, 102-111.

32 Dispas A., Lebrun P., Ziemons E., Marini R., Rozet E., Hubert P. (2014) Evaluation of the quantitative performances of supercritical fluid chromatography: from method development to validation, *J. Chromatogr. A.* (in press).

33 Combes A., El Abdellaoui S., Sarazin C., Vial J., Mejean A., Ploux O., Pichon V. (2013) Validation of the analytical procedure for the determination of the neurotoxin β-N-methylamino-L-alanine in complex environmental samples, *Anal. Chim. Acta.* **771**, 42-49.

13

Modeling of the CO_2 Absorption in a Wetted Wall Column by Piperazine Solutions

Alberto Servia[1,2]*, Nicolas Laloue[1], Julien Grandjean[1], Sabine Rode[2] and Christine Roizard[2]

[1] IFP Energies nouvelles, Rond-point de l'échangeur de Solaize, BP 3, 69360 Solaize - France
[2] LRGP-CNRS Université de Lorraine, 1 rue Grandville, BP 20451, 54001 Nancy Cedex - France
e-mail: alberto.servia@gmail.com

* Corresponding author

Résumé — Modélisation de l'absorption de CO_2 par des solutions de pipérazine dans un film tombant — Des études théoriques et expérimentales sur l'absorption réactive du CO_2 dans des solutions aqueuses de PZ mettant en œuvre un outil expérimental de type film tombant sont présentées. Un modèle rigoureux d'absorption en deux dimensions, prenant en compte les phénomènes cinétique, thermodynamique et hydrodynamique, a été développé pour simuler l'outil expérimental de film tombant. Les principales originalités du modèle, par rapport aux travaux antérieurs, consistent dans la prise en compte de la variation de la concentration en CO_2 de la phase gaz en fonction de la hauteur du réacteur, ainsi que le calcul de l'équilibre gaz-liquide par une approche thermodynamique cohérente.

Un outil expérimental de type film tombant a été spécialement conçu, pour lequel le coefficient de transfert de masse dans la phase gaz a été estimé. Des mesures d'absorption de CO_2 ont été effectuées sur des solutions aqueuses de PZ, vierges et chargées en CO_2, sur la gamme 298-331 K, et pour des concentrations totales en PZ variant de 0,2 à 1 M. Le modèle de réacteur permet de prédire les flux d'absorption avec une précision remarquable de 3,2 % AAD, que ce soit dans les solutions vierges ou chargées. Le gradient de concentration de CO_2 dans la phase gaz ainsi que la réaction de formation du dicarbamate doivent être pris en compte afin de prédire correctement l'absorption du CO_2 dans les solutions aqueuses de PZ chargées en CO_2.

Abstract — Modeling of the CO_2 Absorption in a Wetted Wall Column by Piperazine Solutions — Theoretical and experimental investigations on the reactive absorption of CO_2 in aqueous solutions of PZ using a wetted wall column are presented. A rigorous two dimensional absorption model, accounting for kinetics, hydrodynamics and thermodynamics, has been developed for a wetted wall column. Major innovative features of the model, compared to previous work, are the account on the variation of the gas-side CO_2 concentration over the reactor height as well as the computation of the gas-liquid equilibrium by a thermodynamically consistent approach.

A laboratory-scale wetted wall column was conceived and constructed and the gas-side mass-transfer coefficient was estimated. CO_2 absorption experiments were carried out on unloaded and loaded aqueous solutions of PZ over the range of 298-331 K, and for total PZ concentrations varying from 0.2 to 1 M. The reactor model permitted to predict the absorption fluxes in loaded as well as in unloaded solutions with an excellent accuracy, i.e. 3.2% AAD. In loaded solutions, the gas-side CO_2 concentration gradient, as well as the dicarbamate formation reaction has to be taken into account.

NOMENCLATURE

Chemical species

DEA	DiEthanolAmine
H^+PZCOO^-	Protonated piperazine carbamate
MDEA	N-MethylDiEthanolAmine
MEA	MonoEthanolAmine
PZ	Piperazine
$PZCOO^-$	Piperazine carbamate
$PZ(COO)_2^{2-}$	Piperazine dicarbamate
PZH^+	Protonated piperazine

Others

A	Gas-liquid contact area (m^2)
a	Ratio between the transfer area and the reactor volume (m^{-1})
AAD	Average Absolute Deviation (%)
C^*	Concentration at the gas-liquid interface within the gas phase (mol.m^{-3})
D_h	Hydraulic diameter (m)
D_i	Diffusion coefficient of species "i" (m^2.s^{-1})
E	Enhancement factor adimensional
F	Molar flow (mol.s^{-1})
FEM	Finite Elements Method
G	Gravity acceleration (m.s^{-2})
H	Henry constant (Pa.m^3.mol^{-1})
h	Reactor height (m)
k_G	Gas mass transfer coefficient (mol.Pa^{-1}.m^{-2}.s^{-1})
K_i	Equilibrium constant of reaction i
k_i	Kinetic constant of reaction i (m^3.mol^{-1}.s^{-1})
k_L	Liquid mass transfer coefficient (m.s^{-1})
N	CO$_2$ flux (mol.m^{-2}.s^{-1})
NRTL	Non Random Two Liquid
pH	-log [H$^+$] (adimensional)
pKa	Acid dissociation equilibrium constant adimensional
P	Pressure (Pa)
Q	Volume flow (m^3.s^{-1})
R	Perfect gas law constant (J.mol^{-1}.K^{-1})
r	Radial coordinate (m)
R_i	Reaction rate of chemical reaction i (mol.m^{-3}.s^{-1})
Sh	Sherwood number adimensional
T	Temperature (K)
u	Concentration (mol.m^{-3})

v	Velocity (m.s^{-1})
WWC	Wetted Wall Column
y	Molar fraction
z	Axial coordinate (m)

Greek letters

v_i	Stoichiometric factor associated to species "i" (v_i)
μ	Viscosity (Pa.s)
ρ	Density (kg.m^{-3})
δ	Liquid thickness (m)

Subscripts

app	Apparent
B	Base
G	Gas
L,liq	Liquid
TM	Termolecular
Zw	Zwitterion

Superscripts

eq,*	Equilibrium
in	Inlet
ln	Logarithmic average

INTRODUCTION

Aqueous solutions of alkanolamines are generally used as a solvent for removing acid gases such as CO$_2$ and H$_2$S which can be eventually contained in natural gas, hydrogen or flue gas. MonoEthanolAmine (MEA) is the reference alkanolamine for the CO$_2$ post-combustion capture process while N-MethylDiEthanol-Amine (MDEA) is widely used as solvent for natural gas selective deacidification processes. Even if they present high reaction rates with CO$_2$, the primary and secondary alkanolamines, such as MEA and DiEthAnolamine (DEA), require a high energy consumption in order to be regenerated. Tertiary amines present lower reaction rates with CO$_2$ than primary or secondary amines, however the reaction enthalpy is low, which considerably decreases the required energy to regenerate this type of amine.

The addition of a small quantity of a primary or a secondary alkanolamine (activator) into an aqueous solution of a tertiary alkanolamine strongly increases the reaction rate with CO_2 without significantly modifying the energy to provide for the regeneration of the mixture (Chakravarty et al., 1985). Several studies on the kinetics of CO_2 absorption by aqueous blends of alkanolamines can be found in the literature. PZ has revealed itself as being a high-performance activator compared to the conventional alkanolamines such as the MEA or the DEA. Furthermore, *BASF* commercializes a technology based on the use of a solvent composed by PZ and MDEA (Appl et al., 1982), which illustrates the considerable interest of this cyclic amine. The accurate understanding of the reaction mechanisms between CO_2 and PZ is essential to rigorously investigate the kinetics of CO_2 absorption by MDEA and PZ mixtures.

The aim of this work is to study the reactions between the PZ and its derivatives with CO_2. The studies of the kinetics of CO_2 absorption on unloaded and loaded solutions were conducted to evaluate the reaction rates of CO_2 with PZ and PZCOO$^-$ respectively. The experimental results were interpreted by a rigorous mathematical model coupling all the phenomena occurring within the reactor. This model also accounts for the CO_2 partial pressure evolution in the gas phase in order to test the hypothesis of considering a constant CO_2 partial pressure given by the logarithmic average between the reactor inlet and outlet.

1 KINETICS

1.1 Reaction Mechanism

Two mechanisms are proposed in the literature to explain the chemical interactions existing between an amine and CO_2.

The first mechanism, proposed by Caplow (1968) and reintroduced by Danckwerts (1979), is called Zwitterion mechanism. It was widely used to interpret the kinetic data of aqueous solutions of DEA (Rinker et al., 2000; Littel et al., 1992) and of 2-Amino-2-Methyl-1-Propanol (AMP) (Seo and Hong, 2000). This mechanism consists of two steps. Firstly, the amine provides its free electronic pair to form a chemical bond with the carbon atom of the CO_2 molecule to produce an unstable compound called Zwitterion:

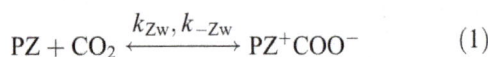

$$PZ + CO_2 \xrightleftharpoons[]{k_{Zw}, k_{-Zw}} PZ^+COO^- \qquad (1)$$

The Zwitterion complex is then deprotonated by any base present in the solution, such as PZ, water,

hydroxide ion, etc., to produce a compound called carbamate:

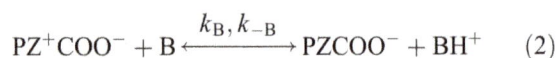

$$PZ^+COO^- + B \xrightleftharpoons[]{k_B, k_{-B}} PZCOO^- + BH^+ \qquad (2)$$

The contribution of each base to the Zwitterion deprotonation depends on its concentration, basicity and steric hindrance.

The CO_2 consumption rate is obtained by assuming the quasi-steady state for the Zwitterion complex and considering that the deprotonation reactions are reversible:

$$r_{CO_2} = \frac{k_{Zw}[PZ][CO_2] \sum_i k_B[B_i] - k_{-Zw} \sum_i k_{-B}[PZCOO^-][B_iH^+]}{k_{-Zw} + \sum_i k_B[B_i]}$$

$$(3)$$

Two limiting cases can be considered for this mechanism. If the deprotonation reaction rate is fast compared to the reverse reaction rate of the Zwitterion formation (k_{-Zw}), the amine partial order is one. The CO_2 rate of consumption is determined using Equation (4), which assumes Zwitterion deprotonation to be irreversible. For instance, this limiting case was verified for MEA, which presents high pKa (9.44 at 298 K, Hamborg and Versteeg, 2009) and no steric hindrance:

$$r_{CO_2} = k_{Zw}[PZ][CO_2] \qquad (4)$$

If the deprotonation path is rate limiting and the Zwitterion deprotonation irreversible, the amine partial order varies between 1 and 2, depending on the degree of contribution of each base within the solution. For example, the reaction rate of CO_2 absorption into aqueous DEA solution was determined using Equation (5), involving water and DEA as bases in the Zwitterion deprotonation step (Rinker et al., 1996):

$$r_{CO_2} = \frac{k_{Zw}[PZ][CO_2] \sum_i k_B[B_i]}{k_{-Zw}} \qquad (5)$$

The second mechanism, called the termolecular mechanism, was proposed by Crooks and Donnellan (1989), and reviewed by da Silva and Svendsen (2004). It considers a simultaneous reaction of the amine, CO_2 and a base to produce the carbamate:

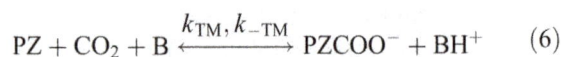

$$PZ + CO_2 + B \xrightleftharpoons[]{k_{TM}, k_{-TM}} PZCOO^- + BH^+ \qquad (6)$$

The CO_2 consumption rate is given by the following expression, assuming the reaction irreversibility:

$$r_{CO_2} = \sum_i k_{TM}[B_i][PZ][CO_2] - \sum_i k_{-TM}[B_iH^+][PZCOO^-] \tag{7}$$

If water only contributes to the termolecular mechanism and the reaction is considered as being irreversible, Equation (7) simplifies to Equation (4) with $k_{Zw} = k_{TM}[H_2O]$. The concentration of water is generally considered as being constant (Bishnoi and Rochelle, 2000; Samanta and Bandyopadhyay, 2007).

The termolecular mechanism is widely used in the literature to explain the chemical reaction existing between CO_2 and PZ (Bishnoi and Rochelle, 2000; Cullinane, 2005; Samanta and Bandyopadhyay, 2007; Dugas, 2009). The authors usually consider one reaction for each amine-function within the PZ molecule:

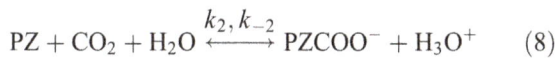

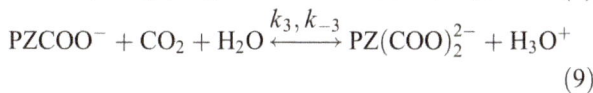

$$PZ + CO_2 + H_2O \xrightleftharpoons{k_2, k_{-2}} PZCOO^- + H_3O^+ \tag{8}$$

$$PZCOO^- + CO_2 + H_2O \xrightleftharpoons{k_3, k_{-3}} PZ(COO)_2^{2-} + H_3O^+ \tag{9}$$

The CO_2 consumption rate is then given by the following equation, assuming that both reactions are first order in PZ and $PZCOO^-$:

$$r_{CO_2} = k_2[PZ][CO_2] - k_{-2}[PZCOO^-][H_3O^+]$$
$$+ k_3[PZCOO^-][CO_2] - k_{-3}\left[PZ(COO)_2^{2-}\right][H_3O^+] \tag{10}$$

The PZ partial order can be estimated by performing CO_2 absorption experiments into PZ unloaded solutions where the CO_2 mass transfer is not limited by the PZ diffusion towards the gas-liquid interface (pseudo-first order regime). Thus, the CO_2 consumption rate is giving by the following expression:

$$r_{CO_2} = k_2[PZ]^\alpha[CO_2] = k_{app}[CO_2] \tag{11}$$

The representation of k_{app} as a function of the PZ concentration allows the determination of the PZ partial order.

1.2 Kinetic Constants

Several authors studied the kinetics between CO_2 and PZ (Bishnoi and Rochelle, 2000; Cullinane, 2005; Derks et al., 2006; Samanta and Bandyopadhyay, 2007; Dugas,

2009; Bindwal et al., 2011). The main features of these studies are shown in Table 1.

Bishnoi and Rochelle (2000) performed experiments on unloaded solutions at temperatures ranging from 298 to 333 K and PZ concentrations of 0.2 and 0.6 M. The experimental data obtained in a wetted wall column were used to estimate the second-order kinetic constant of the reaction between PZ and CO_2 and the PZ partial order. A second set of experiments performed on loaded PZ solutions qualitatively shown that the reaction between $PZCOO^-$ and CO_2 cannot be neglected at these conditions. The experimental data obtained on unloaded solutions were interpreted by a simple model considering that measurements were carried out in the kinetic regime. They observed that the apparent kinetic constant increased linearly with the PZ concentration, suggesting that this reaction is first order in PZ. Moreover, they determined a second-order kinetic constant (k_2) of 53.7 $m^3.mol^{-1}.s^{-1}$ at 298 K, which is significantly higher than the value obtained by Xu et al. (1992) (0.13 $m^3.mol^{-1}.s^{-1}$ at 298 K). Those authors performed experimental tests on loaded solutions, and probably in presence of PZ diffusion limitations towards the gas-liquid interface. The low value of their kinetic constant can be explained by the model developed to interpret the kinetic data which did not account for PZ mass transfer limitations within the liquid phase.

Cullinane (2005) carried out experiments of CO_2 absorption into aqueous PZ solutions by using the same wetted wall column as the one used by Bishnoi and Rochelle (2000). The author proposed a reaction mechanism based on Brönsted theory, which states that the kinetic constant associated to the reactions generating $PZCOO^-$ and $PZ(COO)_2$ (PZ (or $PZCOO^-$), CO_2 and a base: OH^-, H_2O, PZ, CO_3^{2-} and $PZCOO^-$) depends on the pKa of the considered base. The contribution of the hydroxyl ions as a base for catalyzing the chemical reaction between $PZCOO^-$ and CO_2 was neglected since OH^- and $PZCOO^-$ does not coexist within the liquid solution. Three other chemical reactions were also considered to account for the bicarbonate formation in the reaction mechanism (CO_2, H_2O and a base: H_2O, PZ and $PZCOO^-$). The impact of the addition of a neutral salt into the amine solution on the global CO_2 absorption kinetics was also investigated. The reaction rate was found to increase with the ionic strength. The same evolution was observed in the case of the CO_2 absorption into aqueous solutions of PZ and K_2CO_3. Indeed, the presence of K_2CO_3 in the solution increases the bases concentration within the liquid phase (OH^- and CO_3^{2-}) and therefore enhances the reaction rate. Moreover, Cullinane (2005), estimated a

TABLE 1

Literature review on the kinetics study of CO_2 absorption by aqueous PZ solutions

Reference	Experimental device	Loading mol_{CO_2}/mol_{PZ}	[PZ] (M)	T (K)	k_2 (m³.mol⁻¹.s⁻¹)	k_3 (m³.mol⁻¹.s⁻¹)	Kinetic modeling	Mass transfer	Comments
Xu et al., 1992	Disk column	$(0.2393\text{-}2.138) \times 10^3$ mol.m⁻³	0.041-0.21 (mixtures with MDEA)	303-343	0.13 at 298 K	-	Pseudo-first order	Film theory and no k_G (Pure CO_2)	Loaded solutions
Bishnoi and Rochelle, 2000	Wetted wall column	0-0.67	0.2 and 0.6	298-333	$53.7\,\text{exp}\left(-\frac{36\,000}{R}\left[\frac{1}{T}-\frac{1}{298}\right]\right)$	-	Pseudo-first order	Film theory	No interpretation of the data taken on loaded solutions
Bishnoi and Rochelle, 2002	Wetted wall column	0.0011-0.625	0.6 (mixture with MDEA 4 M)	295-343	$53.7\,\text{exp}\left(-\frac{36\,000}{R}\left[\frac{1}{T}-\frac{1}{298}\right]\right)$	$47.0\,\text{exp}\left(-\frac{36\,000}{R}\left[\frac{1}{T}-\frac{1}{298}\right]\right)$	Liquid discretization with constant CO_2 partial pressure	Eddy theory (constant P_{CO_2} within the gas)	Complex kinetics considering synergy between both amines
Cullinane, 2005	Wetted wall column	0-0.019	0.45-1.20 m	298 and 333	-	-	Liquid discretization with constant CO_2 partial pressure	Eddy theory (constant P_{CO_2} within the gas)	Second order on PZ for [PZ] > 0.5 M and study of impact from neutral salts and K_2CO_3 addition
Derks et al., 2006	Stirred cell	0	0.6-1.5	293-313	70.0 at 298 K	-	DeCoursey (1974) and Hogendoorn et al. (1997)	Film theory and no k_G (pure CO_2)	Kinetics of reaction between PZH^+ and CO_2 quantified
Samanta and Bandyopadhyay (2007)	Wetted wall column	0	0.2-0.8	298-313	58.0 at 298 K	59.5 at 298 K	Complex model with constant CO_2 partial pressure	Penetration theory (constant P_{CO_2} within the gas)	Kinetics of the reaction between $PZCOO^-$ and CO_2 quantified
Dugas, 2009	Wetted wall column	0.222-0.412	2-12 m m = molality	313-373	-	-	Pseudo-m,n^{th} order corrected with species activity coefficients	Double film theory	Second order on PZ
Bindwal et al., 2011	Stirred cell	0	0.025-0.1	303	25.8 at 303 K	-	Pseudo-first order	Film theory and no impact from k_G verified	Unloaded solutions

partial order of 2 for the PZ for an amine concentration higher than 0.5 M. This result does not agree with the work of Bishnoi and Rochelle (2000) that determined a partial order of 1 for the PZ, based on measurements on 0.2 and 0.6 M PZ solutions.

Derks *et al.* (2006) determined a value for k_2 of $70.0 \ m^3.mol^{-1}.s^{-1}$ at 298 K through experiments carried out in a stirred cell. They also quantified the kinetics of the reaction between the PZH^+ and CO_2 by performing a new set of experimental CO_2 absorption measurements on partially protonated PZ. The kinetic constant associated to this reaction was $0.280 \pm 0.100 \ m^3.mol^{-1}.s^{-1}$ at 298 K, which is in agreement with the Bronsted theory.

Samanta and Bandyopadhyay (2007) developed a mathematical model accounting for mass transfer, kinetics and equilibrium phenomena to estimate the kinetics of the reactions between PZ and piperazine carbamate ($PZCOO^-$) and CO_2. Their experimental data were performed in a wetted wall column. The kinetic constants obtained were in good agreement with those determined by Bishnoi and Rochelle (2000, 2002). The consistency between these values cannot bet justified through the use of the same device since kinetics determination strongly depends on the mathematical model used to interpret the experimental data.

Dugas (2009) performed experiments in the same wetted wall column as the one used by Bishnoi (2000) and Cullinane (2005). This work considered an activity-based reaction mechanism based on the Brönsted theory. Both PZ and $PZCOO^-$ were involved as bases for catalyzing the chemical reactions between PZ and CO_2, and between $PZCOO^-$ and CO_2, implicitly assuming a partial order of 2 for PZ, in agreement with Cullinane (2005).

Finally, Bindwal *et al.* (2011) observed that the second-order kinetic constant increases with the PZ concentration. They determined a value of $25.8 \ m^3.mol^{-1}.s^{-1}$ at 303 K, which is considerably lower than the one obtained by Bishnoi and Rochelle (2000).

Many discrepancies exist concerning the kinetics of CO_2 absorption by aqueous PZ solutions. The results depend on the type of model involved to interpret the experimental measurements as well as on the experimental device. The easiest way to determine kinetics is through the pseudo-first order assumption, allowing the kinetics to be analytically determined. Nevertheless, the use of this simple mass transfer model is only suitable for a specific and narrow range of experimental conditions. Its use can lead to large errors if mass transfer limitations within the liquid phase are involved (Xu *et al.*, 1992). The kinetics of the CO_2 absorption in a wide range of operating conditions can only be determined by accurately describing all phenomena occurring within the reactor. Besides, even if the gas mass transfer resistance represents a non negligible part of the total mass transfer resistance, all studies from the literature consider a constant CO_2 partial pressure to describe the CO_2 mass transfer, which can lead to errors in the CO_2 flux determination.

2 MODELING SECTION

2.1 Chemical Reactions

Two types of chemical reactions were considered in the reactor model: equilibrium reactions and reactions limited by kinetics. The chemical reactions involving a single transfer of proton were considered instantaneous (Cullinane, 2005; Samanta and Bandyopadhyay, 2007). They were described by their equilibrium constant:

Water dissociation

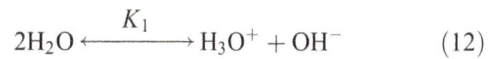

$$2H_2O \xleftrightarrow{K_1} H_3O^+ + OH^- \tag{12}$$

PZ protonation

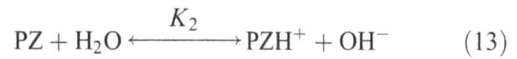

$$PZ + H_2O \xleftrightarrow{K_2} PZH^+ + OH^- \tag{13}$$

$PZCOO^-$ protonation

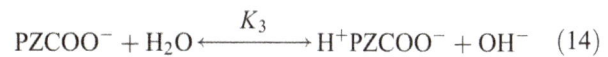

$$PZCOO^- + H_2O \xleftrightarrow{K_3} H^+PZCOO^- + OH^- \tag{14}$$

Carbonate formation

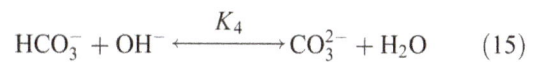

$$HCO_3^- + OH^- \xleftrightarrow{K_4} CO_3^{2-} + H_2O \tag{15}$$

The carbonic acid equilibrium reaction was not considered in the system due to the high pH of the aqueous alkanolamine solutions. The equilibrium constants associated to each reaction were given by the thermodynamic model which is described later in this paper.

Beyond the equilibrated reactions, the chemical reactions involving CO_2 were considered to be kinetically controlled:

Bicarbonate formation

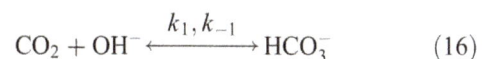

$$CO_2 + OH^- \xleftrightarrow{k_1, k_{-1}} HCO_3^- \tag{16}$$

Piperazine carbamate formation

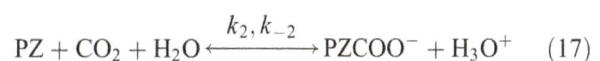

$$PZ + CO_2 + H_2O \xleftrightarrow{k_2, k_{-2}} PZCOO^- + H_3O^+ \tag{17}$$

Piperazine dicarbamate formation

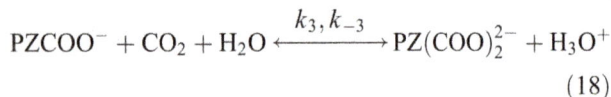

$$PZCOO^- + CO_2 + H_2O \xrightleftharpoons[k_{-3}]{k_3, k_{-3}} PZ(COO)_2^{2-} + H_3O^+$$

$$(18)$$

The CO_2 hydrolysis was neglected as its reaction rate is low compared to the other chemical reactions considered in the kinetic network (Bishnoi and Rochelle, 2000; Samanta and Bandyopadhyay, 2007). The kinetic constant for the bicarbonate formation was obtained from the paper of Pinsent *et al.* (1956) while the kinetic constants of the reactions between piperazine, and piperazine carbamate with CO_2 were taken from the papers of Bishnoi and Rochelle (2000, 2002), respectively. The kinetic constant expression associated to the chemical reaction between the $PZCOO^-$ and CO_2 was given by the following equation.

$$k_2 = 47.0 \exp\left(-\frac{36\,000}{R}\left[\frac{1}{T} - \frac{1}{298}\right]\right) \quad (19)$$

All reactions were considered to be reversible in this work.

2.2 Thermodynamics

The thermodynamic model used in this work was provided by ASPEN Plus. It allowed to compute the concentration of each species within the liquid phase as well as the CO_2 equilibrium vapour pressure. The activity coefficients of all the species within the liquid phase were taken into account through the electrolyte NRTL approach while the Redlich-Kwong-Soave state equation was used to determine the gas phase deviation from the ideal state (Bishnoi and Rochelle, 2000; Cullinane, 2005). The thermodynamic model was validated by comparison with vapour-liquid equilibrium data from literature (Bishnoi and Rochelle, 2000; Hilliard, 2005). Figure 1 is plotted at a fixed temperature of 313 K whereas Figure 2 represents the equilibrium at 2 different temperatures (333 K and 343 K). The equilibrium pressure obviously varies with temperature as well as with loading (*Fig. 3*). Consequently, it is the temperature that makes the difference, not the PZ overall concentration.

The $P_{CO_2}^*$ also increases with temperature and remains almost independent of PZ concentration (*Fig. 1*, *2*). The AAD between experimental and modelled CO_2 vapour pressure was 19%.

The solubility was calculated by the ratio between the CO_2 equilibrium vapour pressure and the CO_2 liquid concentration provided by a flash calculation in ASPEN Plus (*Fig. 3*). The calculated values were in good agreement with the solubilities determined through the N_2O

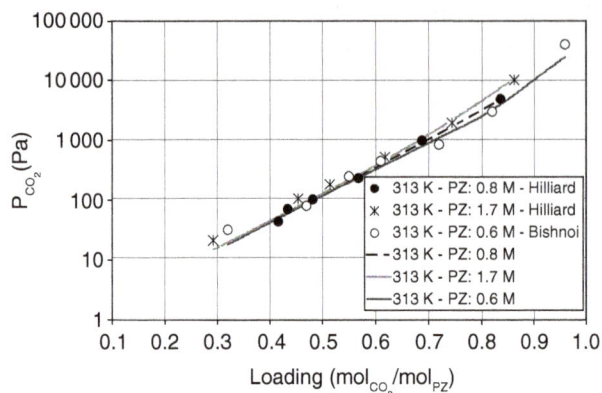

Figure 1

Vapour-liquid equilibrium CO_2 partial pressure as a function of the solution loading at 313 K; symbols: literature data; lines: model calculations.

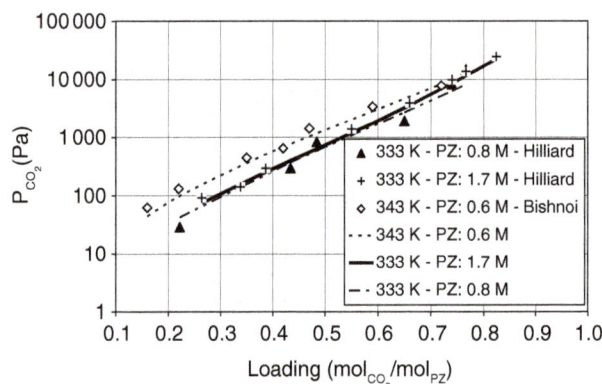

Figure 2

Vapour-liquid equilibrium CO_2 partial pressure as a function of the solution loading at 333 and 343 K; symbols: literature data; lines: model calculations.

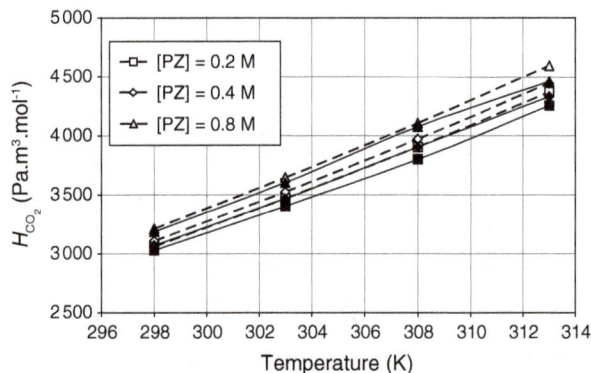

Figure 3

Solubility of CO_2 as a function of temperature at different PZ concentrations. Empty symbols: computed values; filled symbols: N_2O analogy.

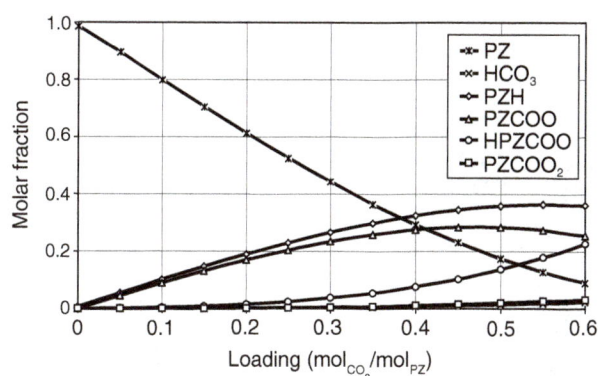

Figure 4

Speciation estimated by the ASPEN Plus thermodynamic model for a solution of 1 M of PZ at 298 K.

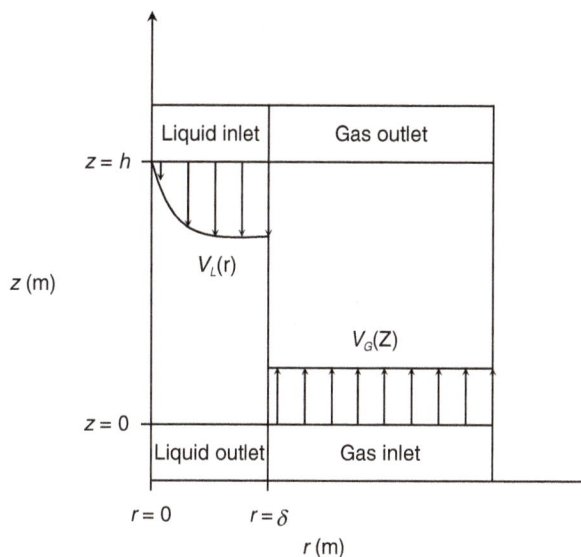

Figure 5

Schematic representation of the reactor geometry.

analogy semi-empirical approach by Samanta et al. (2007), with an AAD of 1.8%. Consequently, the use of the N_2O analogy to determine CO_2 solubility would be possible in this case, since both the solubilities given by the thermodynamic model and by N_2O analogy are similar in the range of tested temperatures. However, the difference between values computed using ASPEN Plus and N_2O analogy increases with temperature which shows that the use of the N_2O analogy at higher temperatures leads to more discrepancy.

The thermodynamic model described in this section was used to determine the solution loading corresponding to the highest concentration of $PZCOO^-$, in order to evaluate its reaction with CO_2. The thermodynamic model predicted a maximum of $PZCOO^-$ concentration at a loading of approximately 0.5 at 298 K, which is in agreement with Bishnoi and Rochelle (2000) (*Fig. 4*). The maximum of $PZCOO^-$ concentration was comparable at 333 K (data not shown).

2.3 Hydrodynamics

The liquid phase velocity profile was determined using the Navier-Stokes equation for an incompressible fluid associated with specific boundary conditions. The velocity profile was considered to be fully developed at the reactor inlet. Hydrodynamic calculations performed in Fluent, not shown here, supported the validity of this assumption:

$$0 = \mu_{\text{liq}} \left[\frac{\partial^2 v_L(r)}{\partial r^2} + \frac{1}{r} \frac{\partial v_L(r)}{\partial r} \right] - \rho_{\text{liq}} g \qquad (20)$$

$$v_L(r = 0) = 0 \qquad (21)$$

$$\frac{\partial v_L(r = \delta)}{\partial r} = 0 \qquad (22)$$

where δ represents the liquid film thickness. The radial (r) and axial (z) coordinates are illustrated in Figure 5.

The gas phase velocity was obtained by performing a mass balance on the nitrogen (N_2). A plug-flow model was used in order to describe the gas phase flow:

$$v_G(z) = \frac{Q_G^{\text{in}} \left(1 - y_{CO_2}^{\text{in}} \right) P}{A \left(P - C_{CO_2}^G RT \right)} \qquad (23)$$

where Q_G^{in} and $y_{CO_2}^{\text{in}}$ represent the total gas flow and the CO_2 molar fraction at the reactor inlet, respectively. The total pressure P was supposed constant within the reactor.

2.4 Reactor Model

A 2D stationary model was developed using COMSOL software to predict the absorption flux of CO_2 into aqueous solutions of PZ within a wetted wall column. This model couples hydrodynamics, mass transfer, chemical reactions and gas-liquid equilibrium. One of the model originalities is the description of the CO_2 partial pressure variation within the gas phase, instead of considering a constant CO_2 partial pressure given by the logarithmic average between the reactor inlet and outlet.

The concentration profile of each chemical species within the liquid phase, $u_i(r,z)$, was obtained by the simultaneous resolution of the mass balance for each compound, the electroneutrality condition and the equilibrium constants associated to the instantaneous-considered proton-transfer reactions.

The species concentrations were renamed as follows in order to simplify the presentation of the algebraic-differential equation system:

CO_2 - u_1, PZ - u_2, H_3O^+ - u_3, OH^- - u_4, PZH^+ - u_5, $PZCOO^-$ - u_6, H^+PZCOO^- - u_7, $PZ(COO)_2^{2-}$ - u_8, HCO_3^- - u_9 and CO_3^{2-} - u_{10}.

– CO_2 mass balance:

$$0 = D_{u_1}\left(\frac{\partial^2 u_1}{\partial r^2} + \frac{1}{r}\frac{\partial u_1}{\partial r} + \frac{\partial^2 u_1}{\partial z^2}\right) - v_L(r)\frac{\partial u_1}{\partial z} - (R_1 + R_2 + R_3) \quad (24)$$

Figure 6
Gas phase distributor.

– global PZ mass balance:

$$0 = D_{u_2}\left(\frac{\partial^2 u_2}{\partial r^2} + \frac{1}{r}\frac{\partial u_2}{\partial r} + \frac{\partial^2 u_2}{\partial z^2}\right) + D_{u_5}\left(\frac{\partial^2 u_5}{\partial r^2} + \frac{1}{r}\frac{\partial u_5}{\partial r} + \frac{\partial^2 u_5}{\partial z^2}\right) + D_{u_6}\left(\frac{\partial^2 u_6}{\partial r^2} + \frac{1}{r}\frac{\partial u_6}{\partial r} + \frac{\partial^2 u_6}{\partial z^2}\right) + D_{u_7}\left(\frac{\partial^2 u_7}{\partial r^2} + \frac{1}{r}\frac{\partial u_7}{\partial r} + \frac{\partial^2 u_7}{\partial z^2}\right) + D_{u_8}\left(\frac{\partial^2 u_8}{\partial r^2} + \frac{1}{r}\frac{\partial u_8}{\partial r} + \frac{\partial^2 u_8}{\partial z^2}\right) - v_L(r)\left[\frac{\partial u_2}{\partial z} + \frac{\partial u_5}{\partial z} + \frac{\partial u_6}{\partial z} + \frac{\partial u_7}{\partial z} + \frac{\partial u_8}{\partial z}\right] \quad (25)$$

– global $PZCOO^-$ mass balance:

$$0 = D_{u_6}\left(\frac{\partial^2 u_6}{\partial r^2} + \frac{1}{r}\frac{\partial u_6}{\partial r} + \frac{\partial^2 u_6}{\partial z^2}\right) + D_{u_7}\left(\frac{\partial^2 u_7}{\partial r^2} + \frac{1}{r}\frac{\partial u_7}{\partial r} + \frac{\partial^2 u_7}{\partial z^2}\right) + v_L(r)\left[\frac{\partial u_6}{\partial z} + \frac{\partial u_7}{\partial z}\right] + R_2 - R_3 \quad (26)$$

– global $PZ(COO)_2^{2-}$ mass balance:

$$0 = D_{u_8}\left(\frac{\partial^2 u_8}{\partial r^2} + \frac{1}{r}\frac{\partial u_8}{\partial r} + \frac{\partial^2 u_8}{\partial z^2}\right) + v_L(r)\frac{\partial u_8}{\partial z} + R_3 \quad (27)$$

– global carbon mass balance:

$$0 = D_{u_1}\left(\frac{\partial^2 u_1}{\partial r^2} + \frac{1}{r}\frac{\partial u_1}{\partial r} + \frac{\partial^2 u_1}{\partial z^2}\right) + D_{u_6}\left(\frac{\partial^2 u_6}{\partial r^2} + \frac{1}{r}\frac{\partial u_6}{\partial r} + \frac{\partial^2 u_6}{\partial z^2}\right) + D_{u_7}\left(\frac{\partial^2 u_7}{\partial r^2} + \frac{1}{r}\frac{\partial u_7}{\partial r} + \frac{\partial^2 u_7}{\partial z^2}\right) + 2 \times D_{u_8}\left(\frac{\partial^2 u_8}{\partial r^2} + \frac{1}{r}\frac{\partial u_8}{\partial r} + \frac{\partial^2 u_8}{\partial z^2}\right) + D_{u_9}\left(\frac{\partial^2 u_9}{\partial r^2} + \frac{1}{r}\frac{\partial u_9}{\partial r} + \frac{\partial^2 u_9}{\partial z^2}\right) + D_{u_{10}}\left(\frac{\partial^2 u_{10}}{\partial r^2} + \frac{1}{r}\frac{\partial u_{10}}{\partial r} + \frac{\partial^2 u_{10}}{\partial z^2}\right) - v_L(r)\left[\frac{\partial u_1}{\partial z} + \frac{\partial u_6}{\partial z} + \frac{\partial u_7}{\partial z} + 2 \times \frac{\partial u_8}{\partial z} + \frac{\partial u_9}{\partial z} + \frac{\partial u_{10}}{\partial z}\right] \quad (28)$$

– electroneutrality balance:

$$0 = D_{u_3}\left(\frac{\partial^2 u_3}{\partial r^2} + \frac{1}{r}\frac{\partial u_3}{\partial r} + \frac{\partial^2 u_3}{\partial z^2}\right) + D_{u_5}\left(\frac{\partial^2 u_5}{\partial r^2} + \frac{1}{r}\frac{\partial u_5}{\partial r} + \frac{\partial^2 u_5}{\partial z^2}\right) - D_{u_4}\left(\frac{\partial^2 u_4}{\partial r^2} + \frac{1}{r}\frac{\partial u_4}{\partial r} + \frac{\partial^2 u_4}{\partial z^2}\right) - D_{u_6}\left(\frac{\partial^2 u_6}{\partial r^2} + \frac{1}{r}\frac{\partial u_6}{\partial r} + \frac{\partial^2 u_6}{\partial z^2}\right) - 2 \times D_{u_8}\left(\frac{\partial^2 u_8}{\partial r^2} + \frac{1}{r}\frac{\partial u_8}{\partial r} + \frac{\partial^2 u_8}{\partial z^2}\right) - D_{u_9}\left(\frac{\partial^2 u_9}{\partial r^2} + \frac{1}{r}\frac{\partial u_9}{\partial r} + \frac{\partial^2 u_9}{\partial z^2}\right) - 2 \times D_{u_{10}}\left(\frac{\partial^2 u_{10}}{\partial r^2} + \frac{1}{r}\frac{\partial u_{10}}{\partial r} + \frac{\partial^2 u_{10}}{\partial z^2}\right) - v_L(r)\left[\frac{\partial u_3}{\partial z} - \frac{\partial u_4}{\partial z} + \frac{\partial u_5}{\partial z} - \frac{\partial u_6}{\partial z} - 2 \times \frac{\partial u_8}{\partial z} - \frac{\partial u_9}{\partial z} - 2 \times \frac{\partial u_{10}}{\partial z}\right] \quad (29)$$

– K_1: $2H_2O \leftrightarrow H_3O^+ + OH^-$

$$K_1 = u_3 \times u_4 \quad (30)$$

– K_2: $PZ + H_2O \leftrightarrow PZH^+ + OH^-$

$$K_2 = \frac{u_5 \times u_4}{u_2} \quad (31)$$

– K_3: $PZCOO^- + H_2O \leftrightarrow H^+PZCOO^- + OH^-$

$$K_3 = \frac{u_7 \times u_4}{u_6} \quad (32)$$

– K_4: $CO_3^{2-} + H_2O \leftrightarrow HCO_3^- + OH^-$

$$K_4 = \frac{u_9 \times u_4}{u_{10}} \quad (33)$$

The kinetics expressions used in the mass balance equations are given as follows:

$$R_1 = k_1[CO_2][OH^-] - \frac{k_1}{K_1}[HCO_3^-] \qquad (34)$$

$$R_2 = k_2[CO_2][PZ] - \frac{k_2}{K_2}[PZCOO^-][H_3O^+] \qquad (35)$$

$$R_3 = k_3[CO_2][PZCOO^-] - \frac{k_3}{K_3}\left[PZ(COO)_2^{2-}\right][H_3O^+] \qquad (36)$$

The partial differential-algebraic system composed by 10 equations was solved using the finite element method (FEM) in COMSOL Multiphysics. It led to the determination of the concentration profiles $C_i(r,z)$ in the liquid phase. Following boundary conditions are applied:

$$u_i(r, z = h) = u_i^{eq} \qquad (37)$$

$$\frac{\partial u_i(r, z = 0)}{\partial z} = 0 \qquad (38)$$

$$\frac{\partial u_i(r = 0, z)}{\partial r} = 0 \qquad (39)$$

$$\frac{\partial u_i(r = \delta, z)}{\partial r} = 0 \quad i \neq CO_2 \qquad (40)$$

$$D_1 \frac{\partial u_1(r = \delta, z)}{\partial r} = k_G(P_{CO_2} - Hu_1(r = \delta, z)) \qquad (41)$$

where h represents the reactor height, H the Henry constant and P_{CO_2} the CO_2 partial pressure in the gas phase.

The concentration of each species at equilibrium conditions were provided by the thermodynamic model described in Section 2.3. The estimated concentrations were used to determine the apparent equilibrium constants of the chemical reactions. The CO_2 Henry constant was determined by the ratio between P_{CO_2} and the molecular CO_2 concentration at equilibrium conditions. This is original, as it is generally estimated in literature using a N_2O analogy.

The use of the same Henry constant for the determination of both interface and liquid compositions, allows the consistency between the global mass transfer driving force defined by the CO_2 partial pressures in the gas and liquid phases and the predicted liquid CO_2 driving force determined at the interface. This is not satisfied when two different solubility values are used.

As CO_2 is absorbed, P_{CO_2} presents a decreasing profile within the reactor. Consequently, the model takes into account the evolution of the CO_2 partial pressure in the gas phase through a one-dimensional plug-flow

model (Eq. 42). The flow can be considered countercurrent since the ratio $h/D_h \gg 1$ and the gas velocity is substantially higher than the liquid velocity. Moreover, the gas distribution has been improved by multiple injection points and the addition of a gas distributor (Fig. 8). Besides, simulations performed with FLUENT software have shown that the flow is essentially countercurrent:

$$0 = \frac{\partial\left(v_G C_{CO_2}^G\right)}{\partial z} + k_G a(P_{CO_2} - Hu_1(r = \delta, z)) \qquad (42)$$

The average CO_2 flux across the gas-liquid interface was determined using the following expression:

$$N_{CO_2} = \frac{\int_{z=h}^{z=0} D_{u_1} \frac{\partial u_1(r=\delta,z)}{\partial r} dz}{h} \qquad (43)$$

2.5 Physicochemical Properties

Properties such as the CO_2 and PZ diffusion coefficients must be known in order to determine the concentration profiles of the different species. The CO_2 diffusion coefficient in water can be obtained through the following correlation (Bishnoi and Rochelle, 2000):

$$D_{CO_2} = 0.02397 \exp\left(\frac{-2\,122.2}{T(K)}\right) \qquad (44)$$

The PZ concentration and the solution loading are not considered in this correlation.

The diffusion coefficients of PZ and of the ionic species were estimated by multiplying the CO_2 diffusion coefficient by 0.7. This is in agreement with the work of Bishnoi (2000), that has shown that the ratio between the diffusion coefficients of the ionic species and CO_2 is comprised between 0.7 and 0.8. Anyway, the ratio considered has not a major influence on the simulation results as the selected experimental operating conditions allow to avoid a significant impact of diffusion limitations of PZ and ionic species in the liquid phase on the CO_2 transfer (Fig. 7).

2.6 Comparison Between Gas Mass Transfer Models

A plug flow reactor model was used in this work to describe the P_{CO_2} evolution within the gas phase. This one-dimensional model was compared to the traditional approach of considering a constant CO_2 partial pressure given by the logarithmic average between the reactor gas inlet and outlet, given by:

$$P_{CO_2}^{ln} = \frac{P_{CO_2}^{inlet} - P_{CO_2}^{outlet}}{\ln \frac{P_{CO_2}^{inlet}}{P_{CO_2}^{outlet}}} \qquad (45)$$

The boundary condition at the gas-liquid interface remaining the same (*Eq. 41*), the only difference between both models is the estimation of P_{CO_2}.

Simulations at two different operating conditions were carried out in order to illustrate the difference between both approaches. The chosen operating conditions and the estimated fluxes obtained are shown in Table 2.

The difference between simulated fluxes given by the two approaches is negligible in unloaded solutions whereas it is of about 16% for a loading of 0.4. Appendix A shows that the results given by both approaches are identical for unloaded solution when

the CO_2 partial pressure at the gas-liquid interface does not change with the reactor height.

2.7 Choice of the CO_2 Partial Pressure

The CO_2 mass transfer in a wetted wall column is a function of the gas mass transfer, the liquid mass transfer, reactions rates, CO_2 solubility, etc. In order to maximize the sensitivity of the model calculations to the values of the kinetic constants, the experimental tests must be carried out in conditions minimizing the resistances due to the PZ mass transfer within the liquid towards the gas-liquid interface. The reactor model was used to identify these conditions.

Figure 7 illustrates the simulated PZ concentration at the gas-liquid interface at $z = 0$ as a function of the CO_2 molar fraction at the gas inlet. Unloaded solutions were considered. Figure 7 shows that PZ is almost depleted at the gas-liquid interface for the high CO_2 partial pressure. For the low CO_2 partial pressure, the PZ depletion at the gas-liquid interface is low, which means that the CO_2 mass transfer is mostly limited by the chemical reactions carried out within the solution. The kinetics of CO_2 absorption can be accurately determined in these conditions.

3 EXPERIMENTAL DEVICE

The experiments were conducted in a wetted wall column at temperatures ranging from 298 to 333 K on unloaded and loaded aqueous solutions of PZ. The solution loading was up to 0.4 mol_{CO_2}/mol_{amine} while the PZ concentrations range was comprised between 0.2-1 M.

The wetted wall column is a suitable equipment to obtain kinetic data of gas-liquid systems presenting high reaction rates due, to the high values of liquid mass transfer coefficients (k_L) associated to this device. The wetted wall column consists of a stainless steel cylinder with a surface area of 36.02 cm^2 (*Fig. 8*). The height and external diameter of the reactor are 9.1 and 1.26 cm, respectively. Within the reactor, the gas phase

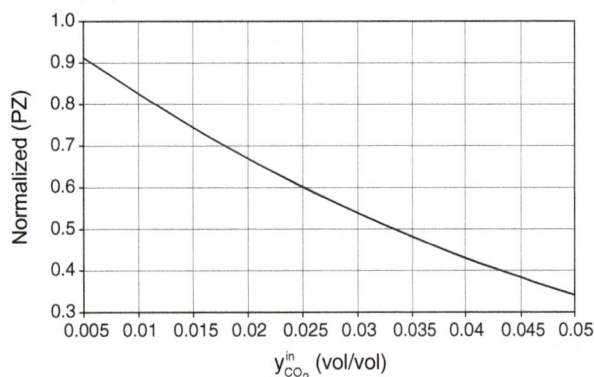

Figure 7

Simulated normalized PZ concentration at the reactor outlet at the gas-liquid interface at 298 K and 0.2 M of PZ (unloaded solution) as a function of CO_2 molar fraction in the gas phase at the gas inlet ($z = 0$).

TABLE 2

CO$_2$ fluxes simulated considering the local CO_2 pressure in the gas phase (model) or a Traditional Approach (TA)
i.e. the logarithmic average of the CO_2 partial pressure

Temperature (K)	[PZ] (M)	Loading (mol_{CO_2}/mol_{PZ})	Flux model ($\times 10^3$) ($mol.m^{-2} s^{-1}$)	Flux TA ($\times 10^3$) ($mol.m^{-2} s^{-1}$)
328	1	0.4	1.12	0.94
319	1	0	1.79	1.78

TA – Traditional Approach.
$Q_L = 16$ L/h, $Q_G = 150$ NL/h, $P = 1.5$ bar, $y_{CO_2}^{in} = 7\,000$ ppm.

flows counter-currently with the liquid that overflows from the inside of a cylinder to form a thin liquid film.

The gas phase, composed of CO_2 and nitrogen (N_2), is water-saturated before being in contact with the liquid in the reactor to prevent from water mass transfer in the reaction zone. The gas phase enters the reactor by 4 injection points. A gas distributor (*Fig. 6*) is located at the reactor inlet, just above the 4 injection points, in order to achieve an efficient gas distribution in the reaction zone. Downstream the reactor, the water contained in the gas is condensed within two consecutives condensers. The water-free gas is finally sent to an infra-red spectrometer that measures in-line the CO_2 gas concentration. The experimental flux is determined by the variation of the CO_2 gas concentration at both the inlet and the outlet of the reactor.

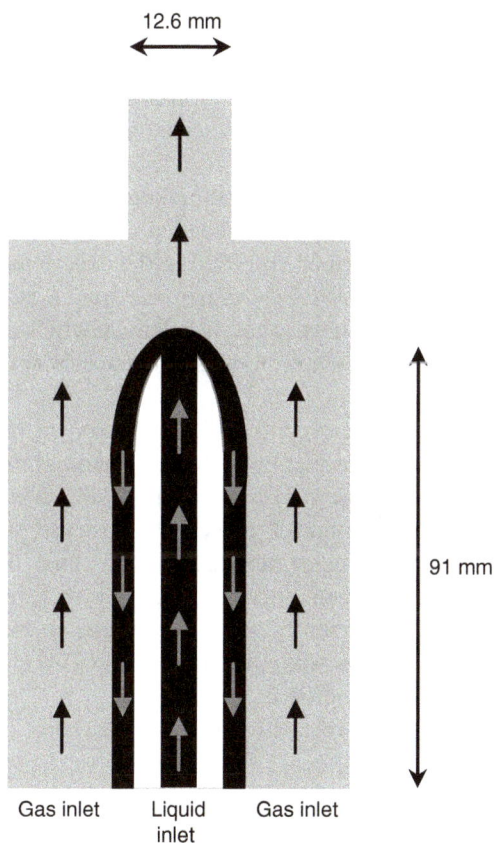

Figure 8

Wetted Wall Column (WWC) scheme.

4 RESULTS AND DISCUSSION

4.1 Gas Phase Mass Transfer Coefficient

The gas-side mass transfer resistance can generally not be neglected in reactive absorption. Hence the correct estimation of the gas-side mass transfer coefficient is crucial to accurately model the overall process. The gas phase mass transfer coefficient was measured by performing experiments of CO_2 absorption in aqueous solutions of MEA. The kinetics and the thermodynamics of MEA are well established in the literature, which justifies the choice of this system to determine the gas side mass transfer coefficient k_G. Besides, this system was used by Pacheco (1998), to estimate mass transfer resistance in gas phase in a similar device. The correlation obtained by Pacheco (1998) was later confirmed by the work of Bishnoi (2000), who performed measurements on a system presenting an instantaneous chemical reaction (SO_2/NaOH). All experiments were performed at a gas flow of 150 L/h and at a constant temperature of 333 K. The absorption tests were carried out on unloaded MEA solutions with overall concentrations ranging from 0.5 to 2 M. The total pressure was set to 1.5 bar. Experiments conducted at different solvent concentrations led to the simultaneous determination of the gas-liquid contact area, A, and of the volumetric gas-side mass transfer coefficient, k_G. A gas-liquid contact area of 38 cm^2 was estimated with an experimental error of approximately 10%. This value is in agreement with the geometric area of 36.02 cm^2. The experimentally determined volumetric mass transfer coefficient k_G, reported in Table 3, was consistent with the solution given by Graetz in a developed mass transfer boundary layer ($Sh_{lim} = 3.66$), however the accuracy was not very high. More details are given in Appendix B. k_G values at 298 and at 319 K were estimated assuming a constant Sherwood number. The estimation of k_G values at another temperature only depends on the CO_2 diffusion coefficient:

$$Sh = \frac{k_G D_h}{D_{CO_2}} = \text{constant} \Rightarrow \frac{k_G(T)}{k_G(T')} = \frac{D_{CO_2}(T)}{D_{CO_2}(T')} \quad (46)$$

The CO_2 diffusion coefficient in N_2 was estimated using the kinetic theory of gases (Poling *et al.*, 2000).

TABLE 3

Estimation of the gas-liquid volumetric mass transfer coefficient at different temperatures

Temperature (K)	298 (Calculated)	319 (Calculated)	333 (Experimental)
k_G ($mol.Pa^{-1}.m^{-2}.s^{-1}$)	6.6×10^{-6}	7.2×10^{-6}	$(7.8 \pm 3.9) \times 10^{-6}$

Resulting estimations of the gas-liquid volumetric mass transfer coefficient are reported in Table 3.

4.2 CO₂ Absorption in Aqueous PZ Solutions

Two set of experimental tests were carried out in order to characterize the kinetics of the reactions between the PZ and the PZCOO⁻ with CO_2. All experimental tests were carried out at constant pressure (1.5 bar) and at a fixed dry CO_2 molar fraction in the gas phase at the reactor inlet (about 7 000 ppm). The liquid and gas flow rates were set to 16 and 150 L/h, respectively. The operating temperature varied between 293 and 331 K.

A large experimental error was expected from the experiments conducted at 298 K since no temperature regulation could be applied. These measurements were performed at ambient temperature, which was comprised between 293 and 298 K.

An average relative gas-side mass transfer-resistance was estimated considering Equation (47):

$$\underbrace{\frac{H}{Ek_L}}_{\text{Liquid phase}} + \underbrace{\frac{1}{k_G}}_{\text{Gas phase}} = \frac{\Delta P_{CO_2}^{\ln}}{N_{CO_2}} \qquad (47)$$

The average relative gas-side mass transfer-resistance was comprised between 18 and 35% as reported in Tables 4 and 5. The high value clearly demonstrates the requirement of a correct estimation of the gas-side mass transfer coefficient for data interpretation.

4.2.1 Unloaded Solutions

CO_2 absorption experiments were conducted at temperatures between 298 and 331 K on unloaded PZ solutions ranging from 0.2 to 1 M. Experimental results and corresponding simulations are reported in Table 4. The simulations were performed considering the experimental temperature and input CO_2 molar fraction.

Model predictions were in good agreement with experimental data, except for the experiment at 297 K in a 1 M PZ solution which might be erroneous. The AAD between the experimental and model data was 3.7%.

The variation of the absorption flux with the total PZ concentration is shown in Figure 9 for three different temperatures. Again, measurements and simulations are shown. The simulations depicted in Figure 9 were performed at the average temperature and CO_2 inlet molar fraction of the measurement series.

The absorption flux increases with the total PZ concentration, as expected, due to the increase of the

reaction rate between CO_2 and the PZ. Curiously, the experimental CO_2 flux is lower at 331 K than at 319 K. This is related to the decrease of the input CO_2 molar fractions at 333 K due to the higher water content within the gas phase at these conditions. The CO_2 solubility decreases as temperature increases, which can also explain the observed evolution of fluxes.

The analysis of the simulated PZ concentration profiles in the liquid film at the reactor outlet (Fig. 10) shows that the PZ depletion at the gas-liquid interface remains moderate in all conditions. The CO_2 mass transfer is thus mainly governed by the CO_2 diffusion and the kinetics of the system.

4.2.2 Loaded Solutions

Experiments were performed in order to study the reaction between PZCOO⁻ and CO_2. The experimental tests were carried out at 298 and 331 K in 1 M PZ solutions and for initial loadings of 0.2, 0.3 and 0.4 mol_{CO_2}/mol_{PZ}. The loadings led to high PZCOO⁻ concentrations without too much modifying the physicochemical properties of the liquid solution.

Experimental results and corresponding simulations are reported in Table 5. As for unloaded solutions, the simulations were performed considering the experimental temperature and input CO_2 molar fraction. Again, model predictions were in very good agreement with simulations, the AAD between model and experimental data being 2.7%.

The influence of the second amine-function on the CO_2 flux has been quantified by performing simulations neglecting the dicarbamate formation, the results being reported in the last column of Table 5. In this case, the model systematically underestimated the CO_2 flux, the average difference between model and experiments being of about 10%. Consequently, the dicarbamate formation has to be taken into account to predict the CO_2 global transfer at these conditions.

The variation of the absorption flux with the solution loading is shown in Figure 11 for the two investigated temperatures. Measurements and simulations are shown, the simulations being performed at the average temperature and CO_2 inlet molar fraction of the measurement series.

At a given temperature, the absorption flux decreases with the solution loading. This can be explained by the fact that the concentration of PZ + PZCOO⁻ decreases with solution loading whereas the CO_2 equilibrium vapour pressure increases. As a result, both the reaction rates and the driving force decrease, leading to a reduction of the CO_2 flux.

TABLE 4

Experimental results of CO_2 absorption into unloaded PZ solutions

Total [PZ]	Temperature	Gas mass transfer resistance	CO_2 gas phase composition		CO_2 flux ($\times 10^3$)	
			Inlet	Outlet	Experimental	Simulated
M	K	%	ppmvol	ppmvol	mol.m^{-2}.s^{-1}	mol.m^{-2}.s^{-1}
0.2	296.6	18	7 085	5 130	1.05	1.07
0.6	297.0	29	7 008	4 147	1.55	1.47
1.0	297.3	37	6 870	3 583	1.82	1.62
0.2	318.8	22	7 211	4 580	1.30	1.20
0.6	319.0	30	7 331	3 890	1.67	1.63
1.0	318.8	35	7 300	3 554	1.83	1.79
0.2	331.4	20	6 905	4 240	1.13	1.16
0.6	331.0	30	6 799	3 306	1.50	1.50
1.0	330.9	35	6 864	2 987	1.65	1.65

Q_L = 16 L/h, Q_G = 150 NL/h, P = 1.5 bar.

TABLE 5

Experimental results of CO_2 absorption into loaded 1 M PZ solutions

Loading	Temperature	Gas mass transfer resistance	CO_2 gas phase composition		CO_2 flux ($\times 10^3$)		
			Inlet	Outlet	Experimental	Simulated	Simulated neglecting dicarbamate formation (*Eq. 18*)
$mol_{CO_2}/$ mol_{PZ}	K	%	ppmvol	ppmvol	mol.m^{-2}.s^{-1}	mol.m^{-2}.s^{-1}	mol.m^{-2}.s^{-1}
0.2	297.8	29	7 280	4 214	1.60	1.57	1.47
0.3	297.0	27	7 002	4 309	1.46	1.44	1.27
0.4	294.7	24	7 310	4 679	1.37	1.38	1.14
0.4	328.1	26	7 005	4 369	1.10	1.11	0.97
0.3	330.3	28	7 005	3 865	1.31	1.36	1.26
0.2	329.8	32	6 843	3 275	1.52	1.50	1.43
0.4	330.5	27	6 921	4 440	1.05	1.04	0.91
0.4	300.0	27	7 135	4 412	1.47	1.39	1.15
0.3	297.7	27	7 103	4 370	1.48	1.46	1.30

Q_L = 16 L/h, Q_G = 150 NL/h, P = 1.5 bar.

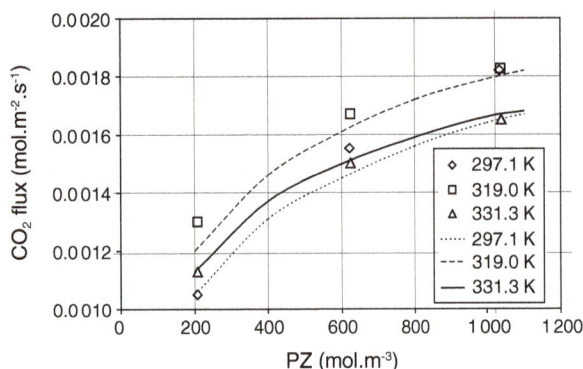

Figure 9

Variation of the absorption flux with total PZ concentration at 297, 319 and 331 K. Symbols: experiments; lines simulations (at the average temperature and CO_2 inlet molar fraction of the experiments).

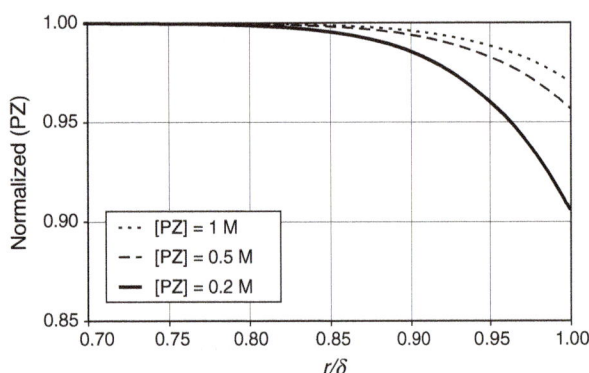

Figure 10

Simulated normalized PZ concentration profiles at the reactor outlet at 331 K.

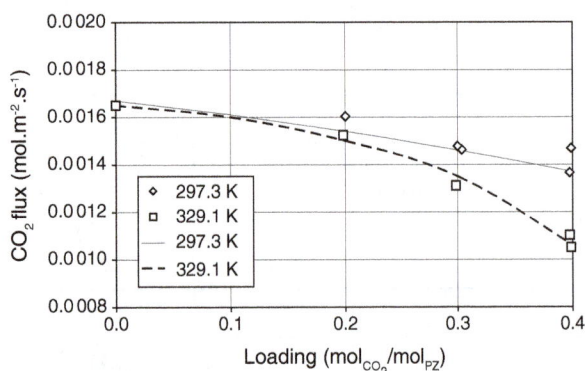

Figure 11

Variation of the absorption flux with solution loading at 297 and 329 K. Symbols: experiments; lines simulations (at the average temperature and CO_2 inlet molar fraction of the experiments).

At a given loading, the absorption flux decreases with increasing operating temperature, the impact being more important at high solution loadings. The effective $PZ + PZCOO^-$ concentration remains almost constant at iso-loading for the two investigated temperatures, but the increase of the CO_2 equilibrium vapour pressure is much more important at 329 K when compared to 298 K. As a result, the mass-transfer driving force decreases with temperature, leading to a decrease of the overall absorption flux.

CONCLUSION AND OUTLOOK

The paper describes theoretical and experimental investigations on the reactive absorption of CO_2 in aqueous solutions of PZ. A rigorous two dimensional absorption model, accounting for kinetics, hydrodynamics and thermodynamics, has been developed for a wetted wall column. The model considers the variation of the CO_2 gas phase concentration over the reactor length, which is more rigorous than previously published work, where average concentrations are considered. Model simulations clearly showed that the gas-phase concentration variation has to be taken into account, especially to assess the kinetics of CO_2 absorption in loaded solutions. The gas-liquid equilibrium was computed using the e-NRTL model, ensuring thus consistency of equations at the gas-liquid interface. The validity of equilibrium calculations has been shown by comparison between model simulations and gas-liquid equilibrium measurement taken from literature.

Model simulations allowed to define accurate operating conditions, where the diffusion of the liquid-side reactants were hardly limiting. However some free PZ depletion was always observed at the gas-liquid interface.

A laboratory-scale wetted wall column was conceived and constructed and the gas-side mass-transfer coefficient was determined experimentally. CO_2 absorption experiments were carried out at different temperatures in the experimental device in loaded as well as in unloaded PZ solutions. The gas-side mass transfer resistance was shown to be responsible of about 30% of the overall mass transfer resistance. Thus the knowledge of the gas-side mass transfer coefficient is crucial in order to correctly interpret absorption measurements.

When applying the kinetic constants published by Bishnoi and Rochelle (2002) the reactor model permits to predict the absorption fluxes with a global AAD of only 3.2% between theory and experiments. It has been shown that in loaded solutions the dicarbamate formation has to be taken into account in order to accurately predict the absorption flux. The model and the experimental device will be used in the future in order to investigate the absorption kinetics in more complex, mixed amine solutions.

REFERENCES

Appl M., Wagner U., Henrici H.J., Kuessner K., Volkamer F., Ernst-Neust N. (1982) Removal of CO_2 and/or H_2S and/or COS from gases containing these constituents, *US Patent* 4336233.

Bindwal A.B., Vaidya P.D., Kenig E.Y. (2011) Kinetics of carbon dioxide removal by aqueous diamines, *Chem. Eng. J.* **169**, 1-3, 144-150.

Bishnoi S. (2000) Carbon Dioxide Absorption and Solution Equilibrium in Piperazine Activated Methyldiethanolamine, *PhD Dissertation*, The University of Texas.

Bishnoi S., Rochelle G.T. (2000) Absorption of carbon dioxide into aqueous piperazine: reactions kinetics, mass transfer and solubility, *Chem. Eng. Sci.* **55**, 22, 5531-5543.

Bishnoi S., Rochelle G.T. (2002) Absorption of carbon dioxide in aqueous piperazine/methyldiethanolamine, *AICHE J.* **48**, 2788-2799.

Caplow M. (1968) Kinetics of carbamate formation and breakdown, *J. Am. Chem. Soc.* **90**, 6795-6803.

Chakravarty T., Phukan U.K., Weiland R.H. (1985) Reaction of Acid Gases with Mixtures of Amines, *Chem. Eng. Prog.* **81**, 32-36.

Crooks J.E., Donnellan J.P. (1989) Kinetics and Mechanism of the Reaction Between Carbon-Dioxide and Amines in Aqueous-Solution, *J. Chem. Soc.-Perkin Transa.* **2**, 331-333.

Cullinane J.T. (2005) Thermodynamics and Kinetics of Aqueous Piperazine with Potassium Carbonate for Carbon Dioxide Absorption, *PhD Dissertation*, The University of Texas.

Danckwerts P.V. (1979) The reaction of CO_2 with ethanolamines, *Chem. Eng. Sci.* **34**, 443-446.

da Silva E.F., Svendsen H.F. (2004) *Ab initio* study of the reaction of carbamate formation from CO_2 and alkanolamines, *Ind. Eng. Chem. Res.* **43**, 3413-3418.

DeCoursey W. (1974) Absorption with chemical reaction: development of a new relation for the Danckwerts model, *Chem. Eng. Sci.* **29**, 1867-1872.

Derks P.W.J., Kleingeld T., van Aken C., Hogendoorn J.A., Versteeg G.F. (2006) Kinetics of absorption of carbon dioxide in aqueous piperazine solutions, *Chem. Eng. Sci.* **61**, 6837-6854.

Dugas R. (2009) Carbon Dioxide Absorption, Desorption, and Diffusion in Aqueous Piperazine and Monoethanolamine, *PhD Dissertation*, The University of Texas.

Hamborg E.S., Versteeg G.F. (2009) Dissociation Constants and Thermodynamic Properties of Amines and Alkanolamines from (293 to 353) K, *J. Chem. Eng. Data* **54**, 1318-1328.

Hilliard M. (2005) Thermodynamics of Aqueous Piperazine/Potassium Carbonate/Carbon Dioxide Characterized by the Electrolyte NRT Model within Aspen Plus®, *MS Thesis*, The University of Texas.

Hogendoorn J., Vas Bhat R., Kuipers J., Van Swaaij W., Versteeg G. (1997) Approximation for the enhancement factor applicable to reversible reactions of finite rate in chemically loaded solutions, *Che. Eng. Sci.* **52**, 4547-4559.

Ko J.J., Tsai T.C., Lin C.Y. (2001) Diffusivity of nitrous oxide in aqueous alkanolamine solutions, *J. Chem. Eng. Data* **46**, 160-165.

Littel R.J., Versteeg G.F., van Swaaij W.P.M. (1992) Kinetics of CO_2 with Primary and Secondary-Amines in Aqueous-Solutions.1. Zwitterion Deprotonation Kinetics for DEA and DIPA in Aqueous Blends of Alkanolamines, *Chem. Eng. Sci.* **47**, 2027-2035.

Pacheco M. (1998) Mass Transfer, Kinetics and Rate-Based Modeling of Reactive Absorption, *PhD Dissertation*, The University of Texas.

Pinsent B.R.W., Pearson L., Roughton F.J.W. (1956) The kinetics of combination of carbon dioxide with hydroxide ions, *Trans. Faraday Soc.* **52**, 1512-1520.

Poling B.R., Prausnitz J.M., O'Connell J.P. (2000) *The Properties of gases and liquids*, McGraw-Hill, Fifth Edition.

Rinker E.B., Ashour S.S., Sandall O.C. (1996) Kinetics and modeling of carbon dioxide absorption into aqueous solutions of diethanolamine, *Ind. Eng. Chem. Res.* **35**, 1107-1114.

Rinker E.B., Ashoun S.S., Sandall O.C. (2000) Absorption of carbon dioxide into aqueous blends of diethanolamine and methyldiethanolamine, *Ind. Eng. Chem. Res.* **39**, 4346-4356.

Samanta A., Roy S., Bandyopadhyay S.S. (2007) Physical Solubility and Diffusivity of N_2O and CO_2 in Aqueous Solutions of Piperazine and (N-Methyldiethanolamine + Piperazine), *J. Chem. Eng. Data* **52**, 1381-1385.

Samanta A., Bandyopadhyay S.S. (2007) Kinetics and modeling of carbon dioxide absorption into aqueous solutions of piperazine, *Chem. Eng. Sci.* **62**, 7312-7319.

Seo D.J., Hong W.H. (2000) Effect of piperazine on the kinetics of carbon dioxide with aqueous solutions of 2-amino-2-methyl-1-propanol, *Ind. Eng. Chem. Res.* **39**, 2062-2067.

Versteeg G.F., Van Dijck L.A.J., Van Swaaij W.P.M. (1996) On the kinetics between CO_2 and alkanolamines both in aqueous and non-aqueous solutions, An overview. *Chemical Engineering Communications* **144**, 113-158.

Xu G.W., Zhang C., Qin S., Wang Y. (1992) Kinetics Study on Absorption of Carbon-Dioxide Into Solutions of Activated Methyldiethanolamine, *Ind. Eng. Chem. Res.* **31**, 921-927.

APPENDIX A

The gas phase material balance, assuming a plug-flow behaviour can be expressed as follows:

$$v_G \frac{\partial C_{CO_2}}{\partial z} = k_G a (C_{CO_2} RT - P^*) \tag{A.1}$$

The integration of Equation (A.1), considering a constant equilibrium partial pressure at the gas-liquid interface gives:

$$\ln \frac{C_{CO_2}^{out} - C^*}{C_{CO_2}^{in} - C^*} = \frac{k_G a}{RT v_G} h \tag{A.2}$$

If a CSTR model is used to perform the gas phase material balance:

$$Q_v^{Gas} \left(C_{CO_2}^{outlet} - C_{CO_2}^{inlet} \right) = \frac{k_G A}{RT} \times B \tag{A.3}$$

Equations (A.2) and (A.3) are identical if B is given by:

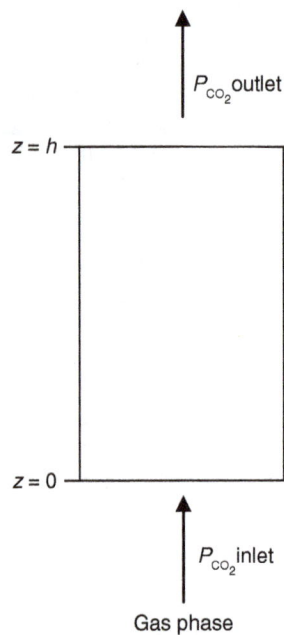

Figure A.1

Representation of the gas phase.

$$B = \frac{C_{CO_2}^{out} - C_{CO_2}^{in}}{\ln \dfrac{C_{CO_2}^{out} - C^*}{C_{CO_2}^{in} - C^*}} \tag{A.4}$$

Consequently, both approaches gives identical results if C^* is constant within the reactor.

APPENDIX B

The mass transfer coefficient in the gas phase, k_G, was determined using CO_2 absorption measurements on MEA solutions at different concentrations. The gas flow was set to 150 L/h for all the experiments. A plug flow model was considered to characterize the gas phase flow, and the double film theory was used to model the mass transfer between the gas and the liquid phase. The CO_2 material balance within the gas phase was then given by:

$$F_{CO_2}|_z - F_{CO_2}|_{z+dz} = \frac{A}{\frac{1}{k_G} + \frac{H}{Ek_L}} P_{CO_2} \tag{B.1}$$

After integration, the following expression is obtained:

$$\ln \frac{y_{CO_2}^{out}}{1 - y_{CO_2}^{out}} + \frac{y_{CO_2}^{out}}{1 - y_{CO_2}^{out}} - \left(\ln \frac{y_{CO_2}^{in}}{1 - y_{CO_2}^{in}} + \frac{y_{CO_2}^{in}}{1 - y_{CO_2}^{in}} \right) = -\frac{\frac{AP}{F_{inert}}}{\frac{1}{k_G} + \frac{H}{Ek_L}} \tag{B.2}$$

Assuming that the experimental tests are carried out in the kinetic regime, the CO_2 mass transfer is not limited by the MEA diffusion towards the gas-liquid interface. Considering that the perfect gas law can be applied, the following equation is obtained. The hypothesis concerning the kinetic regime was verified afterwards:

$$\underbrace{\ln \frac{y_{CO_2}^{out}}{1 - y_{CO_2}^{out}} + \frac{y_{CO_2}^{out}}{1 - y_{CO_2}^{out}} - \left(\ln \frac{y_{CO_2}^{in}}{1 - y_{CO_2}^{in}} + \frac{y_{CO_2}^{in}}{1 - y_{CO_2}^{in}} \right)} = -\frac{\frac{ART}{Q_{inert}}}{\frac{1}{k_G} + \underbrace{\frac{H}{\sqrt{D_{CO_2} k[MEA]}}}} \tag{B.3}$$

After rearrangement:

$$\underbrace{\frac{1}{\ln \frac{y_{CO_2}^{in}}{1 - y_{CO_2}^{in}} + \frac{y_{CO_2}^{in}}{1 - y_{CO_2}^{in}} - \left(\ln \frac{y_{CO_2}^{out}}{1 - y_{CO_2}^{out}} + \frac{y_{CO_2}^{out}}{1 - y_{CO_2}^{out}} \right)}}_{y} = \frac{Q_{inert}}{ART k_G} + \frac{HQ_{inert}}{ART \sqrt{D_{CO_2} k}} \underbrace{\frac{1}{\sqrt{[MEA]}}}_{x} \tag{B.4}$$

A linear regression allows to simultaneously determine the value of the volumetric mass transfer conductance (k_G) and the gas-liquid mass transfer area (A).

The CO_2 solubility in aqueous solutions of MEA was calculated by the correlation provided by Pacheco (1998). The second-order kinetic constant was given by Versteeg et al. (1996) while the CO_2 diffusion coefficient in aqueous solutions of MEA was determined through the N_2O analogy (Ko et al., 2001).

14

Integration of Adaptive Neuro-Fuzzy Inference System, Neural Networks and Geostatistical Methods for Fracture Density Modeling

A. Ja'fari[1]*, A. Kadkhodaie-Ilkhchi[2], Y. Sharghi[1] and M. Ghaedi[3]

1 Mining Engineering Department, Sahand University of Technology, Tabriz - Iran
2 Geology Department, Tabriz University, Tabriz - Iran
3 Chemical Engineering Department, Sharif University of Technology, Tehran - Iran
e-mail: a_jafari@sut.ac.ir - kadkhodaie_ali@tabrizu.ac.ir - ysharghi@sut.ac.ir - m_ghaedi@che.sharif.ir

* Corresponding author

Résumé — Intégration de système d'inférence adaptative de neurone flou (ANFIS), de réseaux de neurones (NN) et de méthodes géostatistiques pour la modélisation de la densité de fractures — Les images de diagraphies fournissent des informations utiles pour l'étude de fractures dans des réservoirs naturellement fracturés. L'inclinaison, l'azimut, l'ouverture et la densité de fractures peuvent être obtenus à partir des images de diagraphies et ces éléments ont une grande importance dans la caractérisation des réservoirs naturellement fracturés. L'imagerie de toutes les parties fracturées des réservoirs d'hydrocarbures et l'interprétation des résultats sont des processus longs et coûteux. Dans cette étude, une méthode améliorée pour faire une corrélation quantitative entre les densités de fractures obtenues à partir des images de diagraphies et de données conventionnelles, a été proposée par l'intégration des différents systèmes d'intelligence artificielle. Pour l'estimation globale de la densité de fractures à partir de données de diagraphies conventionnelles, la méthode proposée combine les résultats obtenus à partir d'algorithmes du système d'inférence adaptative flou de neurones (ANFIS) et du réseau de neurones (NN). Une méthode simple de moyenne a été utilisée pour obtenir un meilleur résultat en combinant les résultats de l'ANFIS et NN. L'algorithme a été appliqué à d'autres puits du champ pour obtenir la densité de fracture. Afin de modéliser la densité de fractures dans le réservoir, nous avons utilisé des algorithmes de simulation et de variographie séquentiels comme la Simulation à Indicateurs Séquentiels (SIS) et la Simulation Gaussienne Tronquée (TGS). L'algorithme global a été appliqué au réservoir d'Asmari de l'un des champs pétroliers du sud-ouest iranien. L'analyse de l'histogramme a été appliquée au contrôle de la qualité des modèles obtenus. Les résultats de cette étude montrent que pour nombre plus élevé de fractures, l'algorithme de faciès TGS fonctionne mieux que le SIS, mais que pour un petit nombre de faciès de fractures les deux algorithmes fournissent des résultats à peu près identiques.

Abstract — Integration of Adaptive Neuro-Fuzzy Inference System (ANFIS), Neural Networks and Geostatistical Methods for Fracture Density Modeling — Image logs provide useful information for fracture study in naturally fractured reservoir. Fracture dip, azimuth, aperture and fracture density can be obtained from image logs and have great importance in naturally fractured reservoir characterization. Imaging all fractured parts of hydrocarbon reservoirs and interpreting the results is expensive and time consuming. In this study, an improved method to make a quantitative correlation between fracture densities obtained from image logs and conventional well log data by integration of different artificial intelligence systems was proposed. The proposed method combines the results of Adaptive Neuro-Fuzzy Inference System (ANFIS) and Neural Networks (NN) algorithms for overall estimation of fracture density from conventional well log data. A simple averaging method was used to obtain a better result by combining results of ANFIS and NN. The algorithm applied on other wells of the field to obtain fracture density. In order to model the fracture density in the reservoir, we used variography and sequential simulation algorithms like Sequential Indicator Simulation (SIS) and Truncated Gaussian Simulation (TGS). The overall algorithm applied to Asmari reservoir one of the SW Iranian oil fields. Histogram analysis applied to control the quality of the obtained models. Results of this study show that for higher number of fracture facies the TGS algorithm works better than SIS but in small number of fracture facies both algorithms provide approximately same results.

INTRODUCTION

The word "fracture" is used as a collective term representing any of a series of discontinuous fractures in rocks such as joints, faults, fractures and/or bedding planes. In naturally fractured reservoir fractures have a significant effect on hydraulic properties of reservoir. When fractures are open, they act as pathways for hydrocarbon production and may even transform a very low permeability reservoir into highly productive zones. When cemented they act as barriers to hydrocarbon flow hindering the motion of hydrocarbon toward the wells (Haller and Porturas, 1998; Khoshbakht *et al.*, 2009). Therefore in modeling fractured reservoirs, understanding fracture properties is very important. Natural fractures can be identified and evaluated by several techniques, with the most common being core analysis, well log analysis and pressure transient testing. Since mid-1980s and introduction of image log tools, the process of fracture detection and characterization of fracture properties; such as dip, dip direction and fracture density; has become less problematic (Serra, 1989; Tokhmchi *et al.*, 2010). As fracture modeling with an inadequate volume of data can lead to misleading interpretations any direct or indirect techniques which increase the knowledge of fracture properties is highly valuable (Tokhmchi *et al.*, 2010).

Earlier attempts to detect natural fractures include the use of sonic waves (Hsu *et al.*, 1987), wavelet transform (Daiguji *et al.*, 1997), core data (Song *et al.*, 1998), seismic data (Behrens *et al.*, 1998), using petrophysical well logs (Tokhmchi *et al.*, 2010).

In this work, we tried to find the best method to predict fracture density from petrophysical log data. NN and ANFIS methods were used for this purpose. ANFIS model was previously done by Ja'fari *et al.* (2012) and the results are available. A same dataset was used for NN to predict the fracture density. A simple averaging method was also used to average both model outputs. The results show that output of this average method has the best correlation with real fracture densities. So this method was used to predict the fracture density in all other wells of the field. Finally the fracture density was modeled in the inter-well region using Sequential Indicator Simulation (SIS) and Truncated Gaussian Simulation (TGS) and histograms were used to quality control of models.

1 METHODOLOGY

In this work, geostatistical and artificial intelligence methods were integrated to create a fracture density model in entire of the field. A brief description and background of each method was described here.

1.1 Adaptive Neuro-Fuzzy Inference System

Neuro-fuzzy modeling is a technique for describing the behavior of a system using fuzzy inference rules within a NN structure. Using a given input/output data set, adaptive Neuro-Fuzzy Inference System (ANFIS) constructs a FIS whose MF parameters are tuned using a back propagation algorithm (Matlab User's Guide, 2007; Labani *et al.*, 2010). So, the FIS could learn from the training data. FL and ANFIS were used by different authors to predict target parameters from a set of input data. Gokceoglu *et al.* (2004) used neuro-fuzzy model for modulus of deformation of jointed rock masses. Kadkhodaie-Ilkhchi *et al.* (2009) used a committee fuzzy inference system to predict petrophysical data from seismic attributes. Labani *et al.* (2010) used a committee machine with intelligent system to predict NMR log parameter from petrophysical log data.

1.2 Neural Network

NN is an intelligent tool for solving non-linear problems. Back propagation, or propagation of error, is a common method of training Artificial Neural Networks (ANN) to learn how to perform a given task. It's a supervised learning method; it means that it requires a set of training data that has the desired output for any given input. The network computes the difference between the calculated output and corresponding desired output from the training data set. The error is then propagated backward through the net and the weights are adjusted during a number of iterations, named epochs. The training stops when the calculated output values best approximate the desired values (Bhatt and Helle, 2002; Labani *et al.*, 2010). NN are used widely in petroleum geoscience to predict different target parameters and also to model a parameter all over a field. FitzGerald (1999) used ANN to predict fracture frequency from petrophysical logs. Ouenes (1999) introduced a new approach in fractured reservoir characterization which uses fuzzy logic and NN. El Ouahed *et al.* (2005) developed a 2D fracture intensity map and fracture network map in a large block of Hassi Messaoud field, using artificial NN and fuzzy logic. Darabi *et al.* (2010) showed the applicability of ANN and fuzzy logic in characterizing Parsi naturally fractured reservoir.

1.3 Variograms

When modeling a reservoir with a pixel-based technique, one has to resort to semivariograms to model the sizes and spatial distributions of the parameter. The user usually is faced to a choice of standard (*e.g.*, spherical, exponential, Gaussian) and nonstandard (*e.g.*, fractal, linear, sinusoidal) variograms. The idea is to model existing data (*e.g.*, core, log) with such variograms until the "best model-fit" provides the variogram structure (often nested of several components) that is used for reservoir modeling. Because data are often complex and the sample semivariogram can rarely be matched accurately, the problem arises on which semivariogram model should be used (Seifert and Jensen, 1999). In this study, spherical, semivariograms were used for our works.

There are many hydrological and geological processes where heuristic considerations suggest that different values of the variable (low and high) possess different variograms. One way of capturing different variograms for high and low values is to use indicator techniques (Journel, 1983, 1993). Indicator semivariograms are similar to traditional semivariograms, except that they are calculated on indicator values rather than the actual value of the variable of interest. Indicator values indicate whether the value of the variable is above (indicator value = 1) or below (indicator value = 0) a particular threshold. The threshold is usually given as the percentile of the univariate distribution of the variable. Since indicator semivariograms can be calculated for a number of different thresholds, they allow for a different spatial structure (*i.e.* semivariogram) at each threshold (Western *et al.*, 1998).

Different authors provided a comprehensive literature on geostatistical methods and their application in petroleum engineering, like Deutsch (2002). Kelkar (2000). Yarus and Chambers (2006) provided an overview for application of geostatistics for reservoir characterization. Gringarten and Deutsch (1999) introduced a methodology for variogram interpretation and modeling for improved reservoir characterization. Liu and Journel (2007) presented a package for geostatistical integration of coarse and fine scale data. Deutsch (2006) illustrated a SIS program for categorical variables with point and block data. Gringarten (1998) offered a computer code for stochastic simulation of fractures in layered systems.

1.4 Sequential Indicator Simulation (SIS)

The SIS method can be used for the stochastic modeling of discrete (*e.g.*, rock types) and continuous attributes (*e.g.*, porosity and permeability). In this method, each attribute to be modeled is described through a binary indicator variable that takes the value "1" if that attribute is encountered at a given location, "0" if not. Each indicator variable is in turn defined by its average frequency and a semivariogram that characterizes the spatial continuity. Following a random path through the three-dimensional grid, individual grid-nodes are simulated, one after another, using constantly updated, thus increasing size and conditioning datasets. The conditioning includes the original data (*e.g.*, well data) and all previously simulated values within a specified neighborhood. This ensures that closely spaced values have the correct short scale correlation. In other words, in this simulation approach, a grid-node is selected randomly and simulated with reference to the original conditioning dataset (*e.g.*, well data). In the next step, another grid-node is selected randomly, and the variable is simulated using the newly generated CCDF (Conditional Cumulative Distribution Function), which is now increased in size by one value. In this way, each node is simulated (Deutsch and Journel, 1992; Seifert and Jensen, 1999).

1.5 Truncated Gaussian Simulation (TGS)

Truncated Gaussian Simulation method has been first designed to provide stochastic images of sedimentary geology, mostly in fluvio-deltaic environments. The basic principle of the method is to replace the handling of the geological description, poorly designed for calculations, by the handling of a Random Function with a multigaussian distribution, for which geostatistical simulations are used routinely. The rock type provided by the simulation depends on the value provided for the Gaussian random function. After definition of thresholds, the rock type is chosen depending between which threshold is the value of the Gaussian random function. The thresholds are given by the proportions of each rock type. These proportions will vary in space, while the properties of the Gaussian random function stay the same (stationary). This method (Matheron *et al.*, 1987) relies on the truncation of a single Gaussian Random Field (GRF) in order to generate realizations of lithofacies. The main feature is the reproduction of the indicator variograms associated with the lithofacies and the hierarchical contact relationship among them. This method is adequate for deposits where the lithofacies exhibit a hierarchical spatial distribution, such as depositional environments or sedimentary formations. The procedure to obtain lithofacies realizations using TGS is described as follows:

– establish the lithofacies proportions and their contact relationships;

– using the truncation rule, perform variography of the lithofacies indicators through the determination of the covariance function of the underlying GRF;

– simulate the GRF at the data locations conditionally to the lithofacies coding. This step is performed using the Gibbs sampler algorithm (Geman and Geman, 1984);

– simulate the GRF at the target locations using the values generated at the previous step as conditioning data;

– truncate the realizations according to the truncation rule.

The TGS here used as an alternative method for fracture facies modeling and the categorical variograms of SIS are used for this method.

2 FIELD DESCRIPTION

This field located in SW of Iran has three major faults. 15 wells were drilled in this field until now. For 12 wells the petrophysical log data are available and 2 wells have image logs. 4 productive zones in fractured Asmari formation exist in this field. Variography and data analysis were done separately for each fracture facies. First, the structural model was built and then fracture modeling was done in the framework of this structural model.

Alavi (2004) collects all lithostratigraphic units of the Zagros fold-thrust belt data and published the perfect description of formations. Asmari (Oligocene to lowermost Miocene), medium-bedded to thick-bedded, locally shelly or oolitic, nummulites-bearing limestones (grainstone, packstone, wackestone) shoaling upward above a thin basal conglomerate from fine-grained (low-energy) deep-marine marly limestone to high-energy shallow-marine skeletal grainstone; composed of a number of sequences; a unconformity-bounded, highly prolific reservoir; interpreted as transgressive-regressive foredeep facies of the proforeland basin (Khoshbakht et al., 2009).

3 ESTIMATING FRACTURE DENSITY

The relationship between fracture densities and well log data including sonic log (DT), deep resistivity log (Rd), neutron porosity log (NPHI) and bulk density log (RHOB) are shown in the crossplots of Figure 1. The data used in Figure 1 are normalized between 0 and 1. As shown, a strong and direct relationship between fracture density and bulk density log is seen (CC = 0.5855). Sonic log shows a strong and inverse relationship with fracture density. The dense rocks have a high potential for fracturing. The increase in density of a rock causes the sonic log to decrease. The neutron and resistivity logs show a weaker relationship with fracture density.

Data scaling is necessary here for two reasons. First, it is desired to account for essential variability in the filtered log data and without some type of scaling process, those logs with the largest original variance would dominate the subsequent analysis. Second, it is desired to have all logs measured in similar units because it will be easier to compare them in the next models and the analysis will not be biased towards those with higher absolute values. In this study, a linear scaling method that maps the maximum log value to 1 and the minimum log value to 0 was used. The linear scaling has the following form:

$$Z_i = \frac{x_i - x_{min}}{x_{max} - x_{min}} \tag{1}$$

where z_i is the scaled value, x_i is the original value, x_{min} is the minimum log value and x_{max} is the maximum log value.

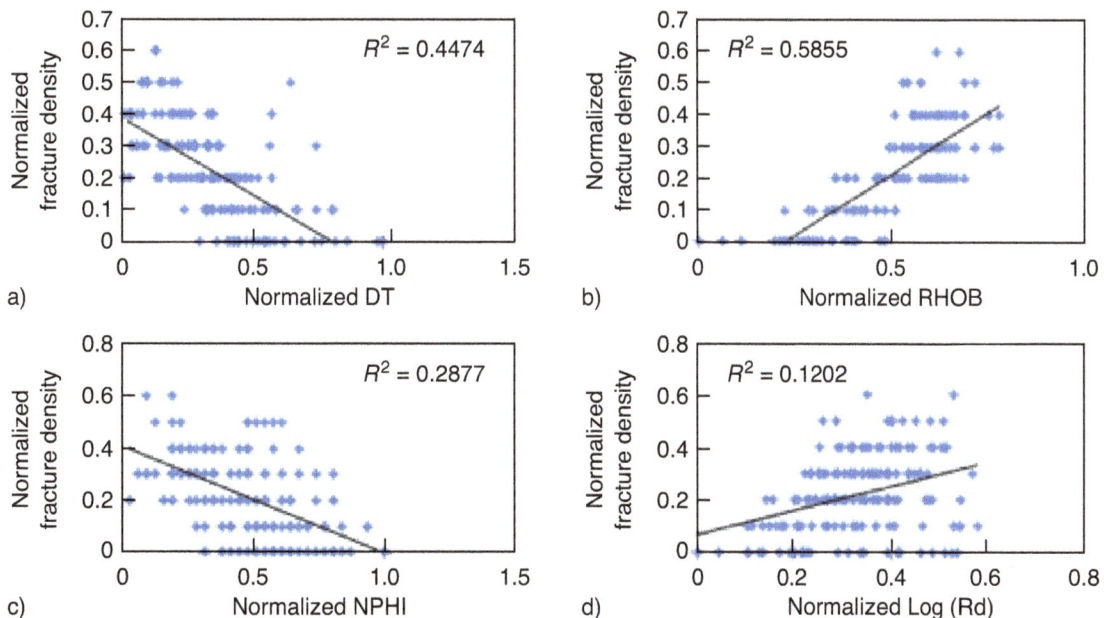

Figure 1

Crossplots showing relationship between fracture density and a) sonic, b) bulk density, c) neutron porosity, d) logarithm of deep resistivity.

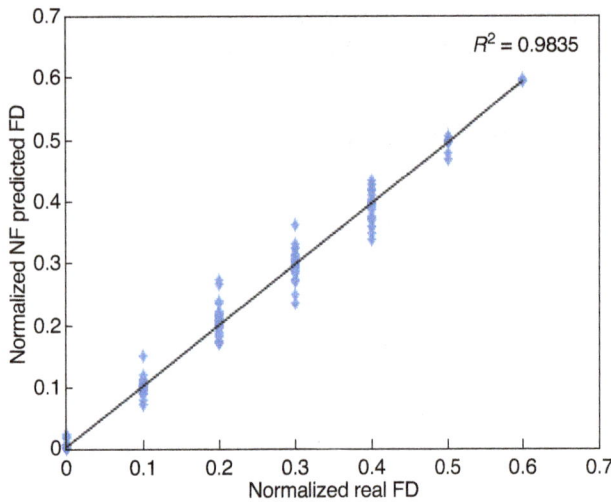

Figure 2

Crossplot of real and NF estimated fracture density.

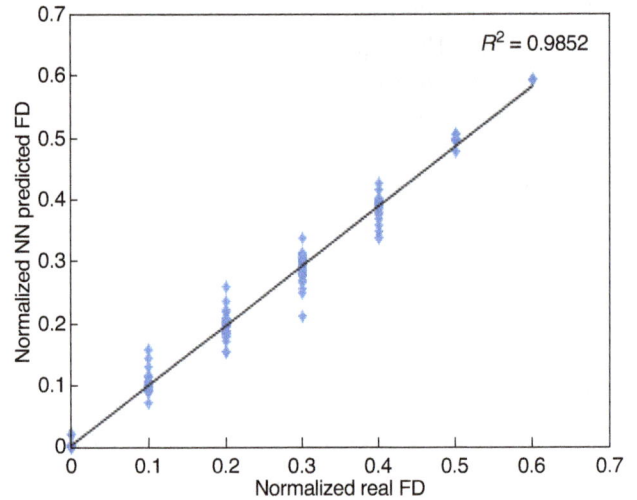

Figure 3

Crossplot of real and NN estimated fracture density.

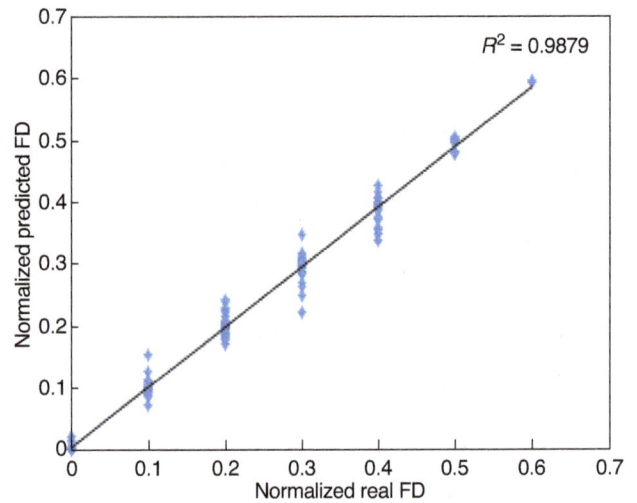

Figure 4

Crossplot of real and estimated average of fracture density.

3.1 NF Model

The formulation between input and output data is performed through a set of fuzzy if-then rules. Normally, fuzzy rules are extracted through a fuzzy clustering process. Subtractive clustering (Chiu, 1994) is one of the effective methods for constructing a fuzzy model. The effectiveness of a fuzzy model is related to the search for the optimal clustering radius, which is a controlling parameter for determining the number of fuzzy if-then rules. Fewer clusters might not cover the entire domain, and more clusters (result in more rules) can complicate the system behavior and may lead to lower performance (Kadkhodaie-Ilkhchi *et al.*, 2010). Depending on the case study, it is necessary to optimize this parameter for construction of fuzzy model. The clustering radius of the system is changed to find the best clustering radius for the system. After selecting the best clustering radius the data are predicted using this radius. Clustering radius of 0.3 was selected as the best radius for our data prediction with low error and high correlation and also very rapid regard to clustering radius of 0.2 and 0.1. The mean square error of the fuzzy model was 3.7×10^{-4} and Figure 2 shows a cross plot of NF predicted and real fracture densities.

3.2 NN Model

A simple feed forward, back propagation error algorithm was used to construct the NN model. A same dataset as ANFIS model was used for the NN model. The learning function of the network was Levenberg-Marquardt algorithm (TrainLM). A same dataset as NF model was used for NN model. Constructed network had a hidden layer with 30 neurons which used Tansig as transfer function and the output layer with one neuron which used Purelin as transfer function. The performance mean square error of the model was 7.82×10^{-4} for fracture density. Figure 3 shows the correlation between predicted and real data in the training dataset. This high correlation shows the ability of the network to find the relationship between input and output data.

We tried to find the best method for fracture density estimation, so we made a simple average of both models outputs and plot the average results *versus* real data, too. Figure 4 shows the result of this cross plot. R^2 of 0.9879 shows this is the best fit model among the above models.

So we used this method to estimate fracture density in other wells of the field. In Figure 5, the value of real and predicted fracture densities are plotted *versus* depth. This plot shows the high similarity of predicted and real fracture density *versus* depth and also the great ability of the proposed method to predict fracture density. The mean square error for this method is 2.8×10^{-4}.

4 MODELING FRACTURE DENSITY

The average fracture density map for the reservoir interval is shown in Figure 6. As shown, towards the northeast and east, the average fracture density increases. The zones with low fracture density are in the western sector of the reservoir around wells 13 and 15.

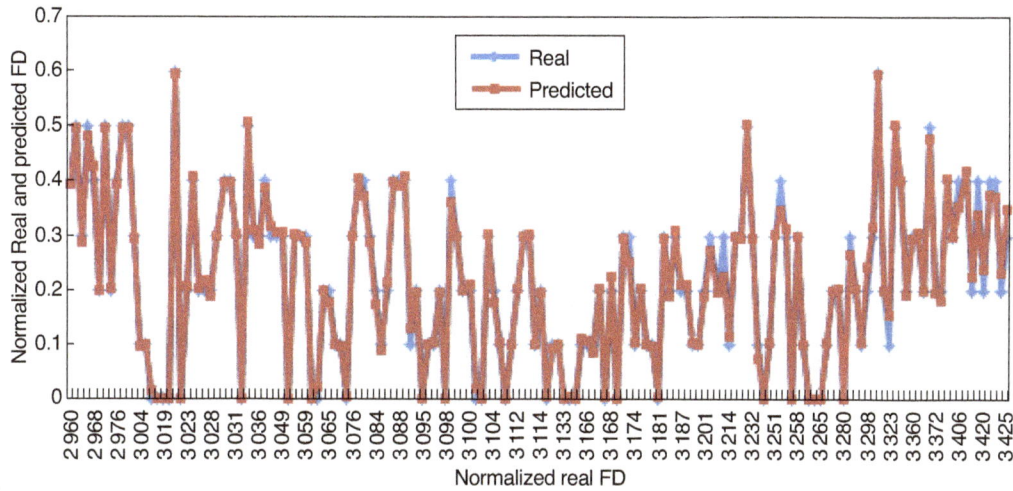

Figure 5

Real and predicted fracture density *versus* depth.

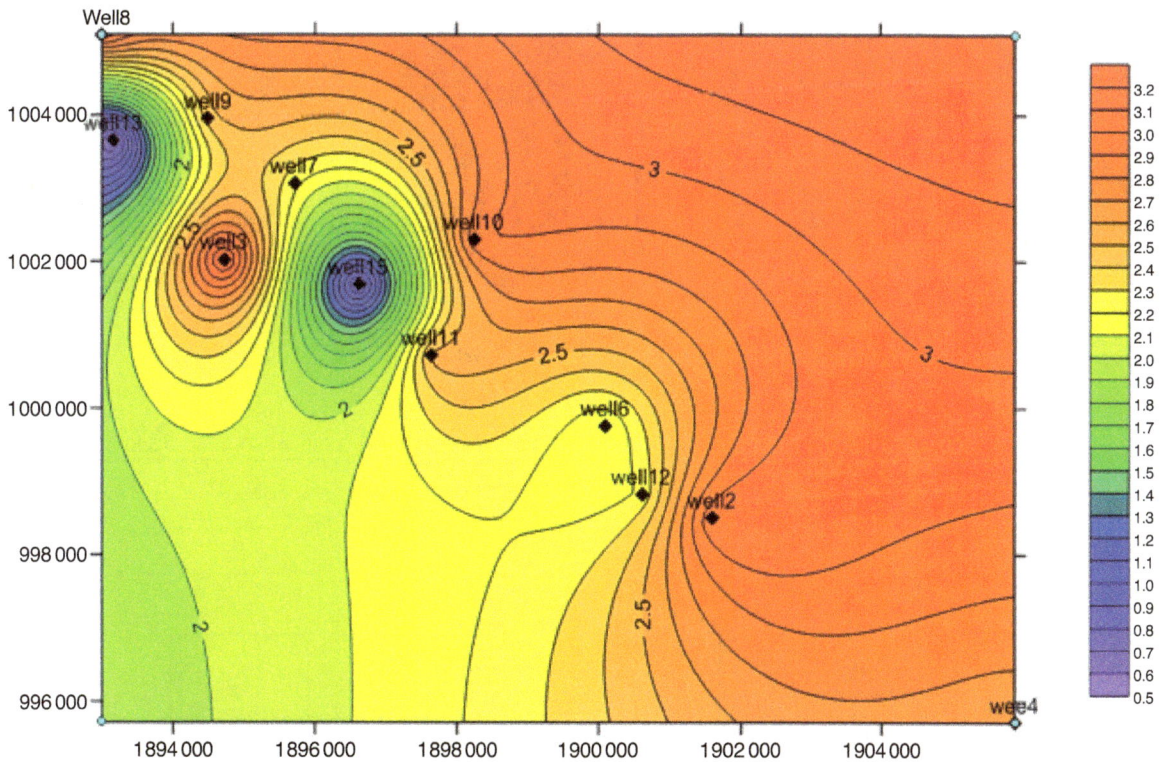

Figure 6

Average fracture density map for reservoir interval.

The result of image logs shows that the number of fractures per meter of length of boreholes changes from zero to eight fractures per meter. This can be seen in the predicted fracture density data, too. Therefore 9 facies for fractures were made and each facies indicates a fracture density. For example, facies zero consists of zero fracture per meter, facies one consists of one fracture per meter and so on. These facies are called fracture facies. Geostatistical simulation variograms were calculated and modeled for each fracture facies, separately. The variograms were calculated in a major direction (where the sample points have the strongest correlation), a minor direction (perpendicular to the major direction) and vertical directions using Petrel software. Variography is composed of three steps: determination of experimental variogram, fitting of a model to this variogram and determination of variogram parameters. At first deferent point pairs with an identical separation distance (lag distance) in an identical azimuth were determined. Then, the mean square errors at this points (or variance according to each lag distance) were calculated. This method is repeated for each lag distance. Finally to obtain an experimental variogram, these variances were plotted *versus* lags. In order to extract the variogram parameters, a model should be fitted to this experimental variogram. Different existing models were tested and finally the model that best fitted the experimental variogram is selected. Before calculating the variograms, the fracture facies proportions in each subzone is determined. Figure 7 demonstrates the proportion of fracture

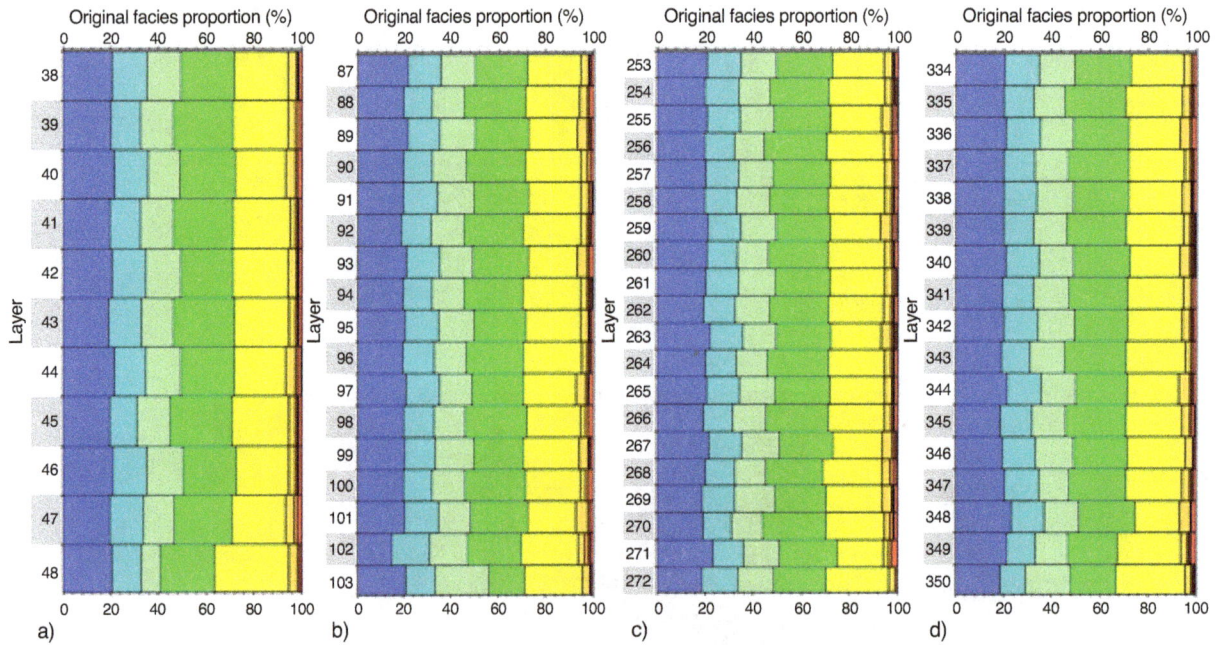

Figure 7

Fracture facies proportion in subzones of 1-4, 2-5, 5-4 and 6-5 are shown in a-d respectively.

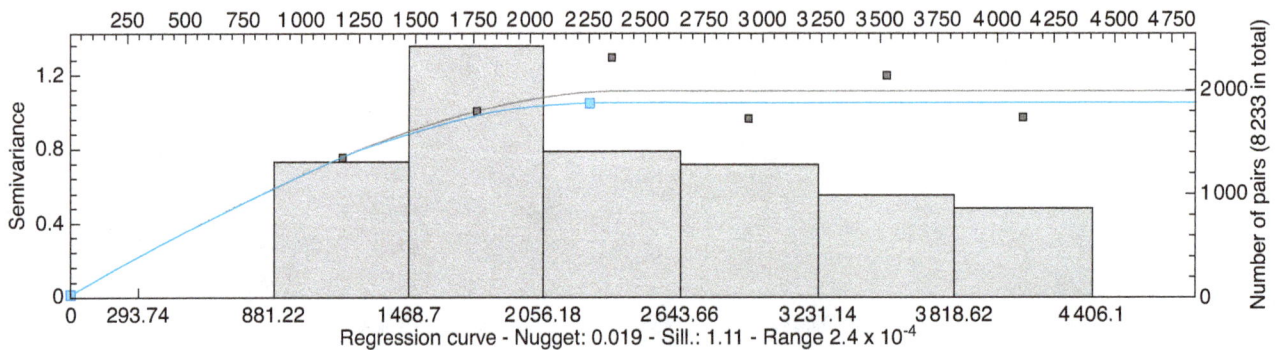

Figure 8

Variogram of two fracture facies in major direction.

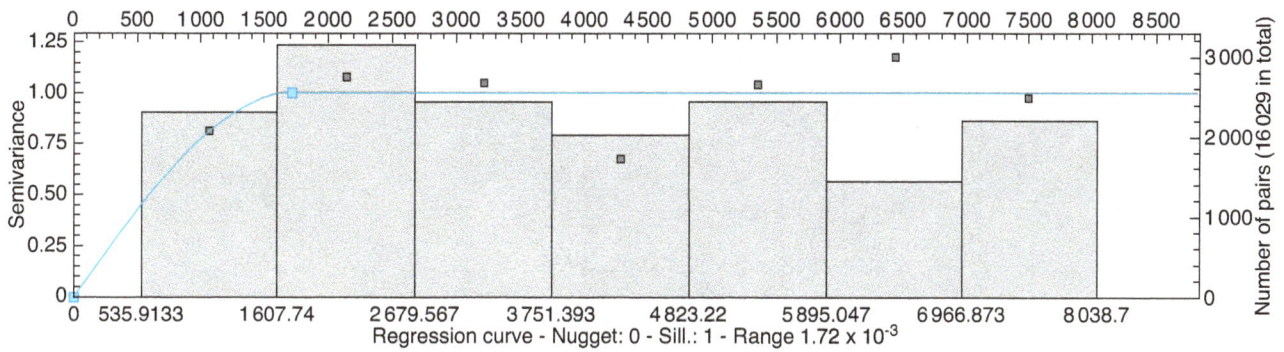

Figure 9

Variogram of three fracture facies in major direction.

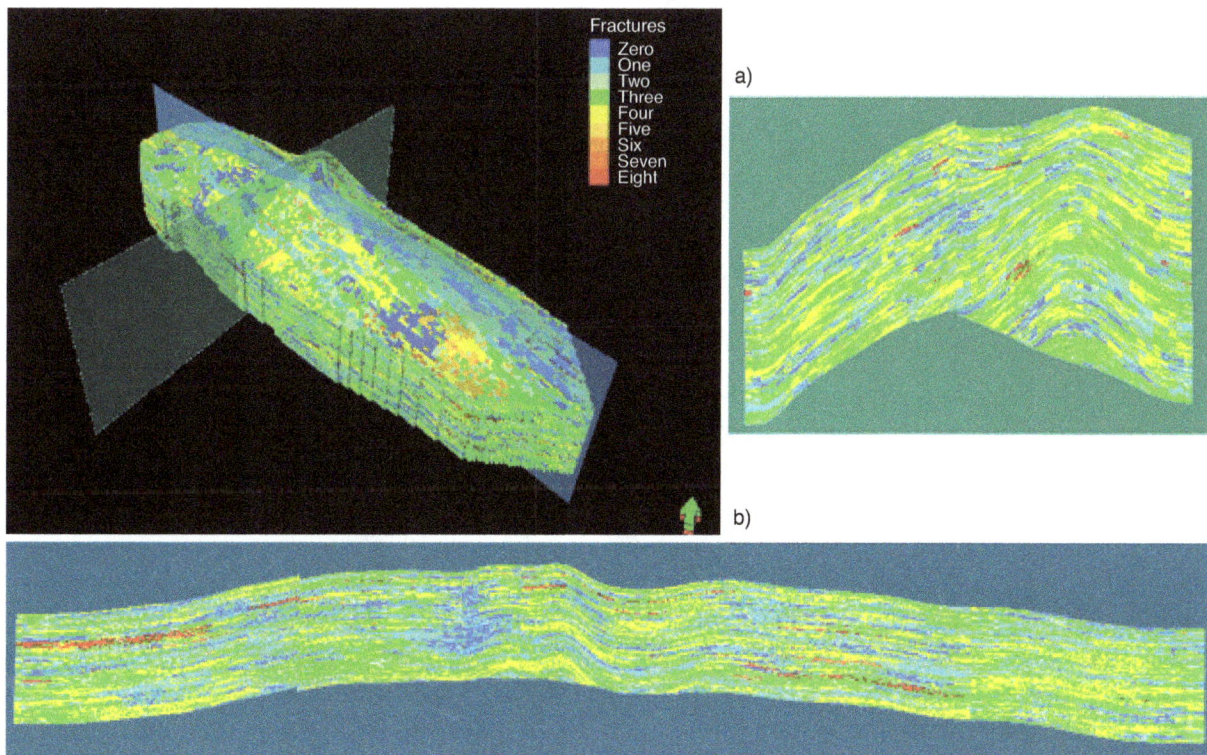

Figure 10

Fracture density model using SIS algorithm. a) Cross section, b) longitudinal section of the model.

facies in the original data of 4 subzones. Figures 8 and 9 show 2 calculated variograms by using spherical modeling. Gray points on the variogram show the semi-variance calculated for each lag distance. There is a histogram in back ground of each variogram. The histograms indicate the number of "point-pairs" used in the variogram calculation. It shows that each point plotted in the variogram is represented by x number of point pairs. The small histograms bars towards a lag distance illustrate that

the variance calculated for that lag distance is not represented by enough point pairs. The variograms have two curves, the gray one which is the best fit for the data and the blue one used for modeling is standardized to sill of one. The parameters of variograms are written below them.

Finally the geostatistical fracture density model was created using variography results. We could generate several models for each parameter; this is considered as one of the advantages

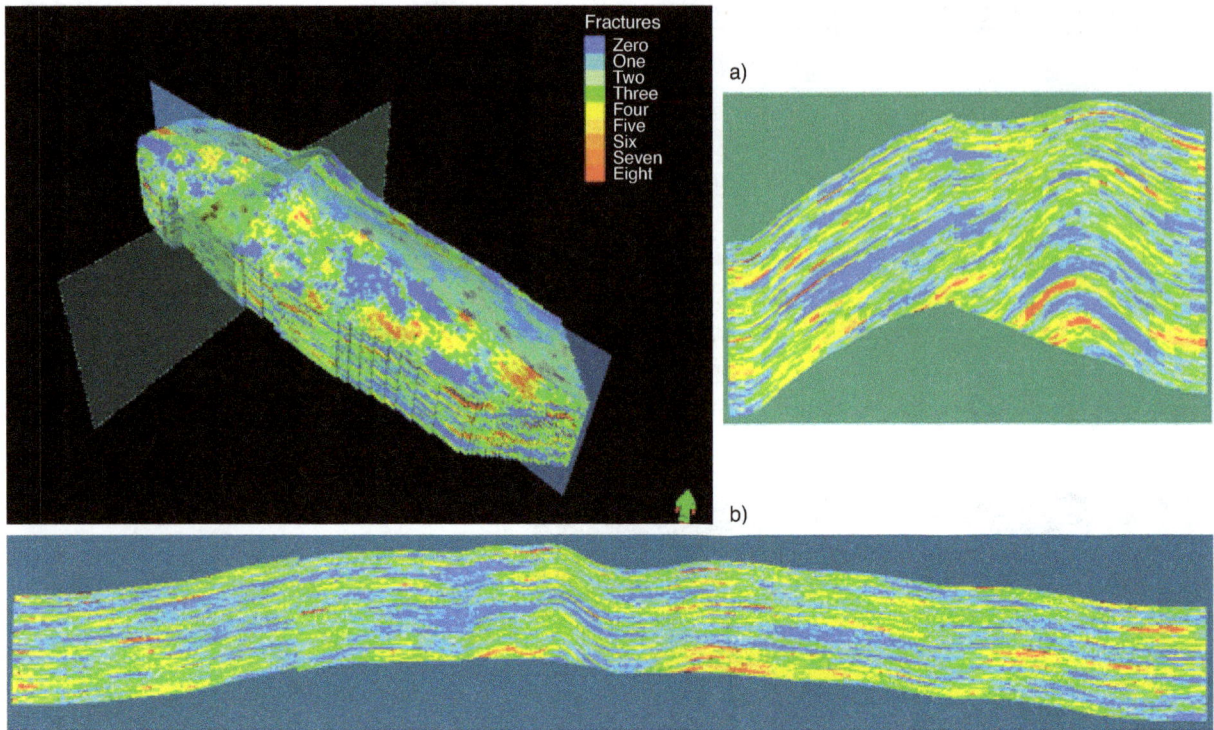

Figure 11

Fracture density model using TGS algorithm. a) Cross section, b) longitudinal section of the model.

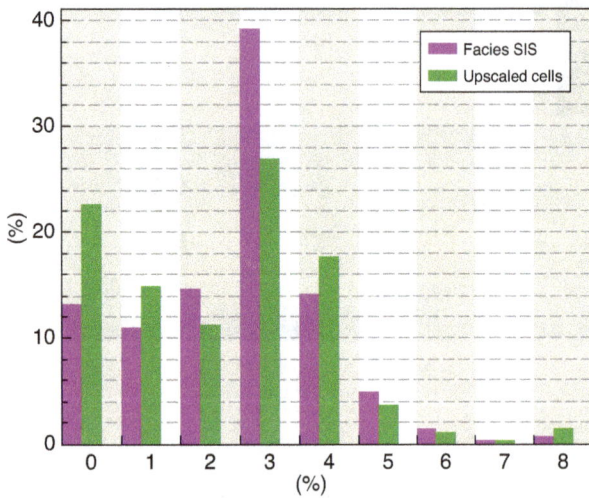

Figure 12

Histogram of upscaled and modeled fracture density using
SIS algorithm.

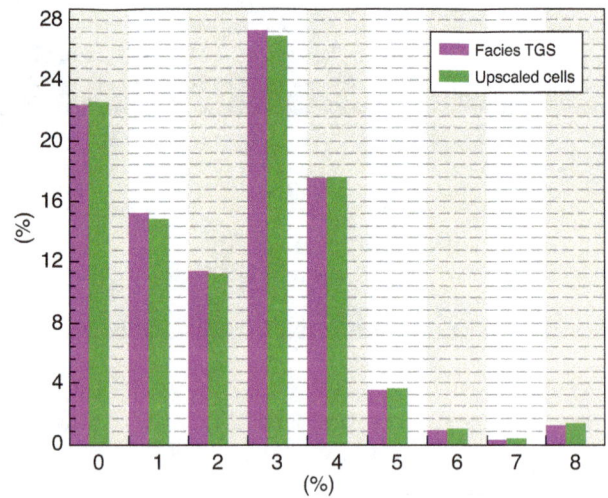

Figure 13

Histogram of upscaled and modeled fracture density using
TGS algorithm.

of stochastic simulation that creates multiple equiprobable
models. However, the probability of each model is equal to
the other one but the characteristics of them are different.
Histograms can be used to control the quality of resulted
models (Petrel User's Guide, 2009).

Figures 10 and 11 show the results of fracture density
modeling, using SIS and TGS algorithms. Figures 12 and 13
show the histograms of upscaled well logs and modeled
properties for fracture density. Similar distribution of histograms
for each model would verify the accuracy of obtained models.

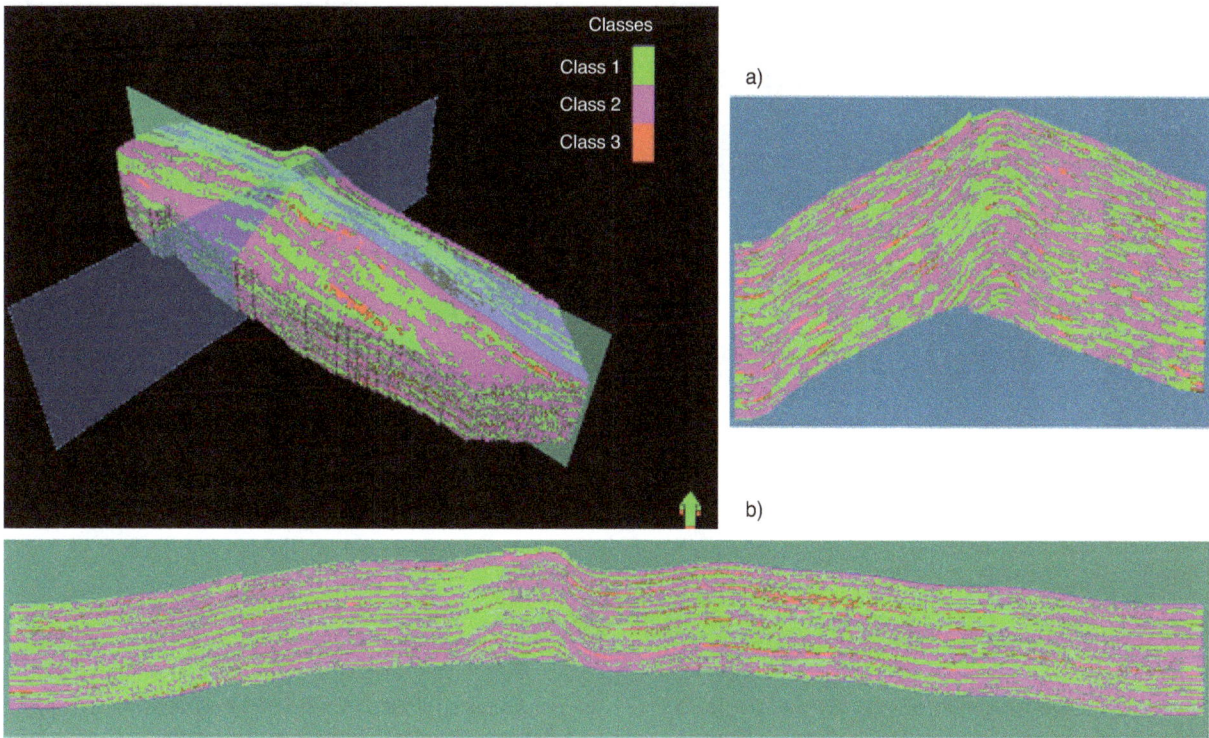

Figure 14

Fracture density model using SIS algorithm. a) Cross section, b) longitudinal section of the model.

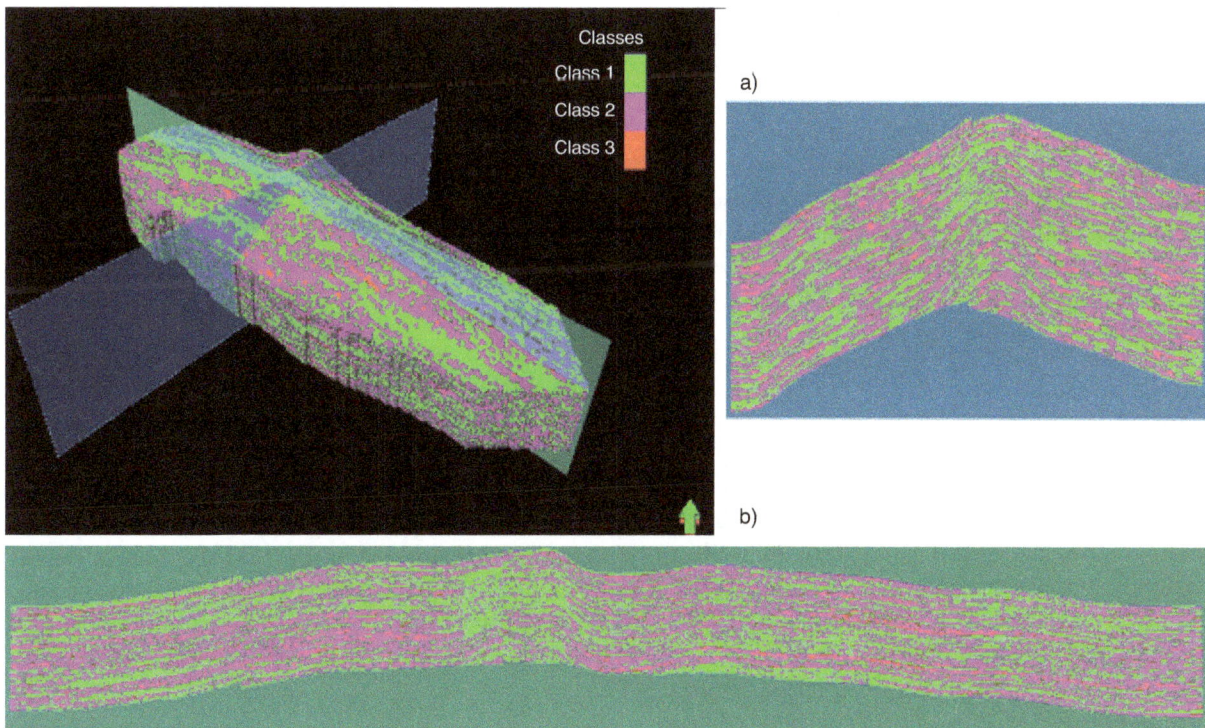

Figure 15

Fracture density model using TGS algorithm. a) Cross section, b) longitudinal section of the model.

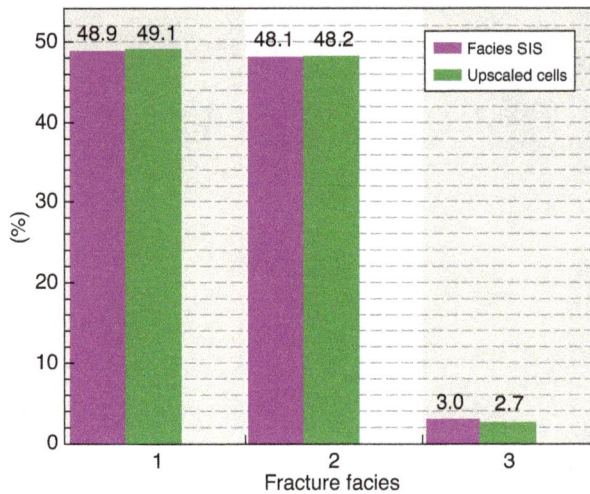

Figure 16

Histogram of upscaled and modeled fracture density using
SIS algorithm.

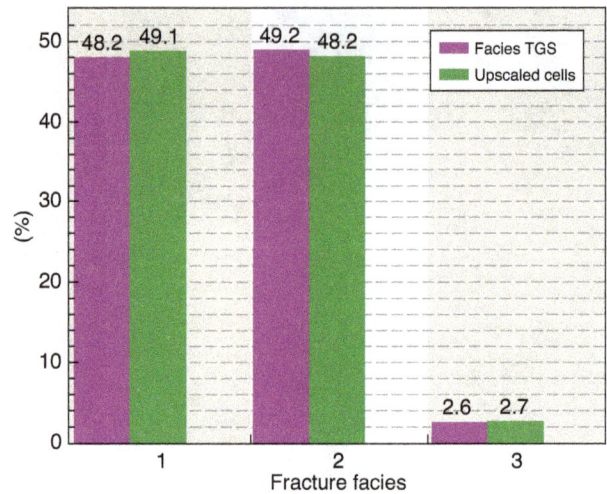

Figure 17

Histogram of upscaled and modeled fracture density using
TGS algorithm.

The results show that histograms of upscaled and modeled fracture density are more similar in TGS algorithm than in SIS algorithm. So in these proposed histograms obtained with TGS work better than the one obtained by SIS.

In another attempt to model fracture densities, the number of fracture facies is reduced by classifying the fracture densities into three classes as class one, two and three containing 0-2 fractures per meter, 3-5 fractures per meter and more than 6 fractures per meter, respectively. Variography analysis and geostatistical modeling of fracture densities were done, using SIS and TGS algorithms. The results of modeling and the histograms of models are shown in Figures 14 to 17. The histograms of modeled and upscaled fracture densities are more similar than the previous modeling results with nine fracture facies. This point verifies that decreasing the number of fracture facies may lead to increase the accuracy of modeling results, when using SIS and TGS algorithms.

CONCLUSIONS

In this study, artificial intelligence tools and geostatistical methods were used to model fracture density in subjected reservoir. The results show that integration of Neural Networks, ANFIS and geostatistical methods could provide very useful data for reservoir characterization and development. The main conclusions of this paper are:

– ANFIS and Neural Networks have a great ability in determining relationships between a series of petrophysical log data and fracture density as target;
– similar results of ANFIS and Neural Networks verify the trueness of the fracture density prediction;

– simple averages of both ANFIS and Neural Networks lead to a better correlation between real and predicted data, so this method can be used for predicting fracture density in this reservoir;
– integration of ANFIS and Neural Networks gives improved results for this dataset and provide the required data for the next step of fracture density modeling;
– the numbers of point-pairs included in variograms are sufficient to obtain reliable results. The nugget effects in all variograms are reasonable. So, we conclude that variograms can determine the spacial variation of fracture data;
– histograms of models show the preference of TGS algorithm to SIS algorithm for this dataset. Even if both models are different in details, histograms show a similar distribution for upscaled and modeled property for both algorithms. This similarity in results verifies the trueness of modeling;
– histograms show that a decrease in the number of fracture facies lead to an increase in the accuracy of obtained models.

ACKNOWLEDGMENTS

The authors would like to express their sincere thanks to the Geology Department of the National Iranian South Oil Company (NISOC especially Mr. Taghavipour and Mr. Heydari-Fard) for their assistance in providing the information and for their technical input to this work.

REFERENCES

Alavi M. (2004) Regional stratigraphy of the Zagross fold-thrust belt of Iran and its proforeland evolution, *Am. J. Sci.* **304**, 1-20.

Behrens R.A., Macleod M.K., Tran T.T., Alimi A.O. (1998) Incorporating seismic attribute maps in 3D reservoir models, *SPE Reserv. Eval.* **1**, 122-126.

Bhatt A., Helle H.B. (2002) Committee neural networks for porosity and permeability prediction from well logs, *Geophys. Prospect.* **50**, 645-660.

Chiu S. (1994) Fuzzy model identification based on cluster estimation, *J. Intelligent Fuzzy Syst.* **2**, 3, 267-278.

Daiguji M., Kudo O., Wada T. (1997) Application of wavelet analysis to fault detection in oil refinery, *Comput. Chem. Eng.* **21**, S1117-S1122 Suppl.

Darabi H., Kavousi H., Moraveji A., Masihi M. (2010) 3D Fracture Modeling in Parsi Oil Feld Using Artificial Intelligence Tools, *J. Petrol. Sci. Eng.* **71**, 67-76.

Deutsch C.V., Journel A.G. (1992) *GSLIB-Geostatistical Software Library and user's guide*, Oxford University Press, Oxford, 340 p.

Deutsch C.V. (2002) *Geostatistical reservoir modeling*, Oxford university press, New York.

Deutsch C.V. (2006) A sequential indicator simulation program for categorical variables with point and block data, *Comput. Geosci.* **32**, 1669-1681.

El Ouahed A.K., Tiab D., Mazouzi A. (2005) Application of artificial intelligence to characterize naturally fractured zones in Hassi Messaoud Oil Field, Algeria, *J. Petrol. Sci. Eng.* **49**, 122-141.

FitzGerald E.M., Bean C.J., Reilly R. (1999) Fracture-frequency prediction from borehole wireline logs using artificial neural networks, *Geophys. Prospect.* **47**, 1031-1044.

Geman S., Geman D. (1984) Stochastic Relaxation, Gibbs Distribution and the Bayesian Restoration of Images, *IEEE Trans. Pattern Anal. Mach. Intell.* **6**, 6, 721-741.

Gokceoglu C., Yesilnacar E., Sonmez H., Kayabasi A. (2004) A neuro-fuzzy model for modulus of joint rock masses, *Comput. Geotechnics* **31**, 375-383.

Gringarten E. (1998) Stochastic simulation of fractures in layered systems, *Comput. Geosci.* **26**, 729-736.

Gringarten E., Deutsch C.V. (1999) Methodology for Variogram Interpretation and Modeling for Improved Reservoir Characterization, *Annual Technical Conference and Exhibition*, Houston, Texas, 3-6 Oct., SPE 56654, 13 p.

Haller D., Porturas F. (1998) How to characterize fractures in reservoirs using borehole and core images: Case studies, *Geol. Soc. London Spec. Publ.* **136**, 249-259.

Hsu K.., Brie A., Plumb R.A. (1987) A new method for fracture identification using array sonic tools, *J. Pet. Technol.*, June, SPE Paper 14397, pp. 677-683.

Ja'fari A., Kadkhodaie-Ilkhchi A., Sharghi A., Ghanavati K. (2012) Fracture density prediction from petrophysical log data using adaptive neuro-fuzzy inference system, *J. Geophys. Eng.* **9**, 105-114.

Journel A.G. (1983) Nonparametric estimation of spatial distributions, *Math. Geol.* **15**, 445-468.

Journel A.G. (1993) Geostatistics: Roadblocks and Challenges, in Soares A. (ed.), *Geostatistics Troia '92*, Kluwer Academic Publishers, Dordrecht, The Netherlands, pp. 213-224.

Kadkhodaie-Ilkhchi A., Rezaee M.R., Rahimpour-Bonab H., Chehrazi A. (2009) Petrophysical data prediction from seismic attributes using committee fuzzy inference system, *Comput. Geosci.* **35**, 2314-2330

Kadkhodaie-Ilkhchi A., Takahashi Monteiro S., Ramos F., Hatherly P. (2010) Rock Recognition from MWD Data: A Comparative Study of Boosting, Neural Networks and Fuzzy Logic, *IEEE Trans. Geosci. Remote Sensing Lett. (GSRL)* **7**, 4, 680-684.

Kelkar M. (2000) Application of Geostatistics for Reservoir Characterization Accomplishments and Challenges, *J. Can. Pet. Technol.* **39**, 25-29.

Khoshbakht F., Memarian H., Mohammadnia M. (2009) Comparison of Asmari, Pabdeh and Gurpi formation's fractures, derived from image log, *J. Petrol. Sci. Eng.* **67**, 65-74.

Labani M.M., Kadkhodaie-Ilkhchi A., Salahshoor K. (2010) Estimation of NMR log parameters from conventional well log data using a committee machine with intelligent systems: A case study from the Iranian part of the South Pars gas field, Persian Gulf Basin, *J. Petrol. Sci. Eng.* **72**, 175-185.

Liu Y., Journel A. (2007) A package for geostatistical integration of coarse and fine scale data, *Comput. Geosci.* **35**, 527-547.

Matheron G., Beucher H., de Fouquet C., Gralli A., Guerillot D., Ravenne C. (1987) Conditional simulation of the geometry of fluvio-deltaic reservoirs, *Proc. SPE, Annual Technical Conference and Exhibition*, Dallas, Texas, 27-30 Sept., SPE 16753, pp. 591-599.

Matlab User's Guide (2007) Matlab CD-ROM, MathWorks, Inc.

Ouenes A. (1999) Practical application of fuzzy logic and neural networks to fractured reservoir characterization, *Comput. Geosci.* **26**, 953-962.

Petrel User's Guide (2009) Petrophysical modeling, CD-ROM, *Schlumberger* Company.

Seifert D., Jensen J.L. (1999) using sequential indicator simulation as a tool in reservoir description: issues and uncertainties, *Math. Geol.* **31**, 527-550.

Serra O. (1989) *Formation MicroScanner image interpretation*, *Schlumberger* Education Services.

Song X., Zhu Y., Liu Q., Chen J., Ren D., Li Y., Wang B., Liao M. (1998) Identification and distribution of natural fractures, *SPE International Oil and Gas Conference and Exhibition in China*, Beijing, China, 2-6 Nov., SPE Paper 50877.

Tokhmchi B., Memarian H., Rezaee, M.R. (2010) Estimation of the fracture density in fractured zones using petrophysical logs, *J. Petrol. Sci. Eng.* **72**, 206-213.

Western W.A., Bloschl G., Grayson R.B. (1998) How well do indicator variograms capture the spatial connectivity of soil moisture? *Hydrol. Process.* **12**, 1851-1868.

Yarus J.M., Chambers R.L. (2006) Practical Geostatistics - An Armchair Overview for Petroleum Reservoir Engineers, (Distinguished Author Series), *J. Petrol. Technol.* **58**, 11, 78-86, SPE 103357.

Permissions

All chapters in this book were first published in OGST, by IFP Energies nouvelles; hereby published with permission under the Creative Commons Attribution License or equivalent. Every chapter published in this book has been scrutinized by our experts. Their significance has been extensively debated. The topics covered herein carry significant findings which will fuel the growth of the discipline. They may even be implemented as practical applications or may be referred to as a beginning point for another development.

The contributors of this book come from diverse backgrounds, making this book a truly international effort. This book will bring forth new frontiers with its revolutionizing research information and detailed analysis of the nascent developments around the world.

We would like to thank all the contributing authors for lending their expertise to make the book truly unique. They have played a crucial role in the development of this book. Without their invaluable contributions this book wouldn't have been possible. They have made vital efforts to compile up to date information on the varied aspects of this subject to make this book a valuable addition to the collection of many professionals and students.

This book was conceptualized with the vision of imparting up-to-date information and advanced data in this field. To ensure the same, a matchless editorial board was set up. Every individual on the board went through rigorous rounds of assessment to prove their worth. After which they invested a large part of their time researching and compiling the most relevant data for our readers.

The editorial board has been involved in producing this book since its inception. They have spent rigorous hours researching and exploring the diverse topics which have resulted in the successful publishing of this book. They have passed on their knowledge of decades through this book. To expedite this challenging task, the publisher supported the team at every step. A small team of assistant editors was also appointed to further simplify the editing procedure and attain best results for the readers.

Apart from the editorial board, the designing team has also invested a significant amount of their time in understanding the subject and creating the most relevant covers. They scrutinized every image to scout for the most suitable representation of the subject and create an appropriate cover for the book.

The publishing team has been an ardent support to the editorial, designing and production team. Their endless efforts to recruit the best for this project, has resulted in the accomplishment of this book. They are a veteran in the field of academics and their pool of knowledge is as vast as their experience in printing. Their expertise and guidance has proved useful at every step. Their uncompromising quality standards have made this book an exceptional effort. Their encouragement from time to time has been an inspiration for everyone.

The publisher and the editorial board hope that this book will prove to be a valuable piece of knowledge for researchers, students, practitioners and scholars across the globe.

List of Contributors

R. Azin
Department of Chemical Engineering, School of Engineering, Persian Gulf University, Bushehr 7516913817 - Iran
Persian Gulf Science and Technology Park, Bushehr - Iran

R. Malakooti
Persian Gulf Science and Technology Park, Bushehr - Iran

A. Helalizadeh
Department of Petroleum Engineering, Petroleum University of Technology, Ahwaz - Iran

M. Zirrahi
Department of Chemical & Petroleum Engineering, Schulich School of Engineering, University of Calgary, Calgary, Alberta, Canada T2N 1N4 - Canada

K. Wojdan
Institute of Heat Engineering, Faculty of Power and Aeronautical Engineering, Warsaw University of Technology, Nowowiejska 21/25 00-665 Warszawa - Poland

B. Ruszczycki
Transition Technologies S.A., Pawia 55, 01-030 Warszawa - Poland

D. Michalk
Statoil Deutschland, Kavernenanlage Etzel, Beim Postweg 2, 26446 Friedeburg - Germany

K. Swirski
Institute of Heat Engineering, Faculty of Power and Aeronautical Engineering, Warsaw University of Technology, Nowowiejska 21/25 00-665 Warszawa - Poland

M. Cancelliere
Politecnico di Torino, Department of Environment, Land and Infrastructure Engineering, 24, Corso Duca Degli Abruzzi, 10129 Torino - Italy

D. Viberti
Politecnico di Torino, Department of Environment, Land and Infrastructure Engineering, 24, Corso Duca Degli Abruzzi, 10129 Torino - Italy

F. Verga
Politecnico di Torino, Department of Environment, Land and Infrastructure Engineering, 24, Corso Duca Degli Abruzzi, 10129 Torino - Italy

Zhiyuan Wang
School of Petroleum Engineering, China University of Petroleum (East China), Qingdao 266555 - China

Baojiang Sun
School of Petroleum Engineering, China University of Petroleum (East China), Qingdao 266555 - China

Ke Ke
SINOPEC Research Institute of Petroleum engineering, Beijing 100101 - China

Romina Digne
IFP Energies nouvelles, Rond-point de l'échangeur de Solaize, BP 3, 69360 Solaize - France

Frédéric Feugnet
IFP Energies nouvelles, Rond-point de l'échangeur de Solaize, BP 3, 69360 Solaize - France

Adrien Gomez
IFP Energies nouvelles, Rond-point de l'échangeur de Solaize, BP 3, 69360 Solaize - France

J. Kittel
IFP Energies nouvelles, Rond-point de l'échangeur de Solaize, BP 3, 69360 Solaize - France

S. Gonzalez
IFP Energies nouvelles, Rond-point de l'échangeur de Solaize, BP 3, 69360 Solaize - France

Marie-Hélène Klopffer
IFP Energies nouvelles, 1-4 avenue de Bois-Préau, 92852 Rueil-Malmaison Cedex - France

Philippe Berne
CEA, LITEN, DTNM, LCSN, 38054 Grenoble - France

Éliane Espuche
Université Lyon 1, CNRS, UMR5223, Ingénierie des Matériaux Polymères, 15 Bd A. Latarjet, 69622 Villeurbanne - France

Emmanuel Richaud
Arts et Métiers ParisTech, CNRS, PIMM UMR 8006, 151 bd de l'Hôpital, 75013 Paris - France

Fatma Djouani
Arts et Métiers ParisTech, CNRS, PIMM UMR 8006, 151 bd de l'Hôpital, 75013 Paris - France

Bruno Fayolle
Arts et Métiers ParisTech, CNRS, PIMM UMR 8006, 151 bd de l'Hôpital, 75013 Paris - France

Jacques Verdu
Arts et Métiers ParisTech, CNRS, PIMM UMR 8006, 151 bd de l'Hôpital, 75013 Paris - France

Bruno Flaconneche
IFP Energies nouvelles, 1-4 avenue de Bois-Préau, 92852 Rueil-Malmaison - France

J. Hermoso
Departamento de Ingeniería Química, Universidad de Huelva, Facultad de Ciencias Experimentales, Campus del Carmen, 21071 Huelva - Spain

F. Martínez-Boza
Departamento de Ingeniería Química, Universidad de Huelva, Facultad de Ciencias Experimentales, Campus del Carmen, 21071 Huelva - Spain

C. Gallegos
Departamento de Ingeniería Química, Universidad de Huelva, Facultad de Ciencias Experimentales, Campus del Carmen, 21071 Huelva - Spain

Merched Azzi
CSIRO Energy Technology, 11 Julius Avenue, 2113 NSW - Australia

Dennys Angove
CSIRO Energy Technology, 11 Julius Avenue, 2113 NSW - Australia

Narendra Dave
CSIRO Energy Technology, 11 Julius Avenue, 2113 NSW - Australia

Stuart Day
CSIRO Energy Technology, 11 Julius Avenue, 2113 NSW - Australia

Thong Do
CSIRO Energy Technology, 11 Julius Avenue, 2113 NSW - Australia

Paul Feron
CSIRO Energy Technology, 11 Julius Avenue, 2113 NSW - Australia

Sunil Sharma
CSIRO Energy Technology, 11 Julius Avenue, 2113 NSW - Australia

Moetaz Attalla
CSIRO Energy Technology, 11 Julius Avenue, 2113 NSW - Australia

Mohammad Abu Zahra
MASDAR Institute, Abu Dhabi

Richard Giot
Laboratoire Géoressources, ENSG-Université de Lorraine, UMR 7359, 2 rue du Doyen Marcel Roubault, TSA 70605, 54518 Vandoeuvre-Lès-Nancy - France

Albert Giraud
Laboratoire Géoressources, ENSG-Université de Lorraine, UMR 7359, 2 rue du Doyen Marcel Roubault, TSA 70605, 54518 Vandoeuvre-Lès-Nancy - France

Christophe Auvray
Laboratoire Géoressources, ENSG-Université de Lorraine, UMR 7359, 2 rue du Doyen Marcel Roubault, TSA 70605, 54518 Vandoeuvre-Lès-Nancy - France

Vincent Cuzuel
LSABM, UMR CBI 8231, ESPCI – CNRS, 10 rue Vauquelin, 75005 Paris - France

Julien Brunet
LSABM, UMR CBI 8231, ESPCI – CNRS, 10 rue Vauquelin, 75005 Paris - France

Aurélien Rey
LSABM, UMR CBI 8231, ESPCI – CNRS, 10 rue Vauquelin, 75005 Paris - France

José Dugay
LSABM, UMR CBI 8231, ESPCI – CNRS, 10 rue Vauquelin, 75005 Paris - France

Jérôme Vial
LSABM, UMR CBI 8231, ESPCI – CNRS, 10 rue Vauquelin, 75005 Paris - France

Valérie Pichon
LSABM, UMR CBI 8231, ESPCI – CNRS, 10 rue Vauquelin, 75005 Paris - France

Pierre-Louis Carrette
IFP Energies nouvelles, Rond-point de l'échangeur de Solaize, BP 3, 69360 Solaize - France

Alberto Servia
IFP Energies nouvelles, Rond-point de l'échangeur de Solaize, BP 3, 69360 Solaize - France
LRGP-CNRS Université de Lorraine, 1 rue Grandville, BP 20451, 54001 Nancy Cedex - France

Nicolas Laloue
IFP Energies nouvelles, Rond-point de l'échangeur de Solaize, BP 3, 69360 Solaize - France

Julien Grandjean
IFP Energies nouvelles, Rond-point de l'échangeur de Solaize, BP 3, 69360 Solaize - France

Sabine Rode
LRGP-CNRS Université de Lorraine, 1 rue Grandville, BP 20451, 54001 Nancy Cedex - France

Christine Roizard
LRGP-CNRS Université de Lorraine, 1 rue Grandville, BP 20451, 54001 Nancy Cedex - France

A. Ja'fari
Mining Engineering Department, Sahand University of Technology, Tabriz - Iran

A. Kadkhodaie-Ilkhchi
Geology Department, Tabriz University, Tabriz - Iran

Y. Sharghi
Mining Engineering Department, Sahand University of Technology, Tabriz - Iran

M. Ghaedi
Chemical Engineering Department, Sharif University of Technology, Tehran - Iran

* 9 7 8 1 6 8 2 8 6 0 7 0 0 *